# ADVANCED MATERIALS, POLYMERS, AND COMPOSITES

*New Research on Properties, Techniques, and Applications*

# ADVANCED MATERIALS, POLYMERS, AND COMPOSITES

*New Research on Properties, Techniques, and Applications*

*Edited by*

**Omari V. Mukbaniani, DSc**

**Tamara Tatrishvili, DSc**

**Marc J. M. Abadie, DSc**

AAP | APPLE ACADEMIC PRESS

First edition published 2022

**Apple Academic Press Inc.**
1265 Goldenrod Circle, NE,
Palm Bay, FL 32905 USA

4164 Lakeshore Road, Burlington,
ON, L7L 1A4 Canada

**CRC Press**
6000 Broken Sound Parkway NW,
Suite 300, Boca Raton, FL 33487-2742 USA

2 Park Square, Milton Park,
Abingdon, Oxon, OX14 4RN UK



**Library and Archives Canada Cataloguing in Publication**

Title: Advanced materials, polymers and composites : new research on properties, techniques, and applications / edited by Omari V. Mukbaniani, DSc, Tamara Tatrishvili, DSc, Marc J.M. Abadie, DSc

Names: Mukbaniani, O. V. (Omar V.), editor. | Tatrishvili, Tamara, editor. | Abadie, Marc J. M., editor.

Description: First edition. | Includes bibliographical references and index.

Identifiers: Canadiana (print) 20210101989 | Canadiana (ebook) 20210101997 | ISBN 9781771889513 (hardcover) | ISBN 9781774638200 (softcover) | ISBN 9781003105015 (ebook)

Subjects: LCSH: Polymers. | LCSH: Polymerization. | LCSH: Composite materials. | LCSH: Nanocomposites (Materials)

Classification: LCC TA455.P58 A35 2021 | DDC 620.1/92—dc23

**Library of Congress Cataloging-in-Publication Data**

Names: Mukbaniani, O. V. (Omar V.), editor. | Tatrishvili, Tamara, editor. | Abadie, Marc J. M., editor.

Title: Advanced materials, polymers, and composites : new research on properties, techniques, and applications / Omari V. Mukbaniani, Tamara Tatrishvili, Marc J.M. Abadie.

Description: First edition. | Palm Bay, FL : Apple Academic Press, 2021. | Includes bibliographical references and index. | Summary: "This new book reviews several domains of polymer science, especially new trends in polymerization synthesis, physical-chemical properties, and inorganic systems. Composites and nanocomposites are also covered in this book, emphasizing nanotechnologies and their impact on the enhancement of physical and mechanical properties of these new materials. Kinetics and simulation are discussed and also considered as promising techniques for achieving chemistry and predicting physical property goals. This book presents a selection of interdisciplinary papers on the state of knowledge of each topic under consideration through a combination of overviews and original unpublished research. Advanced Materials, Polymers, and Composites: New Research on Properties, Techniques, and Applications is addressed to all those working in the field of polymers and composites, including academics, institutes, research centers, as well as engineers working in the industry"-- Provided by publisher.

Identifiers: LCCN 2020058327 (print) | LCCN 2020058328 (ebook) | ISBN 9781771889513 (hardcover) | ISBN 9781774638200 (softcover) | ISBN 9781003105015 (ebook)

Subjects: LCSH: Polymerization. | Polymers.

Classification: LCC TP156.P6 A36 2021 (print) | LCC TP156.P6 (ebook) | DDC 547/.28--dc23

LC record available at https://lccn.loc.gov/2020058327

LC ebook record available at https://lccn.loc.gov/2020058328

ISBN: 978-1-77188-951-3 (hbk)
ISBN: 978-1-77463-820-0 (pbk)
ISBN: 978-1-00310-501-5 (ebk)

# About the Editors

**Omari V. Mukbaniani, DSc**

*Professor, Ivane Javakhishvili Tbilisi State University, Faculty of Exact and Natural Sciences, Department of Chemistry; Chair, Macromolecular Chemistry; Director of the Institute of Macromolecular Chemistry and Polymeric Materials at TSU, Tbilisi, Georgia*

Omari Vasilii Mukbaniani, DSc, is a Professor and Director of the Macromolecular Chemistry, Department of Ivane Javakhishvili Tbilisi State University, Tbilisi, Georgia. He is also the Director of the Institute of Macromolecular Chemistry and Polymeric Materials. For several years he was a member of the advisory board of the *Journal Proceedings of Ivane Javakhishvili Tbilisi State University* (Chemical Series), contributing editor of the journals *Polymer News* and the *Polymers Research Journal*. His research interests include polymer chemistry, polymeric materials, and the chemistry of organosilicon compounds. He is an author of more than 480 publications, 27 books, monographs, and 10 inventions. He created in 2007 the International Caucasian Symposium on Polymers and Advanced Materials, which takes place every other two years in Georgia. The next symposium, ICSP&AM7, will be in July 2021.

**Tamara Tatrishvili, DSc**

*Senior Specialist, Unite of Academic Process Management (Faculty of Exact and Natural Sciences), Ivane Javakhishvili Tbilisi State University; Senior Researcher of the Institute of Macromolecular Chemistry and Polymeric Materials at TSU, Georgia*

Tamara Tatrishvili, PhD, is a Senior Specialist at the Unite of Academic Process Management (Faculty of Exact and Natural Sciences) at Ivane Javakhishvili Tbilisi State University as well as a Head of the Department of the Institute of Macromolecular Chemistry and Polymeric Materials in Tbilisi, Georgia.

**Marc J. M. Abadie, DSc**

*Professor Emeritus, Institute Charles Gerhardt of Montpelier—Aggregates, Interfaces, and Materials for Energy (ICGM-AIME, UMR CNRS 5253), University Montpelier, France*

Professor Marc J. M. Abadie is an Emeritus Professor at the University Montpellier, France. He was the head of the Laboratory of Polymer Science and Advanced Organic Materials-LEMP/MAO. He is currently a "Michael Fam" Visiting Professor at the School of Materials Sciences and Engineering, Nanyang Technological University NTU, Singapore. His present activity concerns high-performance polymers for PEMFCs, composites and nanocomposites, UV/EB coatings, and biomaterials. He has published 11 books and 11 patents. He has advised nearly 95 MS and 52 PhD students with whom he has published over 400 papers. He has more than 40 years of experience in polymer science with 10 years in the industry (IBM, USA; Ministry of Defence, UK; and SNPA/Total, France). In the 1980s, he created the International Symposium on Polyimides and High-Temperature Polymers, a.k.a. STEPI, which takes place every other three years in Montpellier, France. The next symposium, STEPI 12, will be in June 2021.

# Contents

# Contributors

**Marc J. M. Abadie**
Institute Charles Gerhardt Montpellier/AIME CNRS, University of Montpellier, France;
"Petru Poni" Institute of Macromolecular Chemistry, Iasi, Romania,
E-mails: marc.abadie@umontpellier.fr; marc@icmpp.ro

**L. Akhalbedashvili**
Ivane Javakhishvili Tbilisi State University, Alexandre Tvalchrelidze Caucasian Institute of Mineral Resources, Mindeli Str., 11, Tbilisi–0186, Georgia, E-mail: aklali@yahoo.com

**Jimsher Aneli**
Institute of Macromolecular Chemistry and Polymeric Materials, Ivane Javakhishvili Tbilisi State University, Ilia Chavchavadze Blvd. 13, Tbilisi–0179, Georgia; R. Dvali Institute of Machine Mechanics, 10 Mindeli Str. Tbilisi–0186, Georgia, E-mail: janeli@yahoo.com

**Ketevan Archvadze**
TSU Petre Melikishvili Institute of Physical and Organic Chemistry 31, A. Politkovskaia Str., Tbilisi–0186, Georgia

**Olena Astakhova**
Lviv Polytechnic National University, 12, S. Bandery St., 79013 Lviv, Ukraine

**M. Atakay**
Hacettepe University, Department of Chemistry, 06800 Ankara, Turkey

**M. Avaliani**
Iv. Javahishvili Tbilisi State University, Raphiel Agladze Institute of Inorganic Chemistry and Electrochemistry, Mindeli Street 11, Tbilisi–0186, Georgia,
E-mails: avaliani21@hotmail.com; marine.avaliani@tsu.ge

**N. Barbakadze**
Laboratory of Problems of Chemical Ecology, Petre Melikishvili Institute of Physical and Organic Chemistry, Iv. Javakhishvili Tbilisi State University, 31, Ana Politkovskaia Str., Tbilisi–0186, Georgia, Phone: (+995 599) 76-53-15, E-mail: chemicalnatia@yahoo.de

**R. Barberi**
CNR-IPCF, UOS Cosenza, Physics Department, University of Calabria, Rende (Cs), Italy

**N. Barnovi**
Iv. Javahishvili Tbilisi State University, Raphiel Agladze Institute of Inorganic Chemistry and Electrochemistry, Mindeli Street 11, Tbilisi–0186, Georgia

**B. G. Bendeliani**
Ilia Vekua Sukhumi Institute of Physics and Technology, Department of Cryogenic Technique and Technologies, Tbilisi–0186, Georgia

**I. Beshkenadze**
Petre Melikishvili Institute of Physical and Organic Chemistry, Ivane Javakhishvili Tbilisi State University, 31 A. Politkovskaia str. Tbilisi 0186 Georgia

**Michael Bratychak**
Lviv Polytechnic National University, 12, S. Bandery St., 79013 Lviv, Ukraine,
E-mail: mbratychak@gmail.com

**M. D. Luigi Bruno**
CNR-IPCF, UOS Cosenza, Physics Department, University of Calabria, Rende (Cs), Italy

**E. Çatiker**
Ordu University, Department of Chemistry, 52200 Ordu, Turkey, E-mail: ecatiker@gmail.com

**V. Chagelishvili**
Iv. Javahishvili Tbilisi State University, Raphiel Agladze Institute of Inorganic Chemistry and Electrochemistry, Mindeli Street 11, Tbilisi–0186, Georgia

**R. Chedia**
Laboratory of Problems of Chemical Ecology, Petre Melikishvili Institute of Physical and Organic Chemistry, Iv. Javakhishvili Tbilisi State University, 31, Ana Politkovskaia Str., Tbilisi–0186, Georgia

**Ia Chitrekashvili**
TSU Petre Melikishvili Institute of Physical and Organic Chemistry 31, A. Politkovskaia Str., Tbilisi–0186, Georgia

**K. Chubinidze**
Faculty of Exact and Natural Sciences, Iv. Javakhishvili Tbilisi State University, I. Chavchavadze 1, Tbilisi, Georgia; Institute of Cybernetics of Georgian Technical University, Tbilisi, Georgia,
E-mail: chubinidzeketino@yahoo.com

**M. Chubinidze**
Tbilisi Medical State University, Tbilisi, Georgia

**E. J. Churgulia**
Department of Chemistry, Sokhumi State University, A. Politkovskaya, St., 12, Tbilisi–0186, Georgia

**I. Danylo**
Donetsk National Technical University, Department of Chemical Technologies, 85300 Pokrovsk, Ukraine

**Sevan P. Davtyan**
National Polytechnic University of Armenia, Department of General Chemistry,
and Chemical Processes, Teryan Str., 105, Yerevan–0009, Armenia

**G. N. Dgebuadze**
Ilia Vekua Sukhumi Institute of Physics and Technology, Department of Cryogenic Technique and Technologies, Tbilisi–0186, Georgia

**Nora Dokhturishvili**
TSU Petre Melikishvili Institute of Physical and Organic Chemistry 31, A. Politkovskaia Str., Tbilisi–0186, Georgia

**M. Donadze**
Faculty of Chemistry and Metallurgy, Georgian Technical University, Kostava Ave., 69, 0171, Tbilisi, Georgia

**T. Dundua**
Department of Agricultural Sciences and Biosystems Engineering, Georgian Technical University, 17, D. Guramishvili Str. Tbilisi–0192, Georgia

**D. Dzidziguri**
Faculty of Exact and Natural Sciences, Iv. Javakhishvili Tbilisi State University, I. Chavchavadze 1, Tbilisi, Georgia

**N. Esakia**
Iv. Javahishvili Tbilisi State University, Institute of Exact and Natural Sciences, Department of Chemistry, 0179, I. Chavchavadze Ave 3, Tbilisi, Georgia

**Cătălin Fetecău**
ReForm UDJG Interdisciplinary Research Platform, Center of Excellence Polymer Processing, "Dunărea de Jos" University of Galaţi, Galaţi, Romania

**M. Gabrichidze**
Faculty of Chemistry and Metallurgy, Georgian Technical University, Kostava Ave., 69, 0171, Tbilisi, Georgia

**V. Gabunia**
Laboratory of Problems of Chemical Ecology, Petre Melikishvili Institute of Physical and Organic Chemistry, Iv. Javakhishvili Tbilisi State University, 31, Ana Politkovskaia Str., Tbilisi–0186, Georgia

**V. M. Gabunia**
Ilia Vekua Sukhumi Institute of Physics and Technology, Department of Cryogenic Technique and Technologies, Tbilisi – 0186, Georgia; Petre Melikishvili Institute of Physical and Organic Chemistry of the Iv. Javakhishvili Tbilisi State University, Jikia Str., 5, Tbilisi–0186, Georgia

**N. Gagniashvili**
Ivane Javakhishvili Tbilisi State University, Alexandre Tvalchrelidze Caucasian Institute of Mineral Resources, Mindeli Str., 11, Tbilisi–0186, Georgia

**R. A. Gakhokidze**
Iv. Javakhishvili Tbilisi State University, Department of Bioorganic Chemistry, Tbilisi, Georgia

**Eter Gavashelidze**
TSU Petre Melikishvili Institute of Physical and Organic Chemistry 31, A. Politkovskaia Str., Tbilisi–0186, Georgia

**I. Gejadze**
Iv. Javakhishvili Tbilisi State University, Al. Tvalchrelidze Caucasian Institute of Mineral Resources, 11 Mindeli Street, Tbilisi–0186, Georgia

**Nazi Gelashvili**
TSU Petre Melikishvili Institute of Physical and Organic Chemistry 31, A. Politkovskaia Str., Tbilisi–0186, Georgia

**Bagrat Godibadze**
Iv. Javakhishvili Tbilisi State University, I. Chavchavadze Ave., 1, 0179 Tbilisi, Georgia

**M. Gogaladze**
Petre Melikishvili Institute of Physical and Organic Chemistry, Ivane Javakhishvili Tbilisi State University, 31 A. Politkovskaia, Tbilisi–0186, Georgia

**L. Gogua**
Tbilisi State Medical University, 33, Vaja-Pshavela Ave.0186, Tbilisi, Georgia

**M. Z. Gorgoshidze**
E. Andronikashvili Institute of Physics, Iv. Javakhishvili Tbilisi State University, Tbilisi, Georgia

**G. K. Grigoryan**
Scientific-Technological Center of Organic and Pharmaceutic Chemistry NAS
Republic of Armenia, 0014, Yerevan, 26 Azatutyan Av., Armenia

**N. H. Grigoryan**
Scientific-Technological Center of Organic and Pharmaceutic Chemistry NAS Republic of Armenia, 0014, Yerevan, 26 Azatutyan Av., Armenia

**K. G. Guliyev**
Institute of Polymer Materials of Azerbaijan National Academy of Sciences, Az5004, Sumgait, S. Vurgun Str.124, Azerbaijan, E-mail: ipoma@science.az, kazim_pm@mail.ru

**Ts. D. Gulverdashvili**
Azerbaijan Medical University, Biophysical, and Bioorganic Chemistry (Sub)Department, Az1022, Baku, Bakikhanov Str., 23, Azerbaijan

**Marina Gurgenishvili**
TSU Petre Melikishvili Institute of Physical and Organic Chemistry 31, A. Politkovskaia Str., Tbilisi–0186, Georgia, E-mail: marina.gurgenishvili@yahoo.com

**A. Heimowska**
Department of Industrial Commodity Science and Chemistry, Faculty of Entrepreneurship and Quality Science, Gdynia Maritime University, 83 Morska Str., 81-225 Gdynia, Poland

**A. A. Hovhannisyan**
Scientific-Technological Center of Organic and Pharmaceutic Chemistry NAS Republic of Armenia, 0014, Yerevan, 26 Azatutyan Av., Armenia, E-mail: hovarnos@gmail.com

**A. M. Imangazy**
JSC "Institute of Chemical Sciences after A.B. Bekturov," Almaty, the Republic of Kazakhstan

**S. Jalaghania**
Ivane Javakhishvili Tbilisi State University, Alexandre Tvalchrelidze Caucasian Institute of Mineral Resources, Mindeli Str., 11, Tbilisi–0186, Georgia

**N. Janashvili**
Ivane Javakhishvili Tbilisi State University, Alexandre Tvalchrelidze Caucasian Institute of Mineral Resources, Mindeli Str., 11, Tbilisi–0186, Georgia

**Helena Janik**
Gdansk University of Technology, Chemical Faculty, Polymer Technology Department, Gdansk, Poland, E-mail: helena.janik@pg.edu.pl

**M. Japaridze**
Laboratory of Problems of Chemical Ecology, Petre Melikishvili Institute of Physical and Organic Chemistry, Iv. Javakhishvili Tbilisi State University, 31, Ana Politkovskaia Str., Tbilisi–0186, Georgia

**M. Jastrzębska**
Department of Industrial Commodity Science and Chemistry, Faculty of Entrepreneurship and Quality Science, Gdynia Maritime University, 83 Morska Str., 81–225 Gdynia, Poland,
E-mail: m.jastrzebska@wpit.umg.edu.pl

**T. K. Jumadilov**
JSC "Institute of Chemical Sciences after A.B. Bekturov," Almaty, the Republic of Kazakhstan,
E-mail: jumadilov@mail.ru

**Temur Kantaria**
Institute of Chemistry and Molecular Engineering, Agricultural University of Georgia, Tbilisi, Georgia

**Ramaz Katsarava**
Institute of Chemistry and Molecular Engineering, Agricultural University of Georgia, Tbilisi, Georgia,
E-mail: r.katsarava@agruni.edu.ge

**V. Kaulin**
Donetsk National Technical University, Department of Chemical Technologies, 85300 Pokrovsk, Ukraine

**Armenuhi G. Ketyan**
National Polytechnic University of Armenia, Department of General Chemistry, and Chemical Processes, Teryan Str., 105, Yerevan–0009, Armenia

**N. Klarjeishvili**
Petre Melikishvili Institute of Physical and Organic Chemistry, Ivane Javakhishvili Tbilisi State University, 31 A. Politkovskaia str., Tbilisi 0186, Georgia

**R. G. Kondaurov**
JSC "Institute of Chemical Sciences after A.B. Bekturov," Almaty, the Republic of Kazakhstan

**Alicja Kosakowska**
The Institute of Oceanology of the Polish Academy of Sciences, Sopot, Poland

**Kaloian Koynov**
Max Planck Institute for Polymer Research, Department of Physics of Interfaces, Ackermannweg 10, D-55128 Mainz, Germany

**K. Krasowska**
Department of Industrial Commodity Science and Chemistry, Faculty of Entrepreneurship and Quality Science, Gdynia Maritime University, 83 Morska Str., 81-225 Gdynia, Poland

**I. Krutko**
Donetsk National Technical University, Department of Chemical Technologies, 85300 Pokrovsk, Ukraine, E-mail: poshukdoc@gmail.com

**Justyna Kucińska-Lipka**
Gdansk University of Technology, Chemical Faculty, Polymer Technology Department, Gdansk, Poland

**Nino Kupatadze**
Institute of Chemistry and Molecular Engineering, Agricultural University of Georgia, Tbilisi, Georgia

**R. Kvatashidze**
Ivane Javakhishvili Tbilisi State University, Alexandre Tvalchrelidze Caucasian Institute of Mineral Resources, Mindeli Str., 11, Tbilisi–0186, Georgia

**Riva Liparteliani**
TSU Petre Melikishvili Institute of Physical and Organic Chemistry 31, A. Politkovskaia Str., Tbilisi–0186, Georgia

**T. E. Lobzhanidze**
Ivane Javakhishvili Tbilisi State University, Department of Chemistry, Faculty of Exact and Natural Sciences, 0179 Tbilisi, Georgia

**O. Lomtadze**
Petre Melikishvili Institute of Physical and Organic Chemistry, Ivane Javakhishvili Tbilisi State University, 31 A. Politkovskaia str. Tbilisi 0186, Georgia

**N. Loria**
Ivane Javakhishvili Tbilisi State University, Alexandre Tvalchrelidze Caucasian Institute of Mineral Resources, Mindeli Str., 11, Tbilisi–0186, Georgia

**V. Maisuradze**
Iv. Javakhishvili Tbilisi State University, Al. Tvalchrelidze Caucasian Institute of Mineral Resources, 11 Mindeli Street, Tbilisi–0186, Georgia

**N. Makhaldiani**
Faculty of Chemistry and Metallurgy, Georgian Technical University, Kostava Ave., 69, 0171, Tbilisi, Georgia, E-mail: makhaldianinino@gmail.com

**T. Makharadze**
Ivane Javakhishvili Tbilisi State University, Tbilisi, Georgia, E-mail: makharadze_tako@yahoo.com

**Grigor Mamniashvili**
Andronikashvili Institute of Physics of Ivane Javakhishvili Tbilisi State University, 6, Tamarashvili str., 0177, Tbilisi, Georgia; G. Tsulukidze Mining Institute, 7 E. Mindeli St. Tbilisi–0186, Georgia

**Iulian Manole**
ReForm UDJG Interdisciplinary Research Platform, Center of Excellence Polymer Processing, "Dunărea de Jos" University of Galați, Galați, Romania

**E. Markarashvili**
I. Javakhishvili Tbilisi State University, I. Chavchavadze Ave., 1 Tbilisi–0127, Georgia; Institute of Macromolecular Chemistry and Polymeric Materials, Ivane Javakhishvili University, Ilia Chavchavadze Blvd. 13, Tbilisi–0179, Georgia

**I. R. Metskhvarishvili**
Ilia Vekua Sukhumi Institute of Physics and Technology, Department of Cryogenic Technique and Technologies, Tbilisi–0186, Georgia; Georgian Technical University, Faculty of Informatics and Control Systems, 0175 Tbilisi, Georgia

**M. R. Metskhvarishvili**
Georgian Technical University, Faculty of Informatics and Control Systems, 0175 Tbilisi, Georgia

**E. Meyvaci**
Giresun University, Department of Chemistry, 28200 Giresun, Turkey

**Aram H. Minasyan**
National Polytechnic University of Armenia, Department of General Chemistry, and Chemical Processes, Teryan Str., 105, Yerevan–0009, Armenia

**Anna Mkrtchyan**
Institute of Pharmacy, Yerevan State University, Yerevan, Republic of Armenia

**Omar Mukbaniani**
Faculty of Exact and Natural Sciences, Iv. Javakhishvili Tbilisi State University, I. Chavchavadze Ave., 1 Tbilisi – 0127, Georgia; Institute of Macromolecular Chemistry and Polymeric Materials, Ivane Javakhishvili University, Ilia Chavchavadze Blvd. 13, Tbilisi–0179, Georgia, E-mail: omarimu@yahoo.com

**L. Nadaraia**
Department of Agricultural Sciences and Biosystems Engineering, Georgian Technical University, 17, D. Guramishvili Str. Tbilisi–0192, Georgia

**Levan Nadareishvili**
V. Chavchanidze Institute of Cybernetics, Georgian Technical University, Georgia

**A. G. Nadaryan**
Scientific-Technological Center of Organic and Pharmaceutic Chemistry NAS Republic of Armenia, 0014, Yerevan, 26 Azatutyan Av., Armenia

**M. Nadirashvili**
Iv. Javakhishvili Tbilisi State University, Al. Tvalchrelidze Caucasian Institute of Mineral Resources, 11 Mindeli Street, Tbilisi–0186, Georgia

**N. Nonikashvili**
Laboratory of Problems of Chemical Ecology, Petre Melikishvili Institute of Physical and Organic Chemistry, Iv. Javakhishvili Tbilisi State University, 31, Ana Politkovskaia Str., Tbilisi–0186, Georgia

**M. O. Nutsubidze**
Iv. Javakhishvili State University, Department of Chemistry, I. Chavchavadze Ave., 1, 0179, Tbilisi, Georgia

**Giuli Otinashvili**
Institute of Chemistry and Molecular Engineering, Agricultural University of Georgia, Tbilisi, Georgia

**T. Öztürk**
Giresun University, Department of Chemistry, 28200 Giresun, Turkey

**Givi Papava**
TSU Petre Melikishvili Institute of Physical and Organic Chemistry 31, A. Politkovskaia Str., Tbilisi–0186, Georgia

**Akaki Peikrishvili**
F. Tavadze Institute of Metallurgy and Materials Science, 10 E. Mindeli St. Tbilisi–0186, Georgia, E-mail: akaki.peikrishvili@yahoo.com

**A. Petriashvili**
Tbilisi Medical State University, Tbilisi, Georgia

**G. Petriashvili**
Institute of Cybernetics of Georgian Technical University, Tbilisi, Georgia

**T. Petriashvili**
Iv. Javakhishvili Tbilisi State University, Al. Tvalchrelidze Caucasian Institute of Mineral Resources, 11 Mindeli Street, Tbilisi–0186, Georgia

**Sergey Poghosyan**
Institute of Pharmacy, Yerevan State University, Yerevan, Republic of Armenia

**M. Rutkowska**
Department of Industrial Commodity Science and Chemistry, Faculty of Entrepreneurship and Quality Science, Gdynia Maritime University, 83 Morska Str., 81–225 Gdynia, Poland

**A. I. Sadygova**
Azerbaijan Medical University, Biophysical, and Bioorganic Chemistry (Sub)Department, Az1022, Baku, Bakikhanov Str., 23, Azerbaijan

**Ashot Saghyan**
Institute of Pharmacy, Yerevan State University, Yerevan, Republic of Armenia

**B. Salih**
Hacettepe University, Department of Chemistry, 06800 Ankara, Turkey

**M. P. De Santo**
CNR-IPCF, UOS Cosenza, Physics Department, University of Calabria, Rende (Cs), Italy

**K. Satsyuk**
Donetsk National Technical University, Department of Chemical Technologies, 85300 Pokrovsk, Ukraine

**B. Savaş**
Kafkas University, Kars Vocational School, 36100 Kars, Turkey

**Rita Shahnazarli**
Institute of Polymer Materials of Azerbaijan National Academy of Sciences, S. Vurgun Str., 124, AZ5004, Azerbaijan, E-mail: shahnazarli@mail.ru

**E. Shapakidze**
Iv. Javakhishvili Tbilisi State University, Alexander Tvalchrelidze Caucasian Institute of Mineral Resources, Mindeli Street 11, Tbilisi–0186, Georgia, E-mail: elena.shapakidze@tsu.ge

**Alexander Shengelaya**
G. Tsulukidze Mining Institute, 7 E. Mindeli St., Tbilisi–0186, Georgia

**N. G. Shengelia**
Department of Chemistry, Sokhumi State University, A. Politkovskaya, St., 12, Tbilisi–0186, Georgia

**Mariia Shved**
Lviv Polytechnic National University, 12, S. Bandery St., 79013 Lviv, Ukraine

**V. V. Shvelidze**
Department of Physics, Tbilisi State University, I. Chavchavadze Ave., 1, 0179 Tbilisi, Georgia

**Olena Shyshchak**
Lviv Polytechnic National University, 12, S. Bandery St., 79013 Lviv, Ukraine

**N. N. Sidamonidze**
Iv. Javakhishvili State University, Department of Chemistry, I. Chavchavadze Ave., 1, 0179, Tbilisi, Georgia, E-mail: neli.sidamonidze@tsu.ge

**Maciej Sienkiewicz**
Gdansk University of Technology, Chemical Faculty, Polymer Technology Department, Gdansk, Poland

**L. V. Tabatadze**
Department of Chemistry, Sokhumi State University, A. Politkovskaya, St., 12, Tbilisi–0186, Georgia

**D. B. Tagiyev**
Azerbaijan Medical University, Biophysical, and Bioorganic Chemistry (Sub)Department, Az1022, Baku, Bakikhanov Str. 23, Azerbaijan

**Tamara Tatrishvili**
Faculty of Exact and Natural Sciences, Iv. Javakhishvili Tbilisi State University, I. ChavchavadzeAve., 1 Tbilisi–0127, Georgia; Institute of Macromolecular Chemistry and Polymeric Materials, Ivane Javakhishvili University, Ilia Chavchavadze Blvd. 13, Tbilisi–0179, Georgia, E-mail:omarimu@yahoo.com

**Giorgi Tavadze**
F. Tavadze Institute of Metallurgy and Materials Science, 10 E. Mindeli St., Tbilisi–0186, Georgia

**G. Todradze**
Ivane Javakhishvili Tbilisi State University, Alexandre Tvalchrelidze Caucasian Institute of Mineral Resources, Mindeli Str., 11, Tbilisi–0186, Georgia

**P. L. Toidze**
Georgian Technical University, Department of Chemical and Biological Technologies, Tbilisi, Georgia

**Anahit O. Tonoyan**
National Polytechnic University of Armenia, Department of General Chemistry, and Chemical Processes, Teryan Str., 105, Yerevan–0009, Armenia, E-mail: atonoyan@mail.ru

**David Tugushi**
Institute of Chemistry and Molecular Engineering, Agricultural University of Georgia, Tbilisi, Georgia

**V. Ugrekhelidze**
Department of Agricultural Sciences and Biosystems Engineering, Georgian Technical University, 17, D. Guramishvili Str. Tbilisi–0192, Georgia

**Anahit Z. Varderesyan**
National Polytechnic University of Armenia, Department of General Chemistry, and Chemical Processes, Teryan Str., 105, Yerevan–0009, Armenia

**R. O. Vardiashvili**
Iv. Javakhishvili State University, Department of Chemistry, I. Chavchavadze Ave., 1, 0179, Tbilisi, Georgia

**T. S. Vashakmadze**
Iv. Javakhishvili Tbilisi State University, Vekua Institute of Applied Mathematics, Tbilisi, Georgia

**N. S. Vassilieva-Vashakmadze**
Georgian Academy of Engineering, Tbilisi, Georgia, E-mail: nonavas@rambler.ru

**K. Yavir**
Donetsk National Technical University, Department of Chemical Technologies, 85300 Pokrovsk, Ukraine

**Nino Zavradashvili**
Institute of Chemistry and Molecular Engineering, Agricultural University of Georgia, Tbilisi, Georgia

# Abbreviations

| | |
|---|---|
| AA | α-amino acid |
| AAm | acrylamide |
| AIBN | α,α'-azoisobutyronitrile |
| AlG | allylglycine |
| AlG12 | bis-(allyl-glycine)-1,12-dodecylen |
| AlG6 | bis-(allyl-glycine)-1,6-hexylen |
| a-PHB | amorphous poly[(R,S)-3-hydroxybutyrate] |
| BAPO | bis acyl phosphine oxide |
| BCP | block-copolymers |
| BPM | bitumen-polymer mixtures |
| BS | bentonite special |
| C | coal |
| C/O | carbon to oxygen |
| CIF | coumarone-indene fraction |
| CIR | coumarone-indene resins |
| CIRC | coumarone-indene resins with carboxy groups |
| CIRM | coumarone-indene resins with methacrylic fragments |
| CLCs | cholesteric liquid crystals |
| CMCP | 2-chloromethyl-1-(p-vinyl phenyl)cyclopropane |
| CPs | cationic polymers |
| CTP | coal tar pitch |
| CVD | chemical vapor deposition |
| DCP | dicumyl peroxide |
| DHB | dihydroxybenzoic acid |
| DLS | dynamic light scattering |
| DMA | dimethylacetamide |
| DMF | dimethylformamide |
| DPC | differential photocalorimeter |
| DSC | differential scanning calorimetry |
| EB | electron beam |
| ER | epoxy resin |
| ESR | electron spin resonance |
| FA | fulvic acids |
| FC | field-cooled |

| | |
|---|---|
| FD | fluorescent dye |
| FGM | functional gradient materials |
| FP | frontal polymerization |
| FTIR | Fourier transform infrared spectroscopy |
| GMA | glycidyl methacrylate |
| GO | graphene oxide |
| GPC | gel-permeation chromatography |
| HAP | hydroxyacetophenone |
| HDPE | high density polyethylene |
| HLB | hydrophilic-lipophilic balance |
| HPs | hybrid-polymers |
| HTP | hydrogen transfer polymerization |
| LB | Langmuir-Blodgett |
| LDPE | low-density polyethylene |
| LFCT | light fraction of coal tar |
| MA | maleic anhydride |
| MAA | methacrylic acid |
| MB | masterbatch |
| MC | merocyanine |
| MEK | methyl ethyl ketone |
| MEVA | maleized ethylene-vinyl acetate |
| MF | membrane filtration |
| MFR | melt flow rate |
| MM | molecular mechanics |
| MMA | methyl methacrylate |
| MOR | mordenite |
| MotA | motility protein A |
| MotB | motility protein B |
| MS-Ring | symmetry mismatch |
| $Na_2CO_3$ | sodium carbonate |
| $NaN_3$ | sodium azide |
| NaOH | sodium alkali |
| NIR | near-infrared light |
| NPAAs | non-proteinogenic amino acids |
| NPs | nanoparticles |
| OPC | ordinary Portland cement |
| poly(MBA-b-MMA) | poly(β-methyl β-alanine-b-methyl methacrylate) |
| PBS | poly(butylene succinate) |
| PCL | poly(ε-caprolactone) |

| | |
|---|---|
| PE | polyethylene |
| PEAs | poly(ester amides) |
| PECH | poly(epichlorohydrin) |
| PEGs | polyethylenglycols |
| PGF | powdered graphite foil |
| pGFW | powdered graphite foil wastes |
| PHA | polyhydroxyalkanoates |
| PHB | poly(3-hydroxybutyrate) |
| PHB/V | poly(b-hydroxybutyrate/valerate) |
| PI | photo initiator |
| PLA | poly(D,L-lactide) |
| PLLA | poly(L-lactide) |
| PMBA-diBr | poly(β-methyl β-alanine) |
| PMFCs | polymer matrix fibrous composites |
| PMHS | polymethylhydrosiloxane |
| PMMA | polymethyl methacrylate |
| PNC | polymer nanocomposites |
| PP | polypropylene |
| PRs | petroleum resins |
| PVA | polyvinyl alcohol |
| PVB | polyvinylbutyral |
| PVC | polyvinyl chloride |
| RAFT | reversible addition-fragmentation chain transfer |
| REMs | rare earth metals |
| rGO | reduced graphene oxide |
| ROP | ring-opening polymerization |
| SAcP | solution active polycondensation method |
| SEI | secondary electron images |
| SEM | scanning electron microscopy |
| SPS | spark-plasma sintering technology |
| SRB | selective reflection band |
| SW | seawater |
| Td | decomposition temperatures |
| TEOS | tetraethoxysilane |
| TG/DTA | thermogravimetric and differential thermal analyzer |
| Tg | glass-transition temperatures |
| TGA | thermogravimetric analysis |
| THF | tetrahydrofuran |
| TMPTA | trimethylolpropane triacrylate |

| | |
|---|---|
| TPS | thermoplastic starch |
| UV | ultraviolet |
| UVT | ultraviolet light |
| VOC | volatile organic compounds |
| VOCP | vinyloxycyclopropanes |
| VSM | vibrating sample magnetometer |
| XRD | x-ray diffraction |
| ZFC | zero-field-cooled |

# Preface

Polymers, a word that we hear about it a lot, are very vital, and one cannot imagine life without them. Polymers, a large class of materials, consist of many small molecules, named monomers, that are linked together to form long chains and are used in a lot of products and goods that we use in daily life.

There were relatively few materials available for the manufacture of the article needed for a civilized life. Steel, glass, wood, stone, brick, and concrete for most of the construction, and cotton, wood, jute, and a few other agricultural products for clothing or fabric manufacture were used.

The rapid increase in demand for manufactured products introduces new materials. These new materials are polymers, and their impact on the present way of life is almost incalculable. Products made from polymers are all around us: clothing made from synthetic fibers, polyethylene (PE) cups, fiberglass, nylon bearings, plastic bags, polymer-based paints, epoxy glue, polyurethane foam cushion, silicone heart valves, and Teflon-coated cookware.

In modern technologies, new organic materials and tools have been developed to fulfill the strong demand of innovative chemical structures. For the last three decades, increasing need in the high technology industries (space, micro, and nanoelectronics, membranes, fuel cells, etc.) has been the driving force for the development of new polymeric systems that combine thermal stability with specific functional properties, and also others such as lightweight, high corrosion resistance, good wear properties, dimensional stability, low flammability, separation properties, moisture resistance, insulating properties, and ability to be transformed with conventional equipment.

This book reviews several domains of polymer science, especially new trends in polymerization synthesis, physical-chemical properties, and inorganic systems. Composites and nanocomposites are also covered in this book, emphasizing nanotechnologies and their impact on the enhancement of physical and mechanical properties of these new materials. Kinetics and simulation are discussed and also considered as promising techniques for achieving chemistry and predicting physical property goals. This book presents interdisciplinary papers on the state of knowledge of each topic under consideration through a combination of overviews and original unpublished research.

This book is addressed to all those working in the field of polymers and composites, i.e., academics, institutes, research centers, as well as engineers working in the industry.

**—Omar Mukbaniani, DSc**
*Ivane Javakhishvili Tbilisi State University,*
*Faculty of Exact and Natural Sciences,*
*Institute of Chemistry, Department of Macromolecular Chemistry,*
*Director of the Institute of Macromolecular Chemistry and*
*Polymeric Materials at TSU, Georgia*

**—Tamara Tatrishvili, DSc**
*Ivane Javakhishvili Tbilisi State University,*
*Faculty of Exact and Natural Sciences,*
*Institute of Chemistry, Senior Specialist at the Office of*
*Academic Process Management;*
*Head of the Department of the Institute of Macromolecular*
*Chemistry and Polymeric Materials at TSU, Georgia*

**—Marc J. M. Abadie, DSc**
*Professor Emeritus, Doctor Honoris Causa*
*Institut Charles Gerhardt de Montpellier-Agrégats, Interfaces et*
*Matériaux pour l'Energie (IGCM AIME UMR CNRS 5253)*
*STEPI General Chairman, Expert près la Cour d'Appel*
"Michael Fam" Visiting Professor @ NTU/MSE, Singapore

# PART I
# Composites and Nanomaterials

## CHAPTER 1

# Correlation Between Chemical Structure and Photoreactivity in UV Curing Formulation

MARC J. M. ABADIE,[1,2] IULIAN MANOLE,[3] and CĂTĂLIN FETECĂU[3]

[1]*Institute Charles Gerhardt Montpellier/AIME CNRS, University of Montpellier, France, E-mails: marc.abadie@umontpellier.fr; marc@icmpp.ro*

[2]*"Petru Poni" Institute of Macromolecular Chemistry, Iasi, Romania*

[3]*ReForm UDJG Interdisciplinary Research Platform, Center of Excellence Polymer Processing, "Dunărea de Jos" University of Galați, Galați, Romania*

### ABSTRACT

Ultraviolet (UV) curing is a light-induced polymerization of multifunctional monomers/oligomers. It is a very eco-efficient and energy saving crosslinking method that has been extensively used as a coating for wood, paper, walls, and thin-film technology in microelectronic via a photolithography process.

To get a 3D network, most of the formulation uses multifunctional monomers/oligomers or a mixture of them that crosslink under exposure to UV radiations in a free radical or anionic process, depending on the chemical functionality engaged in the monomers/oligomers formulation.

According to the formulation used, *Mc* and crosslink density *Xc* are determined and correlated to the photoreactivity of the system considered. Some examples are given covering cationic UV curing and radical thermal curing systems as well.

## 1.1 INTRODUCTION

Photochemistry through ultraviolet (UV) or electron beam (EB) radiation has been used for more than 40 years, specifically for coatings, adhesives, inks, electronics, and 3D printing in recent years [1]. Fast curing (usually less than 1 second), excellent film properties, and essentially nonvolatile organic compounds (VOC) are the major benefits among many other advantages compared with conventional solvent-borne thermal curing process, which is under increasing pressure from regulatory agencies to limit the number of solvents emitted into the environment.

In the curing process, UV is generally used to cure coatings 200–500 μm thick. For thicker composite structures such as polymer matrix fibrous composites (PMFCs) used in aerospace and automotive applications, EB is a more suitable technique for curing parts up to 10–20 cm thick, depending on the power of the EB [2]. In recent years, major advances have been made in raw materials and equipment designs that make this unique technology available for more and new industrial applications.

In this chapter, we analyze some formulation based on acrylates, epoxies, and/or vinyl ethers that have been published [3], their respective photosensitivity, and photoreactivity [4] compared to *Mc* and the crosslink density *Xc* of the formulations considered.

## 1.2 UV CURING SYSTEM

### *1.2.1 UV BASIC*

The electromagnetic spectrum (Figure 1.1.) goes from low energy (radio waves) to high energy (electron beam). According to the Planck's equation each radiation is associated to energy: $E = AN.hc/\lambda = 119.705 \times 10^{-6}/\lambda\ kJ/mol$. Among UV rays, only visible rays [400 nm (299 kJ/mol) $\leq \lambda \geq$ 700 nm (171 kJ/mol)] and UVA rays [315 nm (380 kJ/mol) $\leq \lambda \geq$ 400 nm (299 kJ/mol)] require the presence of a photoinitiator to initiate polymerization. Indeed, the energy developed by the radiation is not sufficient to break the C-C or C-O bonds ($\approx$ 350 kJ/mol). However, in the case of UVB [280 nm (427 kJ/mol) $\leq \lambda \geq$ 315 nm (380 kJ/mol)] and UVC [100 nm (1196 kJ/mol) $\leq \lambda \geq$ 280 nm (427 kJ/mol)] and especially for lower wavelengths such as EB, the energy of the radiation developed is higher than the bond energies and therefore the resin can be cured without the presence of a photo initiator.

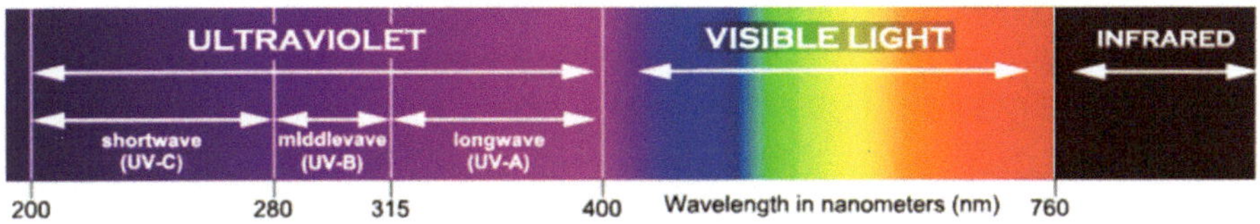


**FIGURE 1.1** The electromagnetic spectrum.

In the polymerization of the double bond, only the π bond is a concern in the propagation reaction. The energy to cut C=C corresponds to π and σ bonds equivalent to $E_{C=C}$ = 615 kJ/mol. To break the σ bond C-C, the energy needed is $E_{C\text{-}C}$ = 347 kJ/mol. Therefore, the energy to provide for opening the π bond will be: (615–347) kJ/mol = 268 kJ/mol, corresponding to a wavelength λ = 446.7 nm. In other words, for λ < 446.7 nm, we need to use a photoinitiator.

Using the Hesse law *vz.* "the bond energy of a reaction ΔH is the total number of bonds broken minus the total number of bonds formed":

$$\Delta H = \Sigma\Delta H_{(\text{bonds broken})} - \Sigma\Delta H_{(\text{bonds formed})} = n(C = C) - 2n(\text{-C-C-})$$
$$= \Delta H = (615 - 2 \times 347)\ \text{kJ/mol} = 79\ \text{kJ/mol}$$

that corresponds to a wavelength λ < 1515.3 nm. According to Hesse law, even in the visible, we could make a polymerization without any photoinitiator, which makes no sense if we consider the time depending. For example, if we leave styrene in the sunlight for a year, it will polymerize alone without a photoinitiator.

### 1.2.2 *RADICAL PHOTOINITIATOR*

A photosensitive formulation for UV curing contains photoinitiator, monomer/oligomer multifunctional, and some additives [4]:

1. Photoinitiator for radical polymerization (acrylates, methacrylates) is classified as [5]:

   **i. Cleavage (Type 1):** HAP (hydroxyacetophenone).

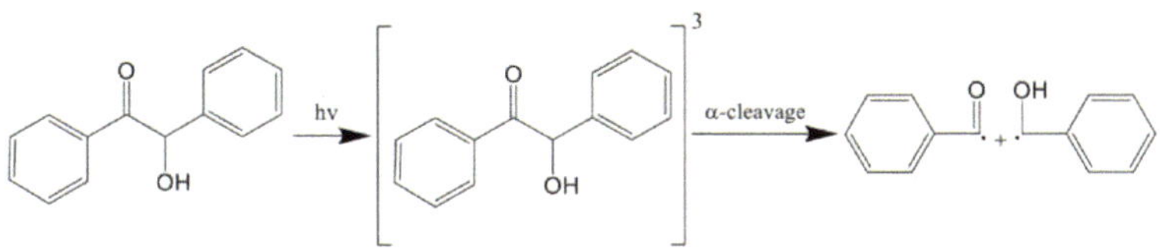


Examples of photoinitiators class 1 are Irgacure® 651, Darocur®1173 (liquid), or bis-acyl phosphine oxide (BAPO) [a mixture of 75 wt.% Darocur®1173 and 25 wt.% BAPO (liquid)], to cite some of them.

ii. **H-Abstraction (Type 2):** BP + $RNH_2$ (benzophenone + amine).

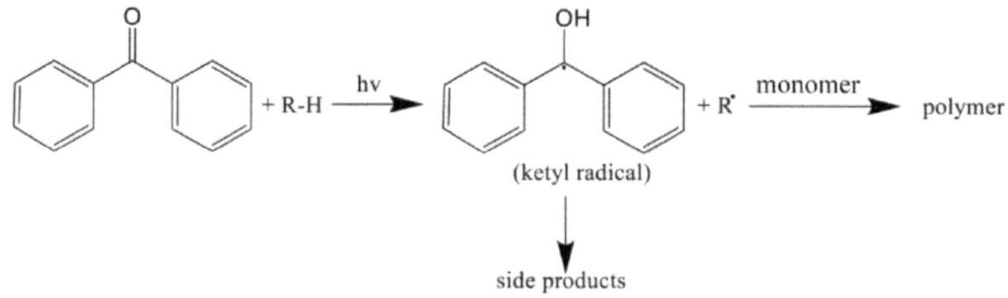


### 1.2.3 *CATIONIC PHOTO INITIATOR*

2. Photoinitiators for cationic polymerization (epoxy and vinyl ether resins), upon exposure to UV light, produce strong acid [6, 7] according to the following scheme:

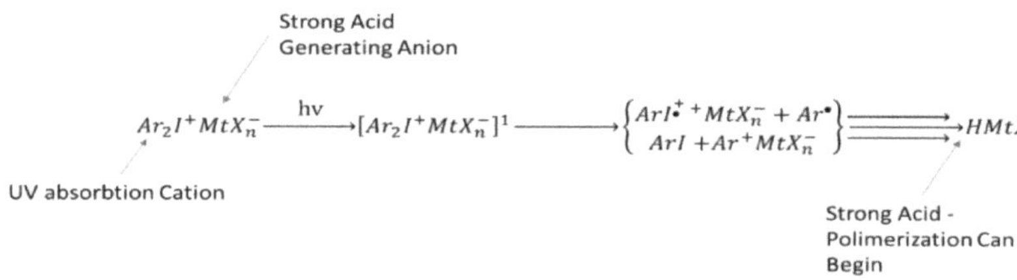


Epoxy polymerize by cationic process [8], initiated by a proton H+ (HMtXn). Sulfonium or iodonium salts exposed to UV produce not only strong acid but also radicals, making the polymerization of acrylates (meth-acrylates) also possible.

The photoinitiator (PI) Cyracure® UVI-6976, largely used, consist of a mixture 40/60 of hexafluoroantimonate of S,S,S,'S'-tetraphenylthiobis (4,1-phenylene) disulfonium and hexafluoroantimonate of diphenyl (4-phenylthiophenyl) sulfonium (CAS no. 89452-32-9 and 71449-78-0) at 50% by weight in propylene carbonate (Figure 1.2).

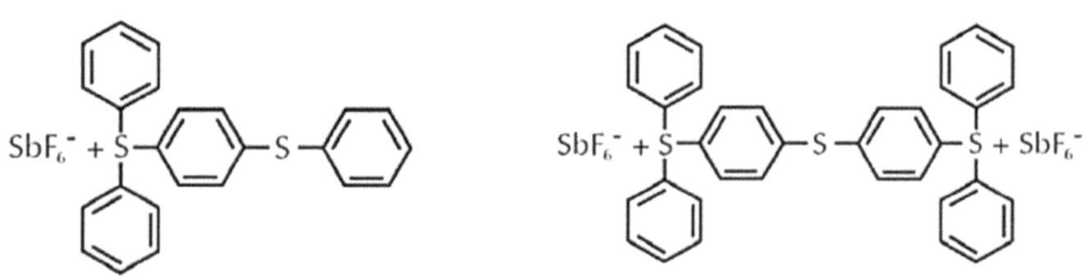

**FIGURE 1.2** Cyracure® UVI 6976.

Note that recently a new cationic photoinitiator has been proposed *viz.* iodonium and sulfonium salt-containing tetrakis (perfluorotbutyloxy) aluminate anion sulfonium/iodonium salt (Figure 1.3) [9, 10].

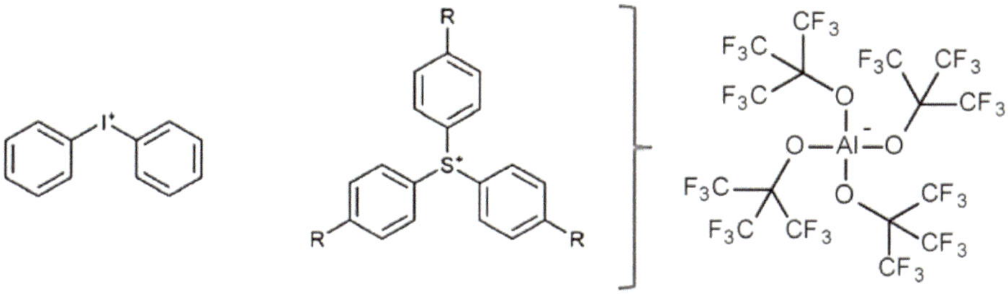

**FIGURE 1.3** Tetrakis aluminate anion.

It should be noted that, while radical polymerizations are sensitive to oxygen (chain peroxidation reaction leading to degradation of the polymer), this is not the case for cationic polymerizations, however moisture-sensitive. This remark is important because it does not require working in the absence of oxygen when using epoxy resins (ERs), but the environment should be dry. Using additive with chlorinated anhydride may improve the UV irradiation stability of the ER. The photopolymerization process is represented in Figure 1.4.

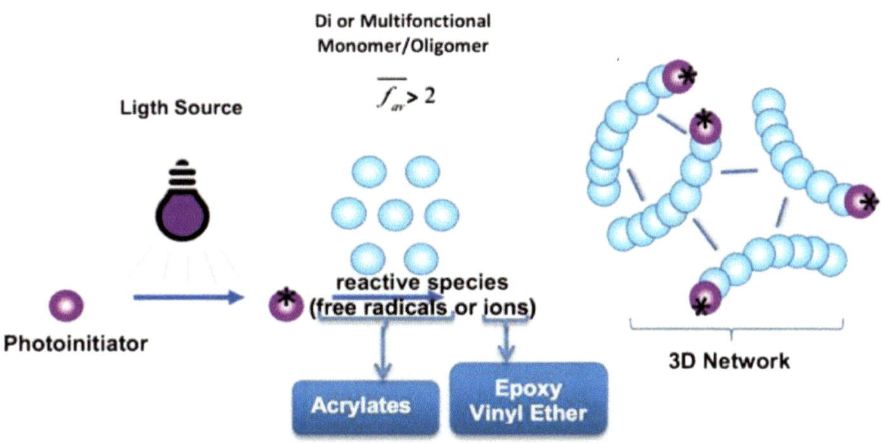


**FIGURE 1.4** Scheme of photopolymerization.

## 1.3 CALCULATION OF *MC* AND *XC*

### *1.3.1 AVERAGE FUNCTIONALITY F*

The main group of polymers belongs to the class of *thermoplastics* and *fibers,* which are linear or branched, uncrosslinked polymers, semi-crystalline, or amorphous, and that can be processed into different shapes by thermal treatment, can be remolded, and are recyclable. Besides other classification, one classification of the polymer types can be done by the increasing degree of crosslinking, whereof the *elastomers* are slightly or moderated crosslinked, occurring very often as rubber-like flexible materials, and *thermosets* or *duromers*, which are highly crosslinked with the formation of the 3D network, amorphous, cannot be remolded and recycled, and available as hard materials.

- Since the networks consist of infinite molecular mass molecules (*Mw* approaches infinity, whereas *Mn* stays finite), virtually insoluble, the characterization of crosslinked polymers cannot be done by molecular mass determination, but rather by dynamic thermal-mechanical analysis technique or by crosslink density and molecular mass between X-links.
- DTMA applies an oscillatory force at a set frequency to the sample and reports changes in stiffness and damping. Young modulus $E$ is determined by deformation under stress [Elastic (Storage) Modulus $E'$ and Viscous (Lost) Modulus $E''$] whereas rigidity and shear modulus G by deformation under shearing stress [$G'$ & $G''$].

The functionality of the monomer $f$ used is defined as the ratio of the number of bonds formed to the number of monomer molecules used during the polymerization process:

$$f = \frac{Number\ of\ Link\ Formed\ per\ Monomer}{Number\ of\ Molecule\ Monomer}$$

If the monomer is monofunctional (having one double bond), the functionality $f = 2$, and the structure of the polymer obtained are linear, therefore a *thermoplastic*. In the case of monomer difunctional, the functionality $f = 4$, or multifunctional, the structure of the polymer is crosslinked, $f > 2$, therefore a *thermoset.*

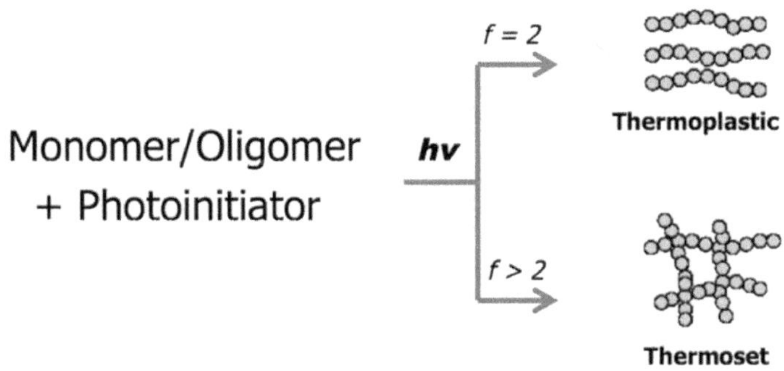


The calculation of the average functionality $f$ of the monomer makes it possible to determine to which category the polymer obtained after polymerization belongs. The average functionality $f_0$ of a mixture of $n_i$ monomer of $f_i$ functionality is given by:

$$f_0 = \frac{n_1 f_1 + n_2 f_2 + \dots\dots + n_i f_i}{n_1 + n_2 + \dots\dots + n_i}$$

For example for monofunctional monomer (acrylate, styrene, epoxy, vinyl ether) $f = 2$ (two possibilities to form a bond), for difunctional monomer (acrylate, styrene, epoxy, vinyl ether) $f = 4$ (four directions to form a link), for trifunctional monomer $f = 6$, etc.

### 1.3.2 MC AND DEGREE OF CROSSLINK XC

The crosslinking density is expressed by $v$ (mol/g-$v \geq 0$) or by the degree of crosslinking $Xc$-$0 \leq Xc \leq 1$. If several components are used, the molar mass $M$ is calculated by the medium molecular mass $M_0$:

$$M_0 = \frac{n_1 M_1 + n_2 M_2 + \dots\dots + n_i M_i}{n_1 + n_2 + \dots\dots + n_i}$$

Resulting in *Mc vz.* molar mass between two crosslinks (thee directions or more) [11]:

$$M_c = \frac{M_0}{f_0 - 2} \quad M_c = \frac{M_0}{p f_0 - 2}$$

where; $p$ is the fraction of double bonds converted ($p \leq 1$).

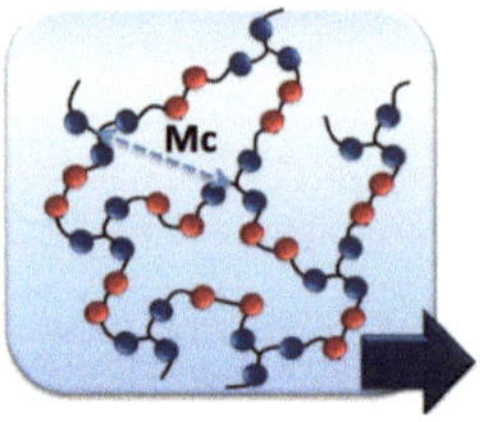


$f_{av} = f_0 = 2 \rightarrow$ *Mc is infinite*
$\rightarrow$ *The polymer is linear*

Then the resulting in a crosslink density *Xc* of:

$$X_c = \frac{1}{M_c}$$

The calculation of the network quantities can also be done according to the theories developed by Macosco and Miller [12]. The prerequisites of this theory are:

- Equal reactivity of all functional groups. It is obvious that this condition is not respected when the reaction is engaged as the increasing viscosity during the reaction will reduce the reactivity of the functional groups.
- The reactivity of the groups being independent from each other.
- Cyclization is not considered.

The network characterizing values can be experimentally determined by DTMA, based on the theory of rubber elasticity [13] where the crosslinking density *Xc* is:

$$X_c = \frac{E'}{3\rho RT}$$

where; E' is the modulus in the rubber elastic region, R is the gas constant, ρ is the density, and T is the absolute temperature.

### 1.3.3 MESH SIZE

The mesh size of a 3D network can be derived from the rubber-elasticity theory using the following equation [14, 15]:

$$\xi = \left(\frac{G' N_A}{RT}\right)^{-\frac{1}{3}}$$

where; G' is the storage modulus in Pa, $N_A$ is the Avogadro constant in $mol^{-1}$, R is the molar gas constant in $J.(mol.K)^{-1}$, and T is the temperature in K.

## 1.4 MATERIALS AND EXPERIMENTAL METHODS

### *1.4.1 RESINS*

We have investigated formulation based on multifunctional ERs UV cured and multifunctional acrylate resins cured thermally for 3D printing applications [3].

#### *1.4.1.1 EPOXY*

Multifunctional epoxy polymers/monomers considered are based on dicyclopentadiene DCPD (Figure 1.5a), on naphthalene (Figures 1.5b and 1.5c), on benzene (Figure 1.5d).

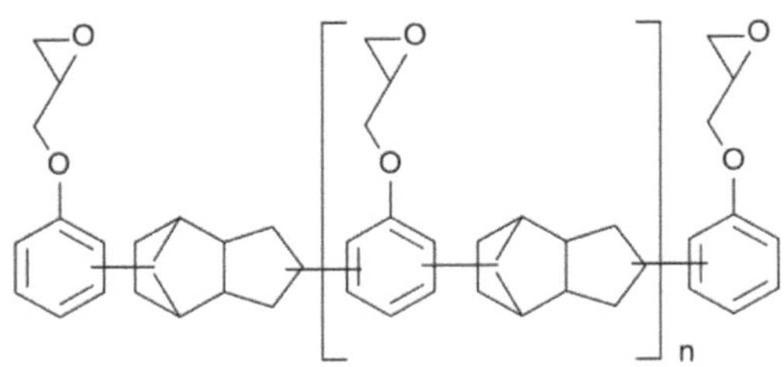


**FIGURE 1.5a** Hepiclon™ HP 720.

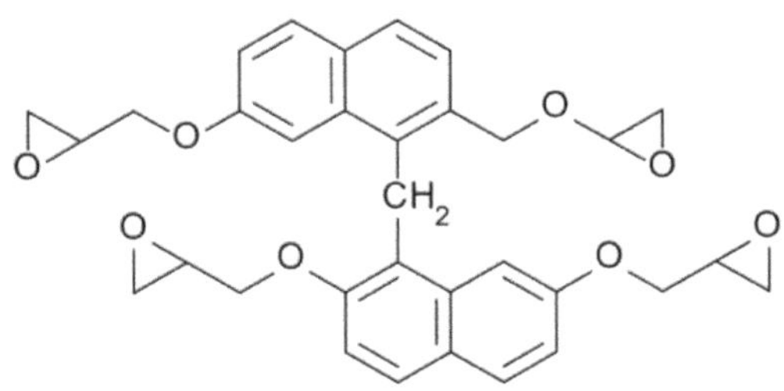


**FIGURE 1.5b** Hepiclon™ HP 4710, *f* = 8.

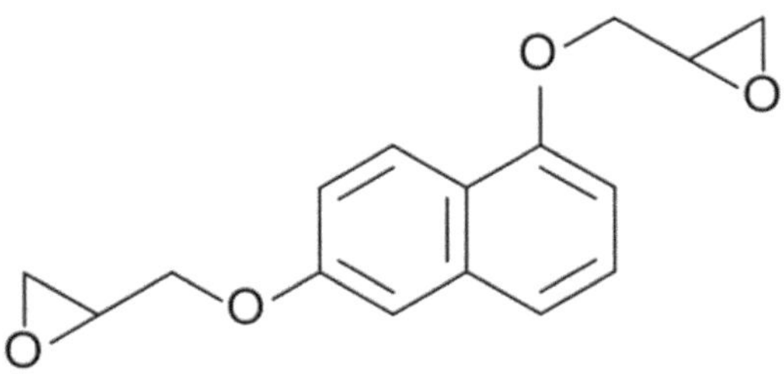


**FIGURE 1.5c** Hepiclon™ HP 4032, *f* = 4.

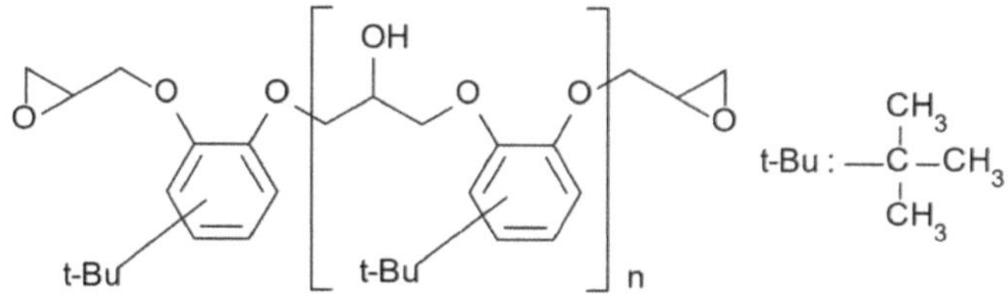


**FIGURE 1.5d** Hepiclon™ HP 820, $f = 4$.

### *1.4.1.2 ACRYLATES*

Two multifunctional acrylates such as trimethylolpropane triacrylate (TMPTA-Figure 1.6a) and ethoxylated (3) trimethylolpropane triacrylate (ETPTA-Figure 1.6b) have been crossling by thermal effect in the presence of benzoyl peroxide BPO.

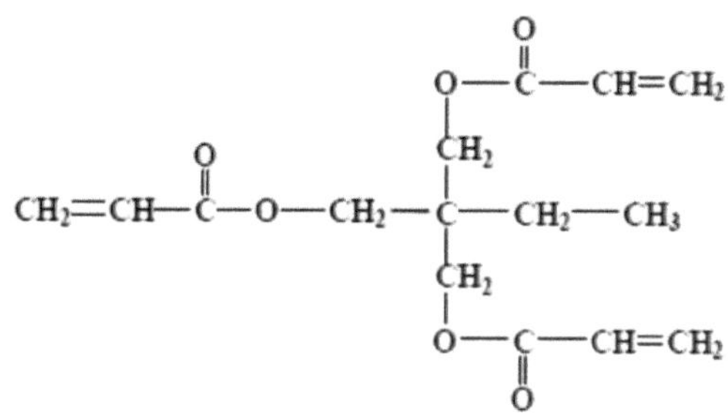

**FIGURE 1.6a** Trimethylolpropane triacrylate, $f = 6$.

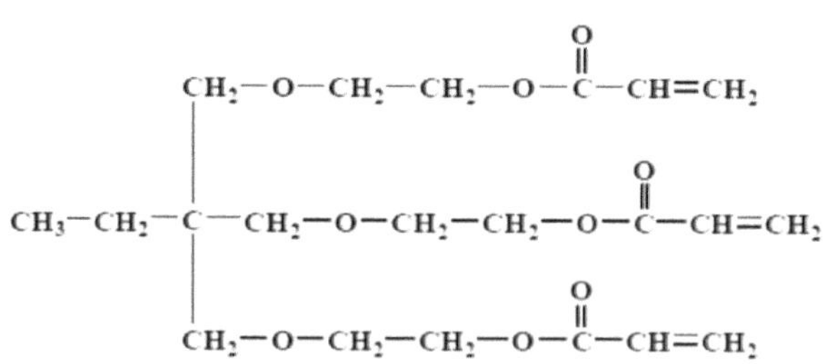

**FIGURE 1.6b** Ethoxylated (3) trimethylolpropane triacrylate, $f = 6$.

## *1.4.2 CO-SOLVENTS*

As all the ERs are solid except Epolam® 515, co-monomers used in all systems will play the role of solvent in view to get liquid formulation and also will participate in the crosslinking reaction once this reaction is initiated. As already mentioned for Epolam® 515 (Figure 1.7a), other liquid epoxies such as Bis(3,4-epoxycyclohexylmthyl) adipate (Figure 1.7b) have been used.

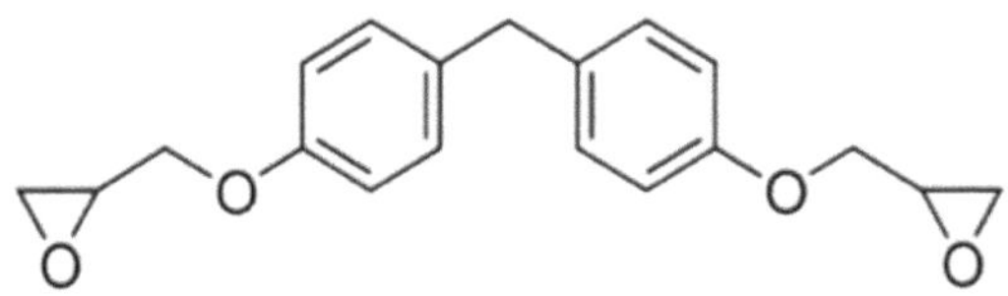


**FIGURE 1.7a** Epolam® 515, *f* = 4.

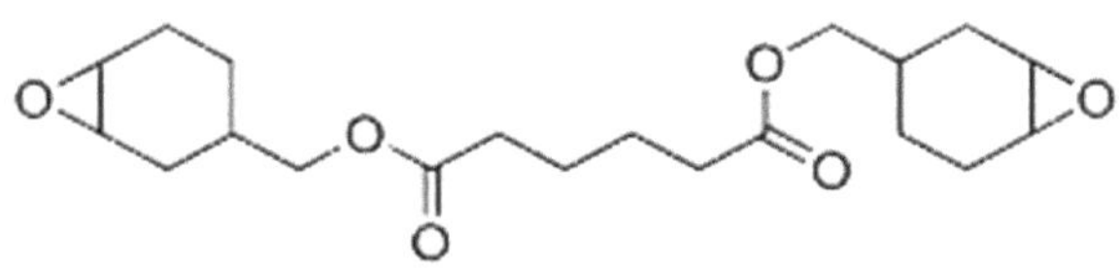


**FIGURE 1.7b** Epoxy adipate Cyracure® UVR 6128, *f* = 4.

We also have considered as a co-reactive solvent, the vinyl ethers (Figure 1.8), that polymerize by the cationic mechanism.

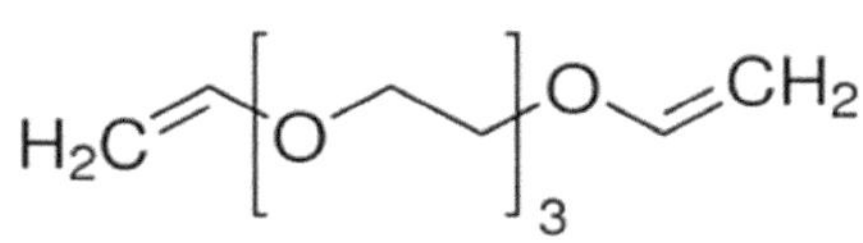


**FIGURE 1.8** Rapid-Cure® DVE-3 tri(ethylene glycol) divinyl ether.

### 1.4.3 *KINETICS*

UV photopolymerization kinetics has been performed by differential scanning photocalorimetry (Figure 1.9), monitoring the photopolymerization that determines the photosensibility of any thin films (Figure 1.10) [16].

Kinetics parameters of the epoxy systems have been studied using a TA instrument #912 based on DSC #2920 model differential photocalorimeter (DPC), equipped with a 200 W high-pressure mercury lamp, giving an optical range from 285 to 440 nm with an intensity of 45 mW/cm$^2$ on the single DSC head (one sample and fully cured reference) [17].

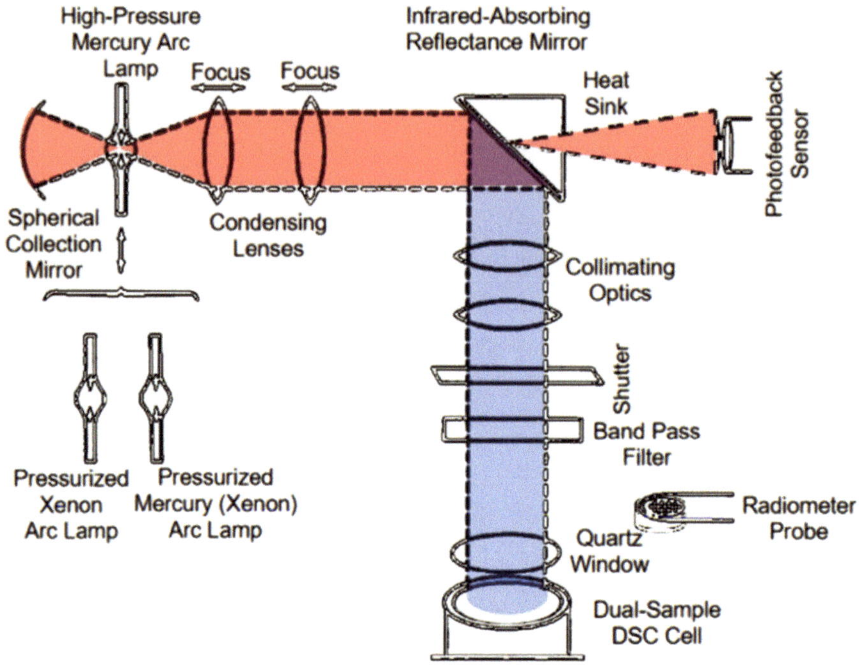


**FIGURE 1.9** Schematic of DPC.

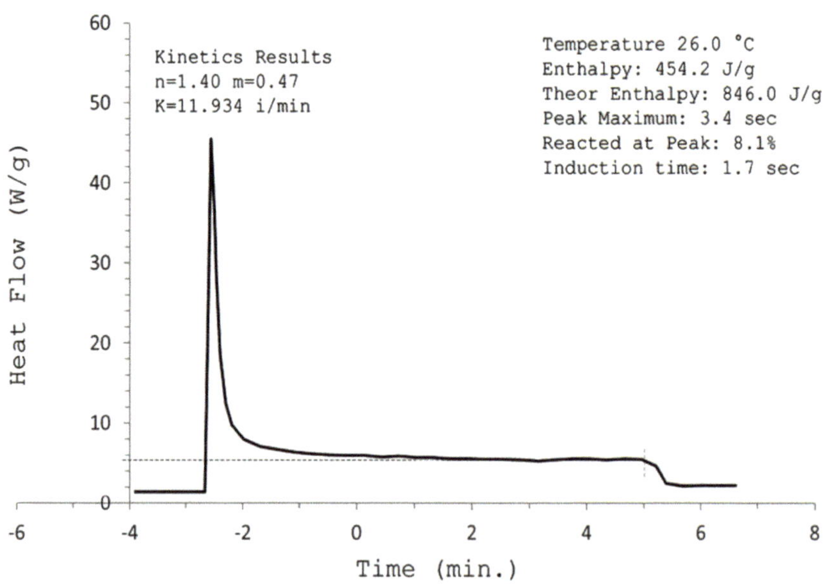


**FIGURE 1.10** DPC photogram.

The pan, filled fully cured under 20 min of the UV exposure sample of resin, was used as a reference. The samples weighting (2.0 ± 0.2) mg were placed in aluminum pans covered by PET film and subjected to 1 min of isothermal conditioning before and 5 minutes after UV exposure.

As any polymerization reaction is an exothermic process, DPC allows doing kinetics according to the DPC curve (Figure 1.10). Calculations of the kinetics of photopolymerization of the considered formulation are based on the general Šesták and Berggren's equation [18].

$$R_{P(T)} = \frac{d\alpha_{(t,T)}}{dt} = k_{(T)}\alpha^m (1-\alpha)^n [-\ln(1-\alpha)]^p$$

where; α is the degree of conversion, *k* is the rate coefficient, *m*, and *n* are the reaction order of initiation and propagation, respectively, and *p* is the order of termination reaction.

In order to simplify this equation, we consider only the outset of the polymerization process. In so doing, the value of *p* in Eq. (1) can be taken as 0. A simplified autocatalytic kinetic equation can thus be obtained, which gives us the following rate equation [19]:

$$R_{P(T)} = \frac{d\alpha_{(t,T)}}{dt} = k_{(T)}\alpha^m (1-\alpha)^n$$

The values of *k* are determined from a ln curve-ln plot of $\frac{d\alpha}{dt}$ *vs.* $[\alpha^{m/n}(1-\alpha)]$:

$$\ln\left(\frac{d\alpha}{dt}\right) = \ln k + n \ln \alpha^{m/n} (1-\alpha)$$

The reactions are conducted at different ratio monomer/co-monomer solvent to determine the rate coefficient *k* of the photopolymerization. Plotting *k* at different temperature allow to reach the activation energy $E_a$ of the considered system for each ratio [3].

$$k_{(T)} = A \exp^{-(\frac{E_a}{RT})}$$

where; *A* is the frequency factor or collision factor, $E_a$ is the activation energy, *R* is the ideal gas constant (8.314 J/mol.K), and *T* is the temperature measured in Kelvin.

## 1.5 RESULTS AND DISCUSSION

### 1.5.1 UV CATIONIC CROSSLINKING REACTION OF EPOXY

Experiments have been performed using the photoinitiator UVI-6976 at a concentration of 3 wt.% with a UV light intensity of 45 $mW/cm^2$.

We have analyzed the following couple epoxy polymer-monomer/co-reactive solvent crosslink by UV radiation:

- Hepiclon™ HP 4710/Epoxy adipate Cyracure® UVR 6128 (Table 1.1).
- Hepiclon™ HP 4032/Epolam® 515 (Table 1.2).
- Hepiclon™ HP 720/DVE di-vinyl ether (Table 1.3).
- Hepiclon™ HP 820/Epolam® 515 (Table 1.2) although Epolam® 515 *is the worst solvent for photoreactivity according to our precedent study* at different ratio polymer or monomer/co-reactive solvent [3].

We have also compared these epoxies to acrylate systems crosslink by thermal effect in the presence of benzoyl peroxide BPO. The ratio polymer/co-reactive solvent given in weight is recalculated in a mole that can be used to determine the values of $f_0$, $M_0$, *Mc*, and *Xc*.

**TABLE 1.1** Hepiclon™ HP 4710/Epoxy Adipate

| **HP 4710/Adipate** | 0/100 | 20/80 | 30/70 | 40/60 |
|---|---|---|---|---|
| **Ea (kJ.mol$^{-1}$)** | 4.0 ± 0.2 | 10.3 ± 1.7 | 21.5 ± 1.5 | 29.4 ± 2.1 |
| $f_0$ | 4 | 4.57 | 4.88 | 5.22 |
| $M_0$ | 366 | 392.85 | 407.64 | 423.95 |
| ***Mc*** | 183 | 153.14 | 141.54 | 131.66 |
| ***Xc*** | 5.5×10$^{-3}$ | 6.5×10$^{-3}$ | 7.1×10$^{-3}$ | 7.6×10$^{-3}$ |

Molar mass of Hepiclon™ HP 4710, Mm = 556 g.mol$^{-1}$, $f_0$ = 8; Epoxy adipate Cyracure® UVR 6128, Mm = 366 g.mol$^{-1}$, $f_0$ = 4.

The monomer Hepiclon™ HP 4710 has a high functionality $f_0$, double of the co-reactive solvent. This induces a greater difference in the values of $f_0$, going from 4.57 for 20/80 to 5.22 for 40/60, which impact *Mc* and make it decreasing. Therefore, the density of crosslinking *Xc* will increase up to the ratio of 40/60 with *Xc* = 7.6×10$^{-3}$. Therefore, when the percentage of the oligomer increases going from 20 wt.% to 40 wt.%, it is normal that the crosslink density increases too due to its high functionality. For high density, we need higher energy for low crosslink density. More compact is the system; more energy has to be provided.

**TABLE 1.2** Hepiclon™ HP 4032/Epolam® 515

| **HP 4032/Epolam® 515** | 0/100 | 40/60 | 60/40 | 90/10 |
|---|---|---|---|---|
| **Ea (kJ.mol$^{-1}$)** | 44.9 ± 3.2 | 37.3 ± 2.5 | 37.4 ± 1.5 | 38.9 ± 2.8 |
| $f_0$ | 4 | 4 | 4 | 4 |
| $M_0$ | 312 | 292.93 | 286.70 | 275.53 |
| ***Mc*** | 156 | 146.47 | 143.35 | 137.77 |
| ***Xc*** | $6.4\times10^{-3}$ | $6.8\times10^{-3}$ | $7.0\times10^{-3}$ | $7.3\times10^{-3}$ |

Molar mass of Hepiclon™ HP 4032, Mm = 272 g.mol$^{-1}$, $f_0 = 4$; for Epolam® 515, Mm = 312 g.mol$^{-1}$, $f_0 = 4$.

In this example, the molar masses values for Hepiclon™ HP 4032 and Epolam® 515 are of the same order of magnitude-272 g/mol and 312 g/mol, respectively, and the same average functionality for both. We do observe that $M_0$ and *Mc* are decreasing. Therefore, the crosslinking density is increasing up to $7.3\times10^{-3}$ for the ratio 90/10, a result that conforms to our expectations.

**TABLE 1.3** Hepiclon™ HP 720/DVE Di-Vinyl Ether

| **HP 720/DVE** | 0/100 | 40/60 | 60/40 | 70/30 |
|---|---|---|---|---|
| **Ea (kJ.mol$^{-1}$)** | 3.6 ± 0.4 | 10.6 ± 1.1 | 22.8 ± 1.8 | 29.4 ± 2.0 |
| $f_0$ | 4 | 4.37 | 4.59 | 4.79 |
| $M_0$ | 202 | 283.62 | 355.42 | 406.94 |
| ***Mc*** | 101 | 119.49 | 137.10 | 145.79 |
| ***Xc*** | $9.9\times10^{-3}$ | $8.4\times10^{-3}$ | $7.3\times10^{-3}$ | $6.9\times10^{-3}$ |

Molar mass of Hepiclon™ HP 720, Mm = 720 g.mol$^{-1}$ for n ≈ 1, $f_0 = 6$; for DVE Di-vinyl ether Mm = 202 g.mol$^{-1}$, $f_0 = 4$.

Considering the energy of activation Ea, we observe that it is increasing when the quantity of Hepiclon™ HP 720 is increasing, which is normal as the activation energy of the co-reactive solvent is low, and its percentage is decreasing. The values of the average functionality $f_0$ are increasing when the concentration of Hepiclon™ HP 720 is increasing as its $f_0$ is higher than the co-reactive solvent. As a result, $M_0$ and *Mc* are increasing, and therefore the crosslinking density decreases due to the higher molar mass of the monomer, 720 g/mol compared to 202 g/mol of the co-reactive solvent. In that case, the degree of crosslink is the highest for the lower percentage of the oligomer that can be explained by the low viscosity and the high reactivity of the co-reactive solvent. Therefore, in this system, the solvent may play a major role in the formation of the network; the higher is the concentration of the co-reactive solvent, the higher is the degree of crosslink. Therefore, in that case, the more expanded the network is, the more difficult is the crosslinking reaction. The highest degree of crosslinking is obtained for the ratio polymer/ co-reactive solvent 40/60 (Table 1.4).

**TABLE 1.4** Hepiclon™ HP 820/Epolam® 515

| **HP 820/Epolam® 515** | 0/100 | 20/80 | 30/70 | 40/60 |
|---|---|---|---|---|
| **Ea ($kJ.mol^{-1}$)** | 44.9 ± 3.2 | 51.1 ± 4.5 | 57.6 ± 2.5 | 65.1 ± 3.1 |
| $f_0$ | 4 | 4 | 4 | 4 |
| $M_0$ | 312 | 326.12 | 383.22 | 414.79 |
| *Mc* | 156 | 178.06 | 191.61 | 207.39 |
| *Xc* | $6.4\times10^{-3}$ | $5.6\times10^{-3}$ | $5.2\times10^{-3}$ | $4.8\times10^{-3}$ |

Molar mass of Hepiclon™ HP 820, Mm = 820 $g.mol^{-1}$ for n ≈ 1, $f_0 = 4$; for Epolam® 515, Mm = 312 $g.mol^{-1}$, $f_0 = 4$.

In that case, as both the monomer-Hepiclon™ HP 820 and the co-reactive solvent-Epolam® 515 have the same $f_0$, the average functionality of the system will not change and $f_0 = 4$ whatever the mixture considered. For the activation energy Ea, $M_0$ and *Mc*, we do observe the same phenomena as Hepiclon™ HP 720. The higher crosslinking degree is obtained for the ratio of 20/80. This surprising result may be explained by the low reactivity of both oligomer, and co-reactive solvent, observed by the high activation energy compare to the precedent system Hepiclon™ HP 720/DVE Di-vinyl ether.

### 1.5.2 THERMAL RADICAL CROSSLINKING REACTION OF MULTIFUNCTIONAL ACRYLATES

We have considered two triacrylates monomers, the trimethylolpropane triacrylate (TMPTA) and the ethoxylated (3) trimethylolpropane triacrylate (ETPTA) polymerized by the benzoyl peroxide for [BPO] = 0.5, 1, 1.5, and 2 wt.% (Table 1.5).

In Table 1.5, one observes that the activation energy $E_a$ of radical polymerization of TMPTA decreases when the concentration of benzoyl peroxide increases; the higher is the concentration of BPO, the easier will be the polymerization as more. The same phenomenon is observed for EPTA.

If we compare the two monomers TMPTA and ETPTA, for the same concentration of BPO, ETPTA polymerizes easier than TMPTA, although the molar mass of ETPTA is higher than TMPTA. This result is due to the fact that the acrylates functions (double bonds) are linked to the core trimethyl propane by two oxygen atoms for ETPTA and only by one oxygen atom for TMPTA. Indeed, the presence of one oxygen will facilitate the flexibility of the chain, thus the mobility of the terminal acrylate function, whereas two oxygen will have a greater effect on the reactivity of the acrylates, leading to a decrease in the activation energy.

**TABLE 1.5** Activation Energy for Thermal Radical Polymerization of TMPTA and ETPTA

| Monomer/BPO | Ea (kJ/mol) |
|---|---|
| 0.5 wt.% PBO | |
| TMPTA | 120.55 |
| ETPTA | 79.42 |
| 1 wt.% BPO | |
| TMPTA | 118.68 |
| ETPTA | 63.07 |
| 1.5 wt.% BPO | |
| TMPTA | 90.90 |
| ETPTA | 57.96 |
| 2 wt.% BPO | |
| TMPTA | 46.15 |
| ETPTA | 35.94 |

As there is no reactive co-solvent in the system, the monomers are used alone, *Mc*, and degree of crosslinking are directly obtained. From Table 1.6, we observe that the density of crosslinking is higher for TMPTA than ETPTA due to the lower molar mass of TMPTA compare to ETPTA. Regarding the activation energy and the degree of crosslinking *Xc*, the more dense or compact the system is, the higher will be the activation energy. It is what we observe.

**TABLE 1.6** *Mc* and Crosslink Density *Xc* for Thermal Radical Polymerization of TMPTA and ETPTA

| Monomer | $M_m$ | Ea kJ/mol (2 wt.% BPO) | $f_0$ | *Mc* | *Xc* |
|---|---|---|---|---|---|
| TMPTA | 296 | 46.15 | 6 | 74 | $1.35\times10^{-2}$ |
| ETPTA | 438 | 35.94 | 6 | 109.5 | $0.91\times10^{-2}$ |

If we compare thermal curing with UV curing, we observe that the activation energy is generally much higher for the thermal effect that it is for UV radiation, due to the energy needed by heating as radiation curing is performed at room temperature.

## 1.6 CONCLUSION

We studied the photoreactivity of different monomers/co-reactive solvents-multi epoxy monomers (sensitive to a cationic polymerization mechanism) or multi acrylates systems (sensitive to a radical polymerization mechanism) and multi-epoxy or vinyl ethers by considering their activation energy Ea. We have correlated the activation energy with *Mc* and the degree of crosslink *Xc* of the network.

As follows from the results, the co-reactive solvent plays an essential role in the photoreactivity of a formulation, particularly the DVE di-vinyl ether because of its low viscosity, but also due to its chemical structure (3O and methylene group) that makes the molecule flexible and therefore more reactive.

For the correlation between Ea and $f_0$, *Mc*, and *Xc*, it is strongly dependent on these latter factors; we have had conflicting results, depending on the systems considered.

1. Thus, when Ea increases, the degree of crosslinking is increasing; the denser the network is, the higher is the activation energy (Tables 1.1 and 1.2), which makes sense as the system is based on high functionality $f_0$ and marked difference in molar mass of the constituents.
2. We get the inverse results (Tables 1.3 and 1.4), both Ea and *Xc* increase, which means that *Mc* is higher for the lowest ratio monomer/co-reactive solvent. In that case and surprisingly, the higher crosslinking density is obtained for monomer < co-reactive solvent. The higher the network structure is, the lower is the activation energy. We may explain these results by invoking the high reactivity of the co-reactive solvent and/or the very low reactivity of both oligomer and co-reactive solvent.

## KEYWORDS


- **3D structure**
- **crosslink density *Xc***
- **differential photocalorimeter**
- **kinetics**
- **photoreactivity**
- **UV curing**


## REFERENCES


1. Redwood, B., Schöffer, F., & Garret, B., (2017). *The 3D Printing Handbook Technologies Design and Applications* (p. 347). Hardcover Brand, ISBN: 9082748509.
2. Anthony, C., (2015). *BNP Media, PCI Magazine*. Electron beam laboratory systems. https://www.pcimag.com/articles/101173-electron-beam-laboratory-systems (accessed on 19 October 2020).
3. Abadie, M. J. M., Manole, I., & Fetecau, C., (2020). UV 3D printing: an overview of the existing technologies. *Materiale Plastice (Mater. Plast.), 57*(1), 2020, 141–152, https://doi.org/10.37358/MP.20.1.5321.
4. Abadie, M. J. M., & Voytekunas, V., (2004). New trends in UV curing. *Eurasian Chem. Tech. Journal, 6*, 67–77.
5. Fouassier, J. P., & Lalev, J., (2012). *Photo Initiators for Polymer Synthesis: Scope, Reactivity, and Efficiency*. Edt: Wiley-VCH.
6. Crivello, J. V., (1999). The discovery and development of onium salt cationic photoinitiators. *J. Polym. Sci. Part A Polymer Chemistry, 37*(3), 4241–4254.
7. Sangermano, M., Roppolo, I., & Chiappone, A., (2018). New horizons in cationic photopolymerization. *Polymers, 10*, 136. doi: 10.3390/polym10020136.
8. Michaudel, Q., Kottisch, V., & Fors, B. P., (2017). Cationic polymerization: From photoinitiation to photocontrol. *Angew. Chem. Int. Ed., 56*, 9670–9679. doi: 10.1002/anie.201 701 425, https://doi.org/10.1002/anie.201701425.
9. Lalevée, J., Mockbel, H., & Fouassier, J. P., (2015). Recent developments of versatile photo initiating systems for cationic ring-opening polymerization operating at any wavelengths and under low light intensity sources. *Molecules, 20*, 7201–7221. doi: 10.3390/molecules20047201.
10. Gotro, J., (2016). *UV Curing Part Five: Cationic Photopolymerization*. Polymer innovation blog. https://polymerinnovationblog.com/uv-curing-part-five-cationic-photopolymerization/ (accessed on 19 October 2020).
11. Berger, J., & Huntgens, F., (1979). *Angew. Makromol. Chem., 76, 77*, 109.
12. Macosco, C. W., & Miller, D. R., (1976 & 1979). *Macromolecules, 9*, 199, 206.
13. Zosel, A., (1996). In: *Lackund Polymerfilme*. Vicentz Verlag, Hannover.

14. Welzel, P. B., et al., (2011). Modulating biofunctional star PEG heparin hydrogels by varying size and ratio of the constituents. *Polymers, 3*(1), 602–620.
15. Lu, S. X., & Anseth, K. S., (2000). Release behavior of high molecular weight solutes from poly(ethylene glycol)-based degradable networks. *Macromolecules, 33*(7), 2509–2515.
16. Abadie, M. J. M., & Appelt, B. K., (1988). Photocalorimetric study of photosensitive materials. *Bull. Soc. Chim. Fr., 1*, 20–24.
17. Appelt, B. K., & Abadie, M. J. M., (1985). Thermal analysis of photocurable materials. *Polym. Eng. and Sci., 25*(15), 931–933.
18. Šesták, J., & Berggren, G., (1972). The study of the kinetics of mechanism of solid-state reactions at increasing temperature, *Thermochimica Acta, 3*, 1–12.
19. Appelt, B. K., & Abadie, M. J. M., (1987). Kinetic and thermodynamic experiments with a photo-DSC. *Polymer Eng. Sci., 27*, 25.

# CHAPTER 2

# The Modified Natural Zeolites in Ion-Exchange Adsorption of Some Heavy Metals

L. AKHALBEDASHVILI, N. GAGNIASHVILI, S. JALAGHANIA, N. JANASHVILI, R. KVATASHIDZE, G. TODRADZE, and N. LORIA

*Ivane Javakhishvili Tbilisi State University, Alexandre Tvalchrelidze Caucasian Institute of Mineral Resources, Mindeli Str., 11, Tbilisi–0186, Georgia, E-mail: aklali@yahoo.com (L. Akhalbedashvili)*

## ABSTRACT

Advanced urbanization and expanded industry cause the release of the emission of wastes in tremendous quantities into the environment, which causes many problems to the environment and to human health. One of the solutions for solving these problems would be obtaining new and efficient materials such as zeolites as an adsorbent for wastewater treatment.

Natural mordenite (MOR), among other purifiers of waters, take a special place due to its unique adsorption and ion-exchange properties. The high content of MOR allows ensuring its wide application, since in our country there is no synthetic zeolite production and their purchase is quite expensive.

The goal of this work was to study the adsorption properties of natural zeolite MOR from deposits of Georgia in regards to heavy metals Pb, Cu, Mn, Cd, and Zn, modified by changed methodic of processing as one of ways for obtaining adsorbents with advanced characteristics. Adsorption experiments were performed in a static-circulation mode.

For obtaining a maximum efficiency of adsorption the conditions of experiments (contact time, ratio solid: solution, granulation degree) were chosen. The order of adsorption activity of MOR for the ions of divalent metals studied is following $Pb^{2+}$> $Cu^{2+}$> $Zn^{2+}$ > $Mn^{2+}$ >$Cd^{2+}$.

## 2.1 INTRODUCTION

Currently, one of the serious issues of environmental protection is the pollution of atmosphere, surface water, and soil with heavy metals, and the problem of extraction of them from the environment is particularly acute, due to their biological activity, persistence, and toxicity. Many heavy metals such as cadmium, nickel, zinc, lead, and chromium form complexes of various structures and stability and, therefore even in trace amounts pose a threat to human life and health [1–3].

Heavy metals in the form of cations or in the form of hydrated complexes cannot be destroyed easily. Exposure of toxic heavy metals can be caused by contamination of drinking water as a result of the use of lead pipes, higher concentrations of toxicants in the air near sources of industrial emissions, or by ingestion through the food chain [4]. With intensive growth and expansion of production, the concentration of various compounds of heavy metals in natural water sources increases [5, 6].

Among the existing methods for cleaning of wastewater (adsorption, precipitation, membrane filtration (MF), and ion exchange) [7] containing heavy metals, sorption methods occupy an important place, which ensures the most complete extraction of toxic ions, especially from solutions with a low concentration. Although ion exchange is an attractive technology for removing heavy metals from drinking and wastewater, it is uneconomical due to the high cost of reagents for regeneration, the need to remove the resulting compounds, or the possibility of secondary pollution, as well as the deficit of ion-exchanging resins.

The advantage of the adsorption method is economy, high efficiency, and the ability to treat wastewater containing heavy metals, organic pollutants, and dyes [8], as well as the recovery of these substances.

The adsorption process is most justified when the adsorbent has a low cost and is produced in large quantities [9]. Adsorbents based on natural zeolites have many properties that contribute to the adsorption and retention of heavy metals [10]. For wastewater treatment, the most important properties of natural zeolite are cation exchangeability and selectivity with respect to ions [11].

Zeolites are porous aqueous crystalline aluminosilicates. Their infinite framework is formed upon articulation through the common vertices of the $SiO_4$ and $AlO_4$ tetrahedra, which create a negative charge on the framework. The charge of the giant anion is compensated by certain exchangeable cations, such as potassium, sodium, calcium, magnesium, copper, nickel, iron, and others [12]. Their pores have molecular sizes, which ensure selectivity in size and shape during catalysis and adsorption [13]. The adsorption of heavy metals from aqueous solutions by natural zeolites with respect to $Co^{2+}$, $Cu^{2+}$, $Zn^{2+}$, and $Mn^{2+}$ has been studying for many years [14–16].

Mordenite (MOR) has a porous structure consisting of parallel main channels (0.65 × 0.70 nm) and small channels (0.57 × 0.26 nm) oriented along the crystallographic c-axis and intersecting by side channels along the b-axis with dimensions 0, 34 × 0.48 nm. Due to the large number of hydroxyl groups, decationized mordenite (H-MOR) are used in the petrochemical industry as catalysts for cracking, isomerization, and alkylation reactions. The presence of OH groups and the high cation exchange capacity of natural adsorbents contribute to the adsorption and retention of heavy metals [16, 17].

The goal of submitting work was to study the adsorption properties of natural zeolite MOR (deposits of Georgia) in regards to heavy metals Pb, Cu, Mn, Cd, and Zn, modified with methodic, based on alternation of acid, alkali, ion-exchange, and thermal treatment.

## 2.2 MATERIALS AND METHODS

### *2.2.1 TREATMENT OF ZEOLITE*

The samples of natural zeolite MOR from Ratevani deposit with content 0.11 $K_2O$ • 0.42 $Na_2O$ • 0.48 CaO •$Al_2O_3$• 9.5 $SiO_2$ • 5.6 $H_2O$ were grinded and sieved through 300 × 600-m sieves and then dried in an oven at 120 ± 5 C for 8 h. Modified samples were obtained by treating with HCl and KOH solutions of different concentration, and salts solutions of transition metals. The phase composition of samples was determined by x-ray diffraction (XRD). The chemical composition of zeolite samples was determined by the standard analytical methods for silicates, also using instrumental methods [18]. $Al_2O_3$, $Fe_2O_3$, CaO, and MgO were analyzed by volumetric methods, and $SiO_2$ was analyzed with gravimetric one. $Na_2O$ and $K_2O$ were analyzed with flame photometry. The X-ray powder diffraction patterns of polycrystalline samples were carried out in isothermal conditions with an X-ray apparatus DRON-3 by using CuKα-radiation (A = 0.154 nm) and synthetic -$Al_2O_3$ as an internal standard.

### 2.2.2 REAGENTS

All the reagents and standard solutions for AAS used for experiments were of analytical grade and were purchased from Sigma-Aldrich. The concentrations of heavy metals under study were determined by the atomic absorption spectrophotometer AAnalyst-200, from the firm "Perkin-Elmer." Buffer solutions for the calibration of pH-meter at pH = 4, 7, 9, concentrated hydrochloric (36%) and nitric acids (63%) were of high grade and purchased from Sigma and Fisher.

A model solution of copper, manganese, and lead was prepared using their nitrate salts, $Cu(NO_3)_2 \cdot 3H_2O$, $Mn(NO_3)_2 \cdot 4H_2O$, and $Pb(NO_3)_2 \cdot 6H_2O$, respectively, in double-distilled water. The model solutions of Zn(II) and Cd(II) were prepared using corresponding chlorides.

### 2.2.3 ADSORPTION STUDIES

The adsorption of heavy metals on MOR samples was studied in a static-circulation mode. The first experiments were conducted at ratio solid phase: solution, equal 1:20; 1:30 and 1:50. The last ratio was proved more effective; therefore, in the following experiments were used 10 g of adsorbent with 500 ml of solutions with heavy metal ions of different concentrations. The particle size of the sample used was in the range of 50–120 μm. The bottles were shaken in a stirrer for 6 h and solutions containing heavy metals were filtered. The initial concentrations of model solutions with metal ions and filtrated solutions concentrations were determined by AAS (AAnalyst 200 atomic absorption spectrophotometer). The flame type was air-acetylene and absorption wavelengths are $Cu^{2+}$ (324.7 nm), $Zn^{2+}$ (213.9 nm), $Pb^{2+}$ (283.3 nm), $Cd^{2+}$ (228.8 nm) and $Mn^{2+}$ (279.5 nm). The percent adsorption (%) and the coefficient of distribution ($K_d$) were calculated using the formulae stated in [19]:

$$\%\ adsorption = \frac{C_i - C_f}{C_f} \times 100 \tag{1}$$

where; $C_i$ and $C_f$ are the concentrations of the metal ion in initial and final solutions, and respectively:

$$K_d = \frac{\text{amount of metal in adsorbent}}{\text{amount of metal in solution}} \times \frac{V}{m}\ \text{ml/g} \tag{2}$$

where; *V* is the volume of the solution (ml) and *m* is the mass of the adsorbent (g).

## 2.3 RESULTS AND DISCUSSION

The data of the chemical analysis of the treated samples are presented in Table 2.1. This study showed that natural zeolite contains a complement of exchangeable sodium, potassium, and calcium ions.

The Si/Al ratio is 5.56 and indicates a well-arranged distribution of Si/Al in structural positions of the framework. The ratio (Na +K + Ca)/Ca are equal 2.86. So, the initial MOR is enriched with calcium-ion, which does not delete with alkali (see Table 2.1). This sample is low-silica, whereas the samples with high-silica are enriched with potassium, sodium, and magnesium. It ensues from the data of quantitative chemical analysis and thermal processing that MOR is not subjected to great structural changes in the course of acid treatment and is stable approximately to 600°C.

By XRD analysis was found that the natural zeolite contains approximately 55–60% MOR and plagioclase, mica, quartz, montmorillonite, and K-feldspar chlorite minerals (Figure 2.1).

**TABLE 2.1** Chemical Composition of Natural and Modified Mordenites

| Zeolite | Components, % | | | | | | | | |
|---|---|---|---|---|---|---|---|---|---|
| | $SiO_2$ | $Al_2O_3$ | CaO | MgO | $Na_2O$ | $K_2O$ | $Fe_2O_3$ | $TiO_2$ | $P_2O_5$ |
| 1. MOR | 69.98 | 10.79 | 3.55 | 0.29 | 2.00 | 0.80 | 1.77 | 0.12 | 0.19 |
| 2. H-MOR* (1N) | 69.61 | 10.03 | 1.96 | 0.15 | 1.40 | 0.75 | 1.06 | 0.09 | 0.18 |
| 3. H-MOR (2N) | 68.96 | 9.89 | 1.5 | 0.13 | 1.02 | 0.45 | 0.97 | 0.08 | 0.18 |
| 4. H-MOR (3N) | 68.94 | 9.87 | 1.45 | 0.13 | 0.95 | 0.43 | 0.84 | 0.08 | 0.17 |
| 5. MOR-OH** (1N) | 67.98 | 9.09 | 3.45 | 0.18 | 3.15 | 5.04 | 1.72 | 0.11 | 0.19 |
| 6. MOR-OH (2N) | 67.45 | 8.85 | 3.24 | 0.16 | 4.38 | 5.96 | 1.72 | 0.11 | 0.19 |
| 7. MOR-OH (3N) | 66.95 | 6.49 | 3.21 | 0.16 | 5.95 | 6.78 | 1.64 | 0.11 | 0.19 |

H-MOR*–mordenite, decathionized with acid of different normality.
MOR-OH**–mordenite, modified with alkali of different normality.

Dependence the amount of adsorption of $Cu^{2+}$, $Pb^{2+}$, $Zn^{2+}$, $Cd^{2+}$ and $Mn^{2+}$ ions on their concentration in model solutions was investigated at a stationary temperature 25°C in the range of concentration of metal 50–400 mg/l at constancy of other parameters: zeolite granulation degree, ratio adsorbent: solution, stirring speed, time of processing (Figure 2.2). It was found that extent of adsorption for $Cu^{2+}$, $Pb^{2+}$, $Zn^{2+}$, $Cd^{2+}$ and $Mn^{2+}$ with increase in concentration of metal in model solutions decreases. The heavy metal adsorption is described with different mechanisms of ion-exchange processes on cation-exchanged sites of zeolite [14, 20], but not on decationated ones.

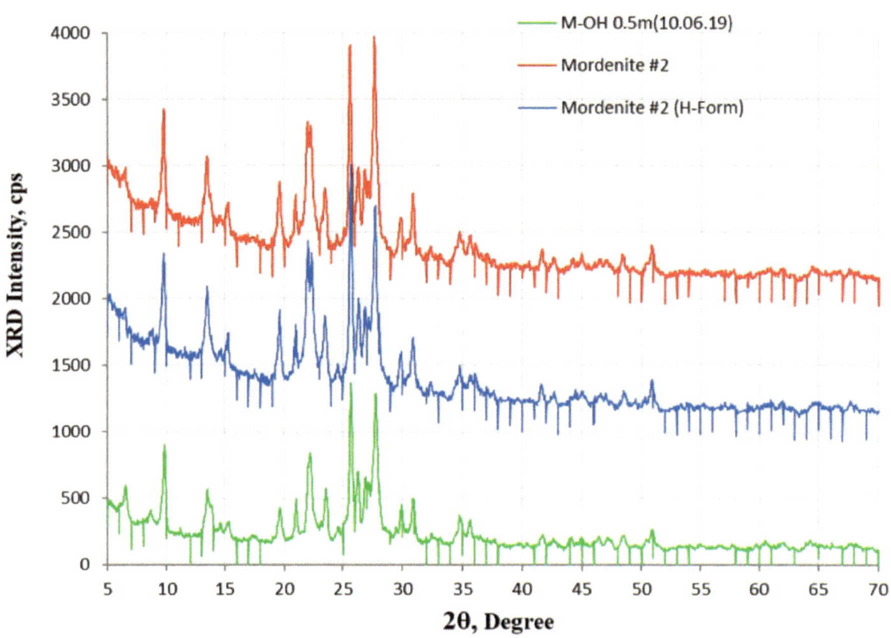


**FIGURE 2.1** X-ray diffraction pattern of natural and modified samples of MOR.

Figure 2.3 shows the dependence of the $K_d$ distribution coefficient on metal ion concentrations. It has been found that $K_d$ values increase with decreasing metal ion content. These results show that as the concentration of the model solution increases during adsorption process energetically more weak sites begin to take part.

Decationated sites take part only in the physical adsorption process, and residual hydroxyl groups in zeolite channels form single-charge Me $(OH)^+$ type complexes with adsorbed transition metal ions. During the ion exchange process, metal ions must transport through the pores of the zeolite as well as through the channels of the lattice, and they must replace exchange cations (mainly sodium and calcium, to a lesser extent potassium). The limiting stage of diffusion of heavy metal ions to adsorption sites is their transportation through smaller diameter channels. In Ref. [21] it was noted that the heat of adsorption of benzene and n-hexane on MOR is 10 kcal/mol higher, which is due to the smaller diameter of intracrystalline channels in MOR and together with rapid and reversible adsorption slow and irreversible adsorption is observed [22]. Especially since cations of heavy metals are present and move in solution in the form of hexa-aquacoplexes with six surrounding water molecules [23]. The amount of adsorbed ions depends on both the charge density of the cations and the diameter of the hydrate cations. Since the metal cation charges are +2, Mn $(OH)_6^{2+}$ complex ions with the largest diameter are adsorbed to the smaller extent and lead ions in the largest amount.

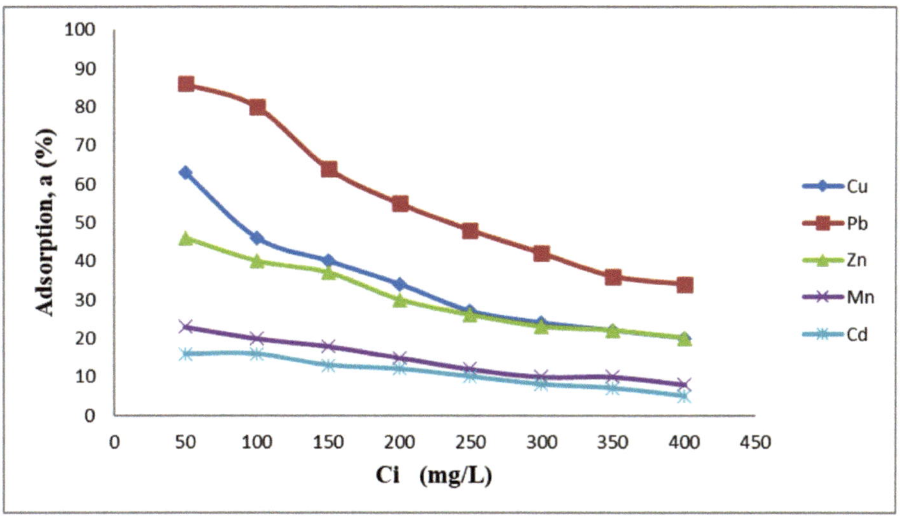


**FIGURE 2.2** Adsorption of metal ions on natural mordenite sample as a function of initial concentration of model solution: m = 10 g, v = 500 ml, pH = 6, holding time-6 h.

So, was estimated the following sequence of adsorption $Pb^{2+} > Cu^{2+} > Zn^{2+} > Mn^{2+} > Cd^{2+}$ for MOR and its acid-treated forms. However, H-MOR samples were less active in the adsorption process. It may be assumed that acid and thermal treatment cause the contraction of the crystal lattice and the difficulty of metal ions adsorption. The change sequence of adsorption was observed for MOR-OH forms: $Pb^{2+} > Zn^{2+} > Mn^{2+} > Cu^{2+} > Cd^{2+}$. MOR removes Pb and Cu from the solutions with metal concentrations below $10^{-4}$ M almost completely. However, MOR in natural and modified forms was not effective in adsorption Cd.

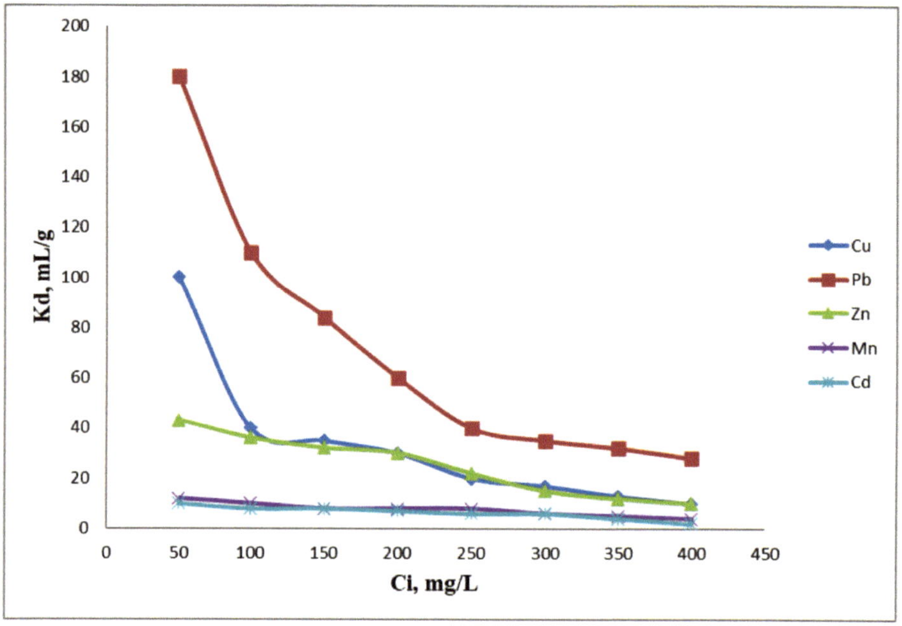


**FIGURE 2.3** Dependence of Kd distribution coefficient on initial concentrations of metal ions at adsorption on MOR: m = 10 g, v = 500 ml, pH = 6, holding time-6 h.

## 2.4 CONCLUSION

The results show that with increasing metal concentrations in the aqueous solution, the energetically more weak sites become involved in adsorption. The heavy metal uptake is attributed to different mechanisms of ion-exchange processes as well as to the physical adsorption. During the ion-exchange process, metal ions have to move through the pores into the zeolite structure, but also through inner channels of the lattice, and they have to replace cations (mainly sodium and potassium), after treating them with acid and alkali. It was established that the adsorption of heavy metals ions on modified samples is carried out from the model solutions according to the physical exchange adsorption mechanism. The following sequence of adsorption activity was estimated for MOR from Georgia (locality Ratevani): $Pb^{2+} > Cu^{2+} > Zn^{2+} > Mn^{2+} > Cd^{2+}$.

## KEYWORDS


- **adsorption**
- **cations**
- **crystal lattice**
- **heavy metals**
- **natural mordenite**
- **x-ray diffraction**


## REFERENCES


1. Demirbas, A., (2008). Heavy metal adsorption onto agro-based waste materials: A review. *Journal of Hazardous Materials, 157*, 220–229.
2. Ceribasis, H. I., & Yetis, U., (2001). Biosorption of Ni(II) and Pb(II) by *Phanaerochaete chrysosporium* from a binary metal system kinetics. *Water SA, 27*(1), 15–20.
3. Hua, M., Zhang, S., Pan, B., Zhang, W., Lv, L., & Zhang, Q., (2012). Heavy metal removal from water/wastewater by nanosized metal oxides: A review. *Journal of Hazardous Materials, 211*, 317–331.
4. Fu, F., & Wang, Q., (2011). Removal of heavy metal ions from wastewaters: A review. *Journal of Environmental Management, 92*, 407–418.
5. Immamuglu, M., & Tekir, O., (2008). Removal of copper (II) and lead (II) ions from aqueous solutions by adsorption on activated carbon from a new precursor hazelnut husks. *Desalination, 228*, 108–113.
6. Kadirvelu, K., Goe, J., & Rajagopal, L., (2008). Sorption of lead, mercury and cadmium ions in multi-component system using carbon aerogel as adsorbent. *Journal of Hazardous Materials, 153*, 502–507.
7. Chingombe, P., Saha, B., & Wakeman, R., (2005). Surface modification and characterization of a coal-based activated carbon. *Carbon, 43*, 3132–3143.
8. Tangjuank, S., Insuk, N., Tontrakoon, J., & Udeye, V., (2009). Adsorption of lead (II) and cadmium (II) ions from aqueous solutions by adsorption on activated carbon prepared from cashew nut shells. *World Academy of Science, Engineering and Technology, 28*, 110–116.
9. Varank, G., Demir, A., Bilgili, M. S., et al., (2014). Equilibrium and kinetic studies on the removal of heavy metal ions with natural low-cost adsorbents. *Environ. Prot. Eng., 40*, 43–61.
10. Turp, S. M., (2018). $Mn^{2+}$ and $Cu^{2+}$ adsorption with a natural adsorbent: Expanded perlite. *Appl. Ecol. Environ. Res., 16*, 5047–5057.
11. Colella, C., (1996). Ion exchange equilibria in zeolites minerals. *Mineral Deposita, 31*, 554–562.
12. Turp, S. M., (2017). Prediction of adsorption efficiencies of Ni(II) in aqueous solutions with perlite via artificial neural networks. *Arch. Environ. Prot., 43*, 26–32.

13. Čejka, J., Centi, G., Perez-Pariente, J., & Roth, W. J., (2012). Zeolite-based materials for novel catalytic applications: Opportunities, perspectives and open problems. *Catal. Today 179*, 2–15.
14. Erdem, E., Karapinar, N., & Donat, R., (2004). The removal of heavy metal cations by natural zeolites. *J. Coll. Inter. Sci., 280*, 309–314.
15. Smirnov, A. D., (1982). Sorption water purification. *L.: Chemistry*, 168.
16. Villasenor, J., Rodriguez, L., & Fernandez, F. J., (2011). Composting domestic sewage sludge with natural zeolites in a rotary drum reactor. *Bioresour. Technol., 102*, 1447–1454.
17. Awasthi, M. K., Wang, M., Pandey, A., et al., (2017). Heterogeneity of zeolite combined with biochar properties as a function of sewage sludge composting and production of nutrient-rich compost. *Waste Manage, 68*, 760–773.
18. Corbin, D. R., Burgess, B. F., Vega, A. J., & Farelee, R. D., (1987). *Anal. Chem., 59*, 2722.
19. Khan, S. A., Rehman, U. R., & Khan, M. A., (1995). *Waste Manage, 15*, 271.
20. Barer, R. M., (1987). *Zeolites and Clay Minerals as Sorbent and Molecular Sieves*. Academic Press, New York.
21. Eberly, P. E., Charles, Jr. N. K., & Alexis, V., (1963). Effect of $SiO_2/Al_2O_3$ ratio on physicochemical properties of mordenite and activity for n-pentane isomerization. *J. Phys. Chem., 67*, 2404.
22. Barrer, R. M., Bultitude, F. W., & Sutherland, J. W., (1957). *Trans. Far. Soc., 53*, 1111.
23. Jama, M. A., & Yücel, H., (1990). The sorption of toxic elements onto natural zeolite, synthetic goethite, and modified powdered block carbon. *Sep. Sci. Technol., 24*(15), 1393.

# CHAPTER 3

# Properties of the Magnetic Polymer Nanocomposites in Magnetic Fields

JIMSHER ANELI[1] and GRIGOR MAMNIASHVILI[2]

[1]*Institute of Machine Mechanics, 10, Mindeli Str., Tbilisi–0186, Georgia, E-mail: janeli@yahoo.com*

[2]*Andronikashvili Institute of Physics of Ivane Javakhishvili Tbilisi State University, 6, Tamarashvili Str., 0177, Tbilisi, Georgia*

## ABSTRACT

Processes of self-assembly were studied in the magnetic polymer carbon nanocomposites doped with cobalt nanoclusters. These processes proceed due to the diffusion of magnetic nanoparticles stimulated by a combined effect of an outer steady magnetic field and heating. The obtained polymer composites are promising for practical applications.

## 3.1 INTRODUCTION

In the last decade, the investigation of such new nanostructure forms of carbon as nanoparticles, nanotubes, and nanowires has become very topical. This is associated with the fact that, due to their sizes and peculiarities of their atomic structure, nanostructural particles reveal such unique physical-mechanical properties that the range of their promising applications covers many areas of activity from microelectronics to medicine.

In recent years, there increased interest in technologies of production of carbon-based materials oriented on the production of doped carbon nanoparticle modifications (nanotubes, nanoclusters, nanowires). This gives scientists and engineers the opportunity of aimed control of unique properties of these materials, which are their natural properties [1].

As a matter of fact, the nanoobject control at the nanometer level using nanoparticles with the aim to arrange them in rows, signatures, and grids is the clue to the production of new functional materials. Hence, in recent years, for obtaining of constructional units of different nanometric sizes, many methods of self-assembling and synthesis were developed. In this connection, the possibility to control perfectly the self-assembling and synthesis processes of nanoparticles is a serious challenge from the point of view of both fundamental and applied investigations.

Based on the fundamental principles, the process of self-assembling requires the existence of interaction between atoms and clusters, as well as of thermodynamic and kinetic driving forces, so that the organization of atoms and clusters for creation of nanosize domain structures should be realized. From this point of view, magnetic nanoparticles deserve a particular interest due to their unique physical-chemical properties and applicability in the new functional material technologies.

Carbon shells provide both the protection of ferromagnetic impurities against aggressive environments and new unique properties for the hydride nanostructures. The self-assembling of magnetic clusters coated with carbon shells represents just such an example that could be used in contemporary materials, for instance, in strong rare-earth free bonded magnets, analytical instruments (nuclear magnetic resonance tomographs), and nanosensors using magnetic one-dimensional nanowires.

Moreover, currently, due to their low toxicity, magnetic carbon nanoparticles are under testing for therapeutic and diagnostic applications.

In recent years the magnetic field was used for the creation of nanoscale materials, which resulted in significant achievements in fabrication of macro- and microstructure synthesized materials possessing unique properties.

In contrast with other existing self-assembling technologies, ordering induced by the magnetic field defines the formation of magnetic nanoparticles in ordered structures with unique properties. Therefore, the area of application of carbon magnetic nanoparticles is quite large. It is enough to name such applications as magnetic fluids, plastic scratch-resist glasses, information storage magnetic media, sensors, biomedicine, etc. It should be noted that, in spite of their broad prospects for multifunctional applications, the carbon nanoparticles doped with ferromagnetic clusters have not been well-researched [2].

In work [3], the electric and magnetic properties measurements were carried out to study the gradient anisotropic conducting and magnetic polymer composites synthesized due to self-assembly processes of nanoparticles under the influence of the elastic forces developed by the stretching of polymer film composites fabricated on the basis of polyvinyl alcohol (PVA) doped with graphite powder and nickel nanoparticles.

In this work, the study of self-assembling properties of carbon nanoparticles doped with cobalt nanoclusters in magnetic polymer nanocomposites (PNC) under the combined influence of magnetic field and heating will be carried out using methods similar ones developed in work [3].

## 3.2 RESULTS AND DISCUSSION

To achieve this goal, we planned experiments with magnetic nanopowders, preparation of filled and unfilled polymer films, and the study of self-organization processes in them. This process was facilitated by heating the films above the glass transition temperature. The particles in conventional composites are essentially immobile in contrast to PNC, particularly above the glass transition temperature $T_g$. The nanoparticle's mobility can affect polymer dynamics resulting in changes in the viscosity modulus, the kinetics of the particle-cluster formation, etc. [3].

The tensile measurements showed that, below $T_g$, the conventional composites and PNCs behave similarly with respect to mechanical properties. However, above $T_g$, the toughness of PNC can increase by order of magnitude with increasing temperature. It was assumed that the mechanism of toughness enhancement is the mobility of nanoparticles. The development of self-healing materials and coatings where nanoparticles migrate towards various defect sites requires a better understanding of the process of nanoparticle diffusion. Heating of PNC above $T_g$ enhances the mobility of polymer chains, which should facilitate the boundary diffusion between polymer interfaces, and this effect should be visualized using the magnetic nanoparticles introduced in the polymer. This process could be improved by applying additional stimuli, in particular a low frequency (AC) magnetic field, a stationary magnetic field, pressure, heating separately or in combinations, etc. Such an impact stimulates self-assembling processes in the prepared films in the result of which one could produce the films "glued" to each other without using other type glues and polymer melting temperature. One of the objectives of this work was the development of a simple technology of production of carbon nanoparticles doped with ferromagnetic clusters and the study of their morphology and composition.

In particular, for production of carbon-based nanopowders and nanocoatings, the method of chemical vapor deposition (CVD) is used mainly along with application of the process of hydrogen reduction of volatile chlorides.

The carbon nanoparticles doped with cobalt magnetic nanoclusters with mean sizes in the range 50–100 nm were synthesized by technology using the combination of pyrolysis of ethanol (and other hydrocarbons), vapor pyrolysis, and the CVD process in the mode of a closed recirculation de cycle with monitored technology parameters.

The developed technological process was realized in the installation, the reactor design and basic units of which provided for the possibility of monitoring of parameters such as the vapor content in reactor zones, catalytic capacity of substrates, partial oxygen pressure (over the range of $10^{-20} \div 10^{-25}$ atm.). This allows carrying out the investigations with the aim to establish the optimal technological parameters for production of finely dispersed carbon nanopowders doped with magnetic nanoclusters. A detailed description of this technology is given in the work [4].

For the preparation of polymer films, polyvinyl butyral (PVB) polymer with low $T_g \sim 45$–55°C was chosen. Polymer PVB is a resin mostly used wherever strong binding, optical clarity, adhesion to many surfaces, toughness, and flexibility is required. As a filler, we used the carbon nanopowder doped with magnetic (Co or Fe) nanoclusters (C/Co) of our production. For a comparative study, we used commercial Co nanopowder with an average diameter of 28 nm (Sun Co., USA). The concentration of used Co-doped carbon nanopowders in the polymeric composite was in the range of 10–50 wt.%, while for the Co nanopowder filled polymer composite concentration was equal to 20 wt.%.

At first, 10% alcohol solution was prepared, and then this solution was poured into Teflon press molds and, after their drying for 48 hours, the films one mm thick were obtained. The filled composites were prepared as follows: magnetic nanopowders were taken in the appropriate proportion (in terms of dry weight), and PVB was mixed with alcohol in a usual way, and then followed the ultrasonic treatment for 10–15 min for the destruction of magnetic nanopowder agglomerates. After thorough mixing for 7–10 min, magnetic polymer composite films similar to unfilled ones were obtained in Teflon press molds. From these films, the circular shape disk samples were cut.

To study the self-assembly processes in these PNC at different concentrations of carbon magnetic nanopowder, we used a simple method from work [5]. In this case, circular samples of the polymer composite (diameter – 28 mm, thickness – 1 mm) were exposed to the magnetic field, which was provided by two attached permanent neodymium magnets and a temperature of 85°C for two hours (Figure 3.1).

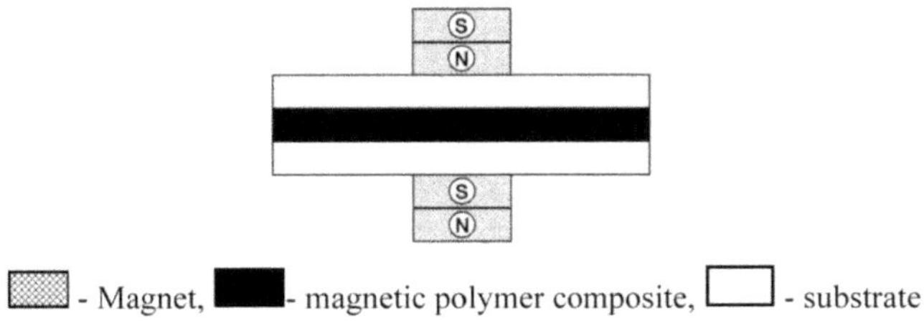


**FIGURE 3.1** Geometry of the samples.

Resulting self-assembly of C/Co nanopowders caused changes in their concentration and modulation of local resistance along the radius of the sample, which was measured by a two-contact method as in our previous work [6] (Figures 3.2–3.7). The resistance was measured between the points spaced 2 mm apart along the radius in all following cases except for Figure 3.3, where the resistance was measured between the sample center and the points 2 mm.

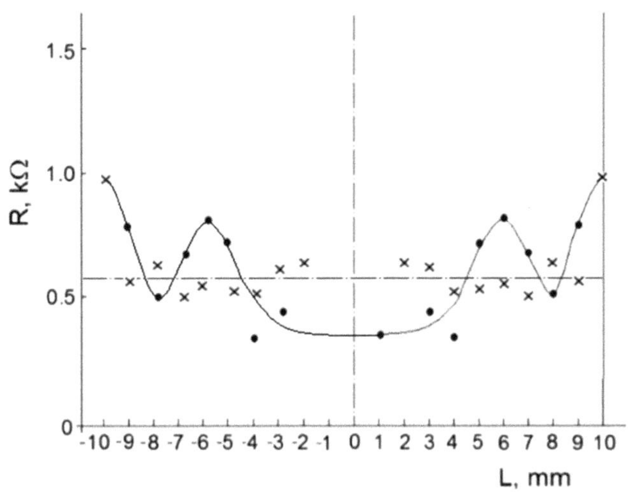


**FIGURE 3.2** Polymer composite C/Co, wt. 30% (× – Sample prepared without magnetic field treatment, • – Sample prepared under combined magnetic field and heat treatment).

Because polymeric composites are magnetic carbon nanocomposites, we were able to study the processes of self-assembling of nanoparticles using radiofrequency resonance magnetic susceptibility measuring devices [5] (Figure 3.8).

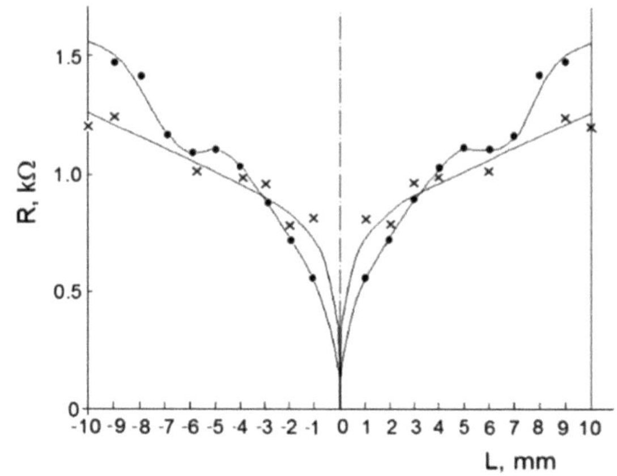


**FIGURE 3.3** Polymer composite C/Co, wt. 30%.

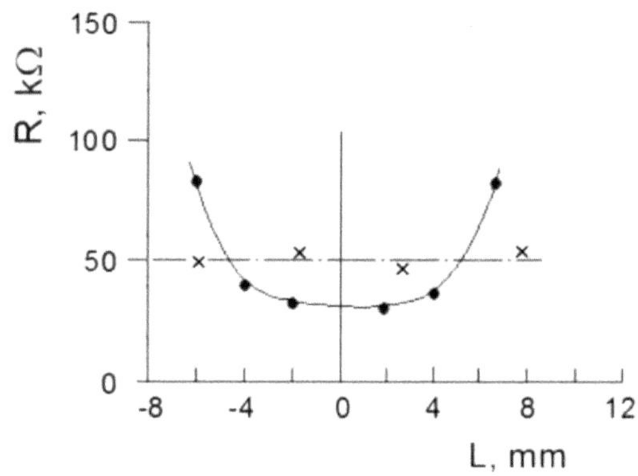


**FIGURE 3.4** Polymer composite C/Co, wt. 50%

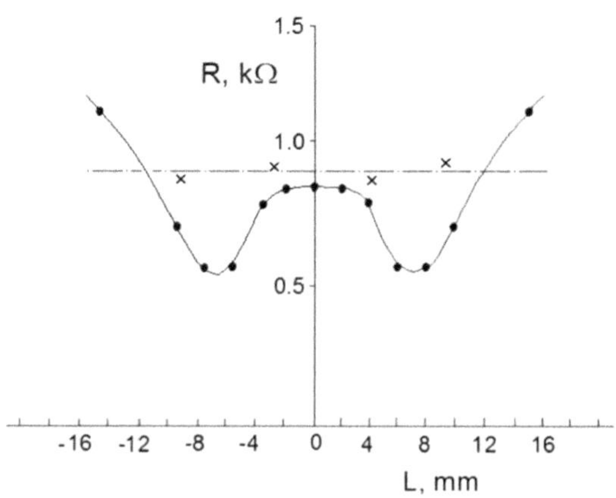


**FIGURE 3.5** Polymer composite C/Co, wt. 20%.

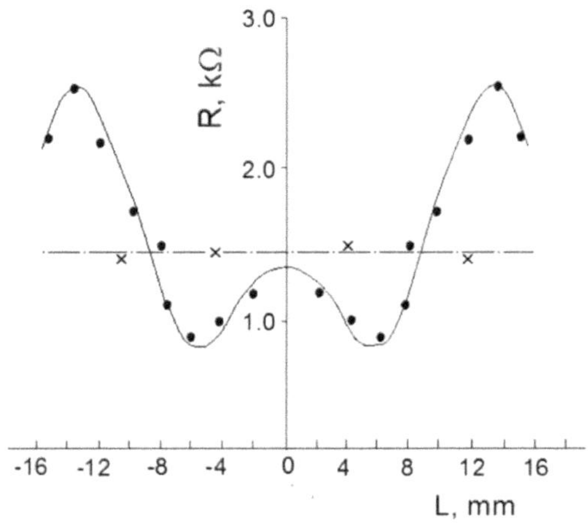


**FIGURE 3.6** Polymer composite C/Co, wt. 15%.

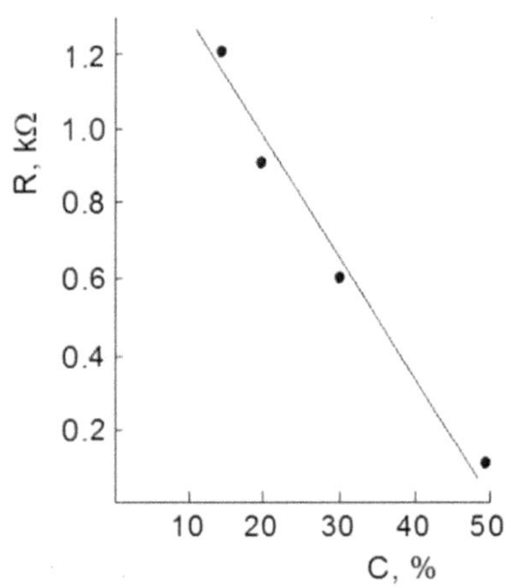


**FIGURE 3.7** Dependence of resistance of initial untreated samples on the nanopowder concentration.

Because polymeric composites are magnetic carbon nanocomposites, we were able to study the processes of self-assembling of nanoparticles using radiofrequency resonance magnetic susceptibility measuring devices [5] (Figure 3.8).

The experimental set-up is presented in Figure 3.8. In the inductive coil of the resonance contour of LC-generator cylindrical tipped ferrite rod is used as a probe. The investigated rectangular shape magnetic polymer composite film is displaced relatively the immovable ferrite tip. The scanning of the film surface is realized along the previously marked disk radius.

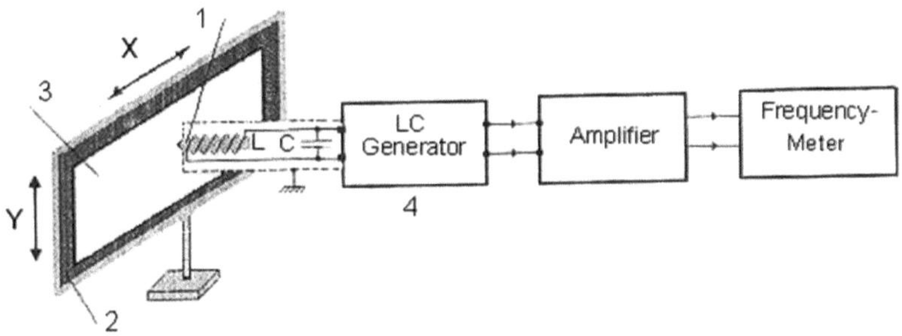


**FIGURE 3.8** Scheme of the measuring of the magnetic characteristics of the polymer films (1 – ferrite probe; 2 – frame; 3 – magnetic polymer nanocomposite film; 4 – LC generator).

The change of magnetic particle concentration causes the inductance change *dL* of the resonance contour of LC-generator resulting in the frequency displacement of LC-generator *df* related with *dL* by relation $df/f = \frac{1}{2}\, dL/L$. This frequency displacement could be precisely measured that stipulates the high sensitivity of the method. At the natural frequency of used LC-generator near ~ 2 MHz the observed range of the frequency change d*f* was about ~1000 Hz at the precision of the frequency measurement ~1 Hz.

For example, we represent the result of self-assembly measurements of Co/C magnetic nanocomposite film in the magnetic field of neodymium magnet (Figure 3.9) one similar to the results obtained during electrical resistance measurements.

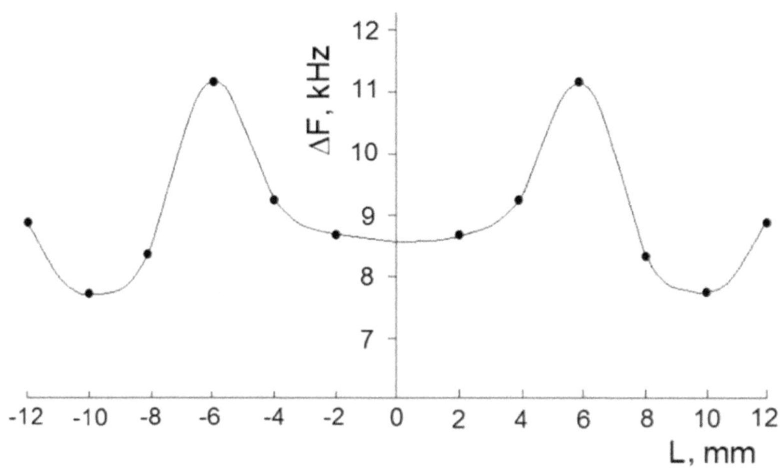


**FIGURE 3.9** Magnetic susceptibility measurements of Co/C 50% polymer nanocomposite along the polymer composite film circular disk diameter.

When we move along the diameter of the disk, the frequency measurements indicate frequency change (frequency interval change of several kHz) associated with the change in the concentration of magnetic nanoparticles in the polymer composite.

Based on the obtained experimental results, it is possible to investigate the self-assembling processes in magnetic nanocomposite polymer films synthesized by the elaborated technology using the carbon magnetic nanopowders fabricated by a technology described in Ref. [4] under combined influence of magnetic field and heating.

## 3.3 CONCLUSION

The self-assembling processes were studied in the magnetic PNC synthesized based on the carbon nanoparticles doped with cobalt nanoclusters, synthesized by a unique CVD technology developed by the authors, under combined effect of magnetic field and heating. These processes took place with the diffusion of magnetic nanoparticles stimulated by the combined effect of outer steady magnetic field and heating. The obtained magnetic PNC have good electrical and adhesive properties and are promising for potential practical applications in magneto-electronics.

## ACKNOWLEDGMENT

The work was supported by the Shota Rustaveli National Science Foundation and STCU Grant #6213.

## KEYWORDS

- **chemical vapor deposition**
- **diffusion**
- **magnetic carbon nanopowders**
- **polymer nanocomposites**
- **resistance**
- **self-organization**

## REFERENCES

1. Jimenez-Contrares, R., (2007). *Nanotechnology Research Developments*. Nova Science Publishers, Inc., New York.
2. King, V. R., (2007). *Nanotechnology Research Advances*. Nova Science Publishers, Inc., New York.
3. Grabowski, C. A., & Mukhopadhyay, A., (2014). *Macromolecules, 47*, 7238.
4. Gegechkori, T., Mamniashvili, G., Kutelia, E., et al., (2015). *J. Magn. Magn. Mater., 373*, 200.
5. Mette, C., Fischer, F., & Dilger, K., (2015). Proceedings of the institution of mechanical engineers. *Part L: Journal of Materials: Design and Applications, 229*(2), 166–172.
6. Aneli, J., Nadareishvili L., & Shamanauri, L., (2016). Electric and Magnetic Properties of Graded Oriented Polymer Composites. *J. Electrical Engineering 4*, 211–219.

# CHAPTER 4

# Physical-Chemical Studies of $M_2^I \cdot M^{II} \cdot L_2 \cdot nH_2O$ Type Heteronuclear Citrates

I. BESHKENADZE,[1] M. GOGALADZE,[1] N. KLARJEISHVILI,[1] L. GOGUA,[2] and O. LOMTADZE[1]

*[1]Petre Melikishvili Institute of Physical and Organic Chemistry, Ivane Javakhishvili Tbilisi State University, 31 A. Politkovskaya, 0186, Tbilisi, Georgia*

*[2]Tbilisi State Medical University, 33, Vaja-Pshavela Ave.0186, Tbilisi, Georgia*

## ABSTRACT

Physical-chemical studies of heteronuclear citrates with the general formula $M^I_2 \cdot M^{II} \cdot L_2 \cdot nH_2O$ has been carried out. For this purpose, the solubility of $M^I_2 \cdot M^{II} \cdot L_2 \cdot nH_2O$ type in various solvents (water, EtOH, acetone, dimethylformamide) is studied, and estimated that they are poorly soluble in all solvents at room temperature and at heating as well.

By thermographic study is proved their thermal stability and thermolysis process. Specifically, the thermal degradation of heteronuclear citrates occurred step-wise: I-water molecule elimination, II-citrate-ion oxidation gradually or simultaneously. Thermolysis final products are: (a) mixture of metal-oxides; (b) mixture of metal-oxides and coal; or (c) metals. For estimate, the character of citrate-ions, deprotonation effect, and their bond type with metals in the $M^I_2 \cdot M^{II} \cdot L_2 \cdot nH_2O$ type chelate compounds were recorded and analyzed their IR spectra. By these studies was established that the hydroxyl-groups are not deprotonated in citrate-ions, but deprotonation occurred to each of the three carboxylic groups of citric acid. Accordingly to COOH group stereochemistry, depending on conditions, citrate-ion could be in the role as bridging carboxylate ligand as well as bi-dentate or tri-dentate bridging carboxylate ligand.

In conclusion, the synthesized compounds are internal type chelate compounds, and as expected, all three deprotonated carboxylic groups of citrate acid form ionic and coordinative bonds with metal atoms.

## 4.1 INTRODUCTION

The deficit of biometals in plant, soil, agricultural animals, and poultry as well is one of the main reasons determining food products' low quantitative and qualitative index. The provision of living organisms with the optimal amount and ratio of biometals is crucial to solve this problem. At the same time, in living organisms, they fulfill their functions as chelate compounds, so filling up the deficit of biometals by such form leads to the sharp increment of their biological activity. This fact can be explained that in the chelate form, biometals have low toxicity, high ability of assimilation, and therefore increased efficiency quality of using of little doses, which in the other hand leads to ecological safety of using biometals by this form.

Inorganic salts are characterizing by high toxicity, low quality of assimilation, and efficiency, which is caused by the formation of the poorly soluble and hardly assimilative compounds in the gastrointestinal tract of animals and poultries of [1–6]. Thus both theoretically and practically is proved the importance of these compounds containing admixes in the feeding of animals and poultries [7–11]. It is well known the role of organic compounds (vitamins, amino acids, organic acids) and microelements containing chelate compounds in the metabolic processes of poultries and animal organisms. Nevertheless, estimation of synthesis methods of oxy-acids, namely citric acid and microelements (Zn, Co, Fe, Mn, Cu) containing chelate compounds, the physical-chemical study of synthesized compounds, and determination of the effectiveness of using them in feeding are still relevant.

## 4.2 MATERIALS AND METHODS

During experiments was studied:

- Qualitative solubility of synthesized compounds in various solvents;
- Thermal stability and thermolysis process;
- Nature of citrate-ion nature, deprotonation effect, and their bond type with metals by spectrophotometric method.

## 4.3 RESULTS AND DISCUSSION

In the Agrarian chemistry, the problem laboratory is continuing research for the creation and testing of a new generation of premixes [12–17]. We have synthesized $M^I_2 \cdot M^{II} \cdot L_2 \cdot nH_2O$ type chelate compounds, by microelemental analysis and melting temperature were proved the composition and identity of synthesized compounds. Their crystalline structure, X-amorphous, and iso-structural orders were revealed by X-ray analysis [18]. In the order of continuing the physical-chemical studies of $M^I_2 \cdot M^{II} \cdot L_2 \cdot nH_2O$ type chelate compounds were determined their qualitative solubility in various solvents (water, EtOH, acetone, dimethylformamide) (Table 4.1).

**TABLE 4.1** Qualitative Solubility of Chelate Compounds

| SL. No. | Compound | Solubility | | | |
|---|---|---|---|---|---|
| | | Water | EtOH | Acetone | DMFA* |
| 1. | $Zn_2 \cdot Mn \cdot L_2 \cdot 4H_2O$ | + t | – | – | – |
| 2. | $Co_2 \cdot Zn \cdot L_2$ | + t | – | – | – |
| 3. | $Cu_2 \cdot Co \cdot L_2 \cdot H_2O$ | + t | – | – | – |
| 4. | $Co_2 \cdot Mn \cdot L_2 \cdot 4H_2O$ | + t | Poor sol.** | Poor sol. | Poor sol. |
| 5. | $Co_2 \cdot Cu \cdot L_2 \cdot 4H_2O$ | + t | Poor sol. | Poor sol. | – |
| 6. | $Fe_2 \cdot Mn \cdot L_2 \cdot 2H_2O$ | + t | – | – | – |
| 7. | $Fe_2 \cdot Zn \cdot L_2 \cdot 2H_2O$ | + t | – | – | – |
| 8. | $Fe_2 \cdot Co \cdot L_2 \cdot 4H_2O$ | + t | Poor sol. | Poor sol. | Poor sol |
| 9. | $Mn_2 \cdot Cu \cdot L_2 \cdot 4H_2O$ | + t | – | – | – |

*DMFA – Dimethylformamide;

**Poor sol. – Poor soluble.

A thermographic investigation has been carried out in order to study the synthesized chelate compounds' thermal stability and sequence of thermolysis process in the presence of air. Hungarian firm F. Paulik, J. Paulik, L. Ergey type derivatograph has been used for investigation. The compounds have been studied under the conditions: TG = 100 mg, T = 700C, DTA = DTG = 1/5, heating rate of samples – 10 Deg/min. All of the thermograms are characterized by several endo- and exo-effects and corresponding effects on the curve (Table 4.2).

For an estimate, the nature of citrate-ions, deprotonation effect and their bond type with metals in the $M^I_2 \cdot M^{II} \cdot L_2 \cdot nH_2O$ type chelate compounds were recorded FTIR spectra in nujol (400–4000 $cm^{-1}$) on the spectrometer

"VARIAAN," CARRY 100. Table 4.3 offers the spectral data of *vibration bands and relevant* appropriations of sodium citrate and synthesized chelate compounds.

**TABLE 4.2** Results of Thermographic Study of Chelate Compounds

| SL. No. | Formula | T (°C) | Weight Loss, % | | Eliminated (Molecule) | Solid Product of Practical Decomposition |
|---|---|---|---|---|---|---|
| | | | Practical | Theoretical | | |
| 1. | $Zn_2{\cdot}Mn\ L_2{\cdot}4H_2O$ | 110 | 6.07 | 6.01 | $2H_2O$ | $Zn_2MnL_2$ |
| | | 400 | 60.60 | 61.35 | 2L | $ZnO+Mn_3O_4$ |
| 2. | $Co_2{\cdot}Zn{\cdot}L_2$ | 265 | 14.61 | 14.61 | 0.43 L | $Co_2ZnL_{1.57}$ |
| | | 300 | 8.10 | 8.00 | 0.2 L | $Co_2ZnL_{1.37}$ |
| | | 410 | 50.60 | 50.94 | 1.37 L | $Co_3O_4$+ZnO+ C |
| 3. | $Cu_2{\cdot}Co{\cdot}L_2{\cdot}H_2O$ | 100 | 2.40 | 3.09 | 1 $H_2O$ | $Cu_2CoL_2$ |
| | | 280 | 60.66 | 060.66 | 2 L | $Cu_2O$+ $Co_3O_4$+ C |
| 4. | $Mn_2{\cdot}Cu{\cdot}L_2{\cdot}4H_2O$ | 85 | 8.30 | 8.66 | 3 $H_2O$ | $Mn_2CuL_2H_2O$ |
| | | 145 | 2.50 | 3.1 | 1 $H_2O$ | $Mn_2CuL_2$ |
| | | 325 | 59.70 | 60.38 | 2 L | $Mn_3O_4$+CuO+ C |
| 5. | $Fe_2\ Mn{\cdot}L_2{\cdot}2H_2O$ | 75 | 0.80 | _ | 0.044 $H_2O$ | $Fe_2MnL_22H_2O$ |
| | | 140 | 5.60 | 6.20 | 2 $H_2O$ | $Fe_2MnL_2$ |
| | | 340 | 70.51 | 69.37 | 2 L | 2Fe + Mn |

## 4.4 DISCUSSION

By the studies of qualitative solubility of synthesized compounds in various solvents (water, EtOH, acetone, dimethylformamide) was estimated that they are soluble in water at a heating time while in the other solvents, they are characterized by low solubility or insolubility at room temperature and heating time as well. As the thermographic study showed, all thermograms are characterized by several endo- and exo-effects and corresponding effects on the curve (Table 4.2).

$Zn_2{\cdot}Mn{\cdot}L_2{\cdot}2H_2O$ compound thermolysis process starts at 110°C on weak endoeffects by 2 mole water molecule elimination (weigh loss is: practical – 6.07%, theoretical – 6.01%). At 400°C, the strong exo-effect corresponds to 2 mol citrate-ion oxidation (Weight loss: practical – 60.6%, theoretical – 61.67%), and the final thermolysis product is the mixture of metal-oxides: ZnO and $Mn_3O_4$.

**TABLE 4.3** IR Spectra of Absorption *Vibration Bands of* $Na_3L$ and $M^{I}_{2}\cdot M^{II}\cdot L_{2}\cdot nH_{2}O$ Type Chelate Compounds ($cm^{-1}$)

| SL. No. | Compound | ν OH ($cm^{-1}$) | ν $COO^-$ ($cm^{-1}$) | Appropriation |
|---|---|---|---|---|
| 1. | $Na_3\cdot C_6H_5O_7$ | 3450 | – | ν OH |
| | | 1240, 1200, 1105, 1125, 1050, 900, 880, 830 | – | δOH |
| | | – | 1580 | $\nu_{as}$ $COO^-$ |
| | | – | 1400–1320 | $\nu_s COO^-$ |
| 2. | $Mn_2\cdot Cu\cdot L_2\cdot 4H_2O$ | 3479–3394 | – | ν OH |
| | | 1288, 1257, 1149, 1072, 933, 887, 848 | – | δOH |
| | | – | 1569 | $\nu_{as}$ $COO^-$ |
| | | – | – | $\nu_s COO^-$ |
| 3. | $Zn_2\cdot Mn\cdot L_2\cdot 4H_2O$ | 3464–3394 | – | ν OH |
| | | 1288, 1257, 1142, 1072, 972, 902, 856, 802 | – | δOH |
| | | – | 1558 | $\nu_{as}$ $COO^-$ |
| | | – | – | $\nu_s COO^-$ |
| 4. | $Fe_2\cdot$ $Mn\cdot L_2\cdot 2H_2O$ | 3417–3370 | – | ν OH |
| | | 1250, 1195, 1134, 1088, 1057, 964, 941, 895, 840 | – | δOH |
| | | – | 1550 | $\nu_{as}$ $COO^-$ |
| | | – | – | $\nu_s COO^-$ |
| 5. | $Co_2\cdot Zn\cdot L_2$ | 3479–3394 | – | ν OH |
| | | 1327, 1288, 1257, 1195, 1072, 933, 902, 864, 802 | – | δOH |
| | | – | 1558 | $\nu_{as}$ $COO^-$ |
| | | – | – | $\nu_s COO^-$ |
| 6. | $Cu_2\cdot Co\cdot L_2\cdot H_2O$ | 3564–3479, 3394 | – | ν OH |
| | | 1326, 1288, 1257, 1195, 1072, 972, 910, 864 | – | δOH |
| | | – | 1558 | $\nu_{as}$ $COO^-$ |
| | | – | – | $\nu_s COO^-$ |

$Co_2 \cdot Zn \cdot L_2$ compound thermolysis process starts at a relatively high temperature (265°C) due to it does not include water molecules. The citrate-ion decomposition in the compound proceeds gradually: I endo-effect (265°C) corresponds to 0.43-mole citrate-ion oxidation (Weight loss: practical – 14.61%, theoretical – 14.61%), II endo-effect (300°C) – 0.2 mole citrate-ion oxidation (Weight loss: practical – 8.10%, theoretical – 8.00%). The other citrate-ion is oxidized at strong exo-effects (400°C) followed by the formation of the mixture of metal-oxides and coal: $Co_3O_4$, ZnO, and C (coal).

$Cu_2 \cdot Co \cdot L_2 \cdot H_2O$ and $Mn_2 \cdot Cu \cdot L_2 \cdot 4H_2O$ compounds thermograms analysis show that at I endo-effects in the temperature range 100–145°C, water molecules are oxidized. The further rise in temperature at strong exo-effects corresponds to citrate-ion oxidation, and in both cases, the final thermolysis products are the mixtures of metal-oxides and coal.

$Fe_2 \cdot Mn \cdot L_2 \cdot 2H_2O$ compound thermolysis process proceeds differently. In particular, I endo-effect at 75°C corresponds to the elimination of moisture (0.8%), which is 0.044 mole of water. At II endo-effect (140°C) occurs oxidation of 2 mole water. At strong effects (340°C), the citrate-ion is oxidized, and the final thermolysis products are metals (Fe, Mn) in contrast to other compounds.

Concerning the spectrophotometric study of heteronuclear citrates, from literature, is well known [19, 20] that in the citric acid ($H_3L$) spectrum are observed the not-dissociative carboxylic group (C = O) valence vibration bands at 1750–1720 $cm^{-1}$ region while the not-dissociative oxy-group valence vibration bands are at 3450 $cm^{-1}$ region and also the other bands connected with oxy-group deformative vibrations (1290, 1205, 1165, 1125, 1050, 925, 880, 865, 825 $cm^{-1}$).

In the sodium citrate ($Na_3L$) spectrum, the citric acid valence vibration bands νCOOH (1750–1720 $cm^{-1}$) are disappeared completely and appeared $COO^-$ ion asymmetric valence vibration band $\nu_{as}$ (1580 $cm^{-1}$) and symmetric valence vibration band $\nu_s$ (1400–1320 $cm^{-1}$). Among them, the $\nu_{as}$ absorption band is more characteristic due to it has a constant value. Besides, in the $Na_3L$ spectrum is also revealed the oxy-group strong valence vibration band νOH at the 3450 $cm^{-1}$ region and also the other deformative vibration bands at the 1290–825 $cm^{-1}$ region. The OH-group appropriate absorption band is the same of $H_4L$ spectrum (3450 $cm^{-1}$). Considering the x-ray diffraction (XRD) analyses is suggested the opinion that the citrate-ion forms the metalocycles and is coordinated by central and terminal carboxylic acid oxygen.

As it is shown in Table 4.3 in the test compounds spectra at the region 1550–1569 $cm^{-1}$ are observed the asymmetric valence bands $\nu_{as}$ appropriated to the citric acid deprotonated carboxylic group $COO^-$ ion. The absorption bands appropriated to $\nu_s COO^-$ at the region 1400–1320 $cm^{-1}$ are covered by the nujol absorption bands.

In the compounds IR spectrum clearly appeared the strong absorption band $\nu_{as}$ (3564–3370 $cm^{-1}$) appropriate to not-dissociative oxy-group and also in the region 1100–1300 $cm^{-1}$ are observed as quantitatively, as well as intensively deformative vibration bands appropriated to δ OH strong absorption bands. All of these results indicate about the formation of chemical bonds in the compounds without the participation of oxy-groups of citrate-ion.

In conclusion, the IR spectral analysis shows that in the $M^I_2 \cdot M^{II} \cdot L_2 \cdot nH_2O$ type chelate compounds, the hydroxy-groups are not deprotonated, but deprotonation occurred to each of the three carboxylic groups of citric acid, which are directly involved in the formation of coordinative bonds as well as ionic bonds with metal ions. According to the COOH group, stereochemistry depending on conditions citrate-ion could be as bridging carboxylate ligand as well as bi-dentate or tri-dentate bridging carboxylate ligand.

The supposed structure of the $M^I_2 \cdot M^{II} \cdot L_2 \cdot nH_2O$ type compounds shows that polyhedrons around $M^I$ atoms are pseudo-octahedrons with double four-member chelate cycles and pairs of water molecules, while around the $M^{II}$ atoms, the pseudo-tetrahedral coordination is realized by oxy-groups of deprotonated carboxylic acid. It can be presented as in Figure 4.1.

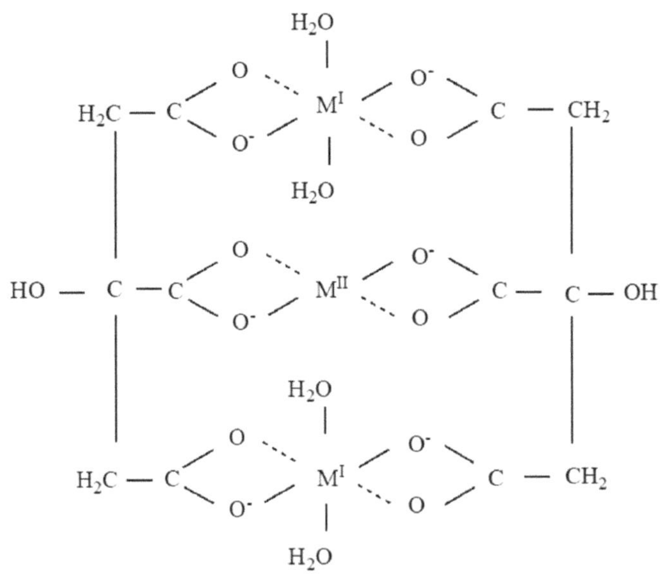


**FIGURE 4.1** Possible configuration of $M^I_2 \cdot M^{II} \cdot L_2 \cdot nH_2O$ type compounds.

## 4.5 CONCLUSION

Basis of carried out experiments it can be concluded:

- $M^{I}_{2}\cdot M^{II}\cdot L_{2}\cdot nH_{2}O$ type heteronuclear chelate citrates are soluble in water at a heating time while in the other solvents, they are characterized by low solubility or insolubility at room temperature and heating time as well.
- Thermal degradation of citrates occurred step-wise: I-water molecule elimination, II-citrate-ion oxidation gradually or simultaneously. Thermolysis final products are (a) mixture of metal-oxides; (b) mixture of metal-oxides and coal; or (c) metals.
- Synthesized compounds are internal type chelate compounds, and as it was expected there is occurred the deprotonation of the three carboxylic groups of both citrate acid and the formation of ionic and coordinative bonds with metal atoms. The structural stability is provided by successfully deployed intra-molecular and inter-molecular hydrogen bonds, which reduce the strain (especially inside of the tetrahedron) inside of four-member chelate cycles leading to chelate coordination of carboxylate groups.

## ACKNOWLEDGMENTS

We thank the Science and Technology Center in Ukraine and Shota Rustaveli National Science Foundation for financial support. The work was implemented with the support of the Science and Technology Center in Ukraine Project Proposal #5461 and Shota Rustaveli National Science Foundation Grant #30/06.

## KEYWORDS

- **chelate**
- **citrate-ion**
- **heteronuclear**
- **physical-chemical**
- **spectrophotometric**
- **thermolysis**

## REFERENCES

1. Loginov, G. P., (2005). Effect of metal chelates with amino acids and protein hydrolysates on the productive functions and metabolic processes in animal body. *Doctor's Thesis, Kazan*, pp. 3–59.
2. Kalashnikov, A. P., Fisinin, V. I., Shchuglov, V. V., Kleimanov N. A., et al., (2003). *Norms and Rations in the Feeding of Agricultural Animal,* Moscow, p. 456.
3. Kochetkova, N. A., (2009). Influence of metal citrates on biochemical indices of tissues and organs of chicken-broilers and the quality of the products. *Abstract of a Thesis, Specialty Biochemistry.*
4. Kochetkova, N. A., Shaposhnikov, A. A., Smirova, A. V., et al., (2009). Investigation of structure of zinc, cobalt, manganese, iron (II) citrates by IR-spectroscopy. *J. Scientific Bulletin of the State University of Belgorod. Series Natural Science, 3*(8), 380–387.
5. Kochetkova, N. A., Shaposhnikov, A. A., Zateev, S. V., et al., (2010). Citrates of biometals in the chicken-broiler's diet. *J. Poultry Raising*. www.webpticeprom.ru (accessed on 19 October 2020).
6. Boiko, I., & Miroshnichenko, I., (2011). Application of manganese citrate in rearing chicken-broilers, *J. Poultry, Poultry Factory,. 11*, 110–117.
7. Lebedev, S. V., Miroshnikov, S. A., Sukhanova, O. N., Rakhmatullin Sh. G., (2009). Method of elevation of productivity of broiler-chickens, Russian Federation, Patent for invention, # 2370095 A23K 1/00.
8. Beshkenadze, I. A., Gogaladze, M. A., Zhorzholiani, N. B., et al., (2013). Synthesis of the chelates continuing amino acids and citric acid for creation of new generation premixes. *J. Annals of Agrarian Science, 11*(2), 84–86.
9. Beshkenadze, I., Gogaladze, M. A., Zhorzholiani, N., et al., (2008). *Chemical Admix to Poultry Fodder* (pp. 4917, 4918). National Center for Intellectual Property of Georgia, Sakpatenti, Tbilisi.
10. Tsintsadze, G., Beshkenadze, I., Zhorzholiani, N., et al., (2006). Physical-chemical investigation of $M^I M^{II} L \cdot nH_2O$ heteronuclear citrates. *J. Proceedings of the Georgian Academy of Sciences, Chemical Series, 32*(3/4), 248–252.
11. El-Said, A. I., Zidan, A. S. A., El-Meliqy, M. S., et al., (2001). Coordination properties of some mixed amino acid metal complexes. *J. Synth. and Rect. Inorg. and Metal-Org. Chem., 31*(4), 633–648.
12. Zhorzholiani, N. B., Beshkenadze, I. A., Gogaladze, M. A., et al., (2014). Synthesis and study heteronuclear citrates. *Journal of Chemistry and Chemical Engineering, 8*(4), 385–390.
13. Beshkenadze, I. A., Zhorzholiani, N. B., Gogaladze, M. A., et al., (2011). *Possibility of Obtaining the New Generation Premixes Containing Bio-Metals and Natural Zeolites* (pp. 94–97). Transactions of P. Melikishvili Institute of Physical and Organic Chemistry.
14. Beshkenadze, I. A., Tsitsishvili, V. G., Gogaladze, M. A., et al., (2012). *Natural Zeolites and Biometal-Containing Composites* (pp. 13–17). Materiali IV Mezhdunarodnoi Konferencii-Sorbenti kak Faktor Kachestva Zhizni i zdorovya, Belgorod (In Russian).
15. Beshkenadze, I. A., Urotadze, S. L., Zhorzholiani, N. B., et al., (2014). Study of physiological activity of bio-coordinated substances and their premixes with natural zeolite-containing composites. *International Scientific Journal Intellectual, 26*, 245–250.

16. Beshkenadze, I. A., Zhorzholiani, N. B., Gogaladze, M. A., et al., (2014). *Chelate-Containing Admixes to Poultry Fodder*. National Center for Intellectual Property of Georgia, Sakpatenti, Tbilisi, U 1787.
17. Beshkenadze, I. A., Zhorzholiani, N. B., Gogaladze, M. A., et al., (2014). *Chemical Admix for Poultry Nutrition*. National Center for Intellectual Property of Georgia, Sakpatenti", Tbilisi, U1800.
18. Beshkenadze, I. A., Zhorzholiani, N. B., Gogaladze, M. A., et al., (2014). Synthesis and study of heteronuclear citrates. *Journal of Chemistry and Chemical Engineering, 8*(4), 385–390.
19. Bellami, L., (1963). Infrared Spectra of Complicated Molecules, M. p. 590.
20. Roelofsen, G. B., & Kantera, I. A., (1972). Citric acid monohidrate $C_6H_8O_7{\cdot}H_2O$. *Crist. Struct. Commun., 1*(1)23–26.

# CHAPTER 5

# Coumarone-Indene Resins with Functional Groups

MARIIA SHVED, OLENA SHYSHCHAK, OLENA ASTAKHOVA, and MICHAEL BRATYCHAK

*Lviv Polytechnic National University, 12, S. Bandery St., 79013 Lviv, Ukraine, E-mail: mbratychak@gmail.com (M. Bratychak)*

## ABSTRACT

Coumarone-indene resins (CIR) with various functional groups have been synthesized on the basis of the light fraction of coal tar (LFCT). The resins with epoxy, carboxy groups, and methacrylic fragments were synthesized. The initial LFCT contains coumarone 5.75 wt.%, indene 44.45 wt.% and styrene 6.73 wt.%. CIR was synthesized *via* initiated polymerization. 2,2'-Azobis(2-methylpropionitrile) and peroxy derivative of dioxydiphenylpropane diglycidyl ether (PO) were used as initiators. To increase the yield of CIR and to introduce corresponding functional groups into its structure, styrene, glycidyl methacrylate (GMA), or methacrylic acid (MAA) were added to LFCT. The effect of initiator amount, reaction temperature, and time on the yield and characteristics of CIR has been examined. The structure of resulting products was confirmed by chemical and spectral methods. It was proposed to use CIR as additives for the production of polymer bitumen compositions.

## 5.1 INTRODUCTION

When building modern roads, petroleum bitumen and polymeric additives are used as components of road materials [1–3]. It is recommended to use elastomers, thermoplastic, and reactive polymers, as well as various resins as polymeric additives [1]. The introduction of the polymeric component into the composition of road bitumen significantly improves the performance characteristics of the coating, in particular, adhesion to mineral fillers, elasticity, and reduces brittleness [4, 5].

Over the past few years, the researchers of the Department of Chemical Technology of Oil and Gas Processing at Lviv Polytechnic National University have been conducting studies on the possibility of using petroleum [6–8], phenol formaldehyde [9], and epoxy [10] resins as polymeric additives. A particular attention should be paid to petroleum resins (PRs), because they are synthesized from liquid by-products obtained after hydrocarbon raw materials pyrolysis intended for ethylene production [11–12]. The C9 fraction of liquid pyrolysis products containing such unsaturated hydrocarbons as styrene, vinyl toluenes, dicyclopentadiene, indene, etc., was used to synthesize PRs with epoxy [13], carboxy [14], and hydroxy [15] groups. PRs with various functional groups are used as additives during oil tar oxidation to bitumen and for the preparation of bitumen-polymer mixtures (BPM) [16]. The joint oxidation of tar and PRs was carried out at 523 K for 3 h with an air volumetric flow rate of 2.5 $min^{-1}$. The amount of PRs in the raw material mixture was 1–10 wt.%. The increase in the amount of PRs with epoxy or carboxy groups increases the softening temperature of bitumen, its ductility, and adhesion [8, 16]. At the same time, the penetration of bitumen, which characterizes its hardness, decreases. The optimum amount of PRs in such a mixture is 5 wt.%.

When preparing BPM, PRs with epoxy, carboxy, and hydroxy groups [16] have been studied as polymer components. As in the previous case, the amount of PRs in the mixture ranged from 1 to 10 wt.%. The introduction of PRs in bitumen makes it possible to significantly increase the adhesion of commercial bitumen to mineral fillers, with the best results achieved in the case of using PRs with carboxy groups in an amount of 5 wt.% [16].

Good results in the creation of BPM were also achieved in the case of using phenol-formaldehyde and epoxy resins (ERs) containing different functional groups [8, 9]. The disadvantage of such mixtures is the expensiveness of resins used. Therefore, the search for the raw materials, which would give the opportunity to obtain cheap resins (polymers), is still relevant.

It is known [17] that a coal tar resin is a by-product of coal coking. As a result of the coal tar rectification, a coumarone-indene fraction (CIF) [18] and a light fraction of coal tar (LFCT) are obtained. Both fractions contain sufficient amounts of indene, coumarone, styrene, and other unsaturated compounds, and therefore they are used to produce the so-called coumarone-indene resins (CIR) which may be used as polymer components for the creation of BPM [19].

Researches on the development of new methods for CIR obtaining and the introduction of various functional groups into their structure are promising because functional groups increase the adhesive properties of the resulting products and expand the possibilities of their use. This work presents the results regarding the obtaining of LFCT-based CIR with functional groups, and the possibility of using these resins as additives to petroleum road bitumen in order to create BPM.

## 5.2 EXPERIMENTAL

### *5.2.1 MATERIALS*

The initial LFCT with the density $d_4^{20} = 0.9702$ was taken from JSC "Zaporizhkoks" (Ukraine). Among its components (Table 5.1) are styrene (6.73 wt.%), coumarone (5.75 wt.%) and indene (44.45 wt.%).

**TABLE 5.1** Components Content in Raw Materials

| Component | Component Content, wt.% |
|---|---|
| Benzene | 0.98 |
| Toluene | 1.23 |
| Ethylbenzene | 0.32 |
| m + p-Xylenes | 8.10 |
| o-Xylene | 2.61 |
| Styrene | 6.73 |
| o-Ethylbenzene | 4.35 |
| Pseudocumene | 7.71 |
| Hydrindane | 2.64 |
| Coumarone | 5.75 |
| Indene | 44.45 |
| Naphthalene | 6.96 |
| b-Methylnaphthalene | 0.31 |
| a-Methylnaphthalene | 0.13 |
| Other hydrocarbons | 7.73 |
| **Total** | **100.0** |

The following compounds were used for CIR synthesis:

- Styrene, which was dried before the experiments with solid alkali and purified by distillation at 323 K under residue pressure of 300–400 Pa. Its characteristics: refractive index $n_D^{20}$ = 1.5471 (literature value $n_D^{20}$ = 1.5468); density $d_4^{20}$ = 0.902.
- Methacrylic acid (MAA) purchased from Aldrich with $d_4^{20}$ = 1.015.
- Glycidyl methacrylate (GMA) purchased from Aldrich (USA) with $d_4^{20}$ = 1.042, $n_D^{20}$ = 1.449.
- Maleic anhydride (MA) purchased from Ukrorgsintez (Kyiv, Ukraine) with melting point 325.8 K and boiling point 475 K.
- 2,2'-azobis(2-methylpropionitrile) in the form of 0.2 M solution in toluene (AMP) purchased from Aldrich was used as the initiator of LFCT co-oligomerization. It was found for the solution $d_4^{20}$ = 0.858, $n_D^{20}$ = 1.495.
- Monoperoxy derivative of diglycidyl ether Bisphenol A (PO) was used as an initiator:

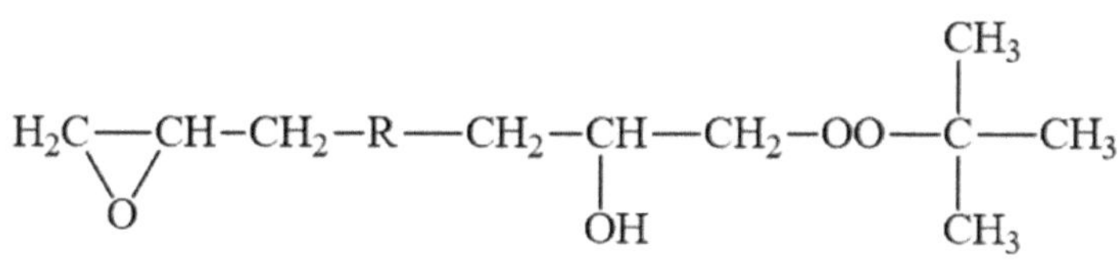


where; R = $–OC_6H_4C(CH_3)_2C_6H_4O–$.

PO was synthesized according to the modified method, taking into account the results presented in [20]. A three-necked reactor equipped with a mechanical stirrer, backflow condenser and thermometer, was loaded with 100 g of Bisphenol A diglycidyl ether and 100 ml of toluene. Separately a mixture was prepared consisting of 84.0 g of 70% aqueous solution of *tert*-butyl hydroperoxide dissolved in 100 ml of toluene, 19.1 g of benzyl triethyl ammonium chloride and 3.9 g of crystalline KOH. The solution of the diglycidyl ether in toluene was heated to 323 K under stirring and a mixture containing hydroxides and catalysts was added to it through a dropping funnel. Then the mixture was kept at 323 K for 6 h, cooled to room temperature, and transferred to a separating funnel. After washing the catalysts, the organic layer was vacuumized at 323–333 K. The residue was re-precipitated in petroleum ether and dried under vacuum to a constant weight. 120.8 g of the resulted product with a molecular weight 420 g/mol, active oxygen content 2.5% and content of epoxy groups 8.4% were obtained.

- Toluene purchased from Aldrich.
- Petroleum ether was a fraction with a boiling range of 313–343 K and $d_4^{20} = 0.650$.
- Methyl methacrylate (MMA) purchased from VWR Prolabo Chemicals (CAS: 80-62-6) with $d_4^{20} = 0.935$. Apart from MMA the styrene was used as an additional monomer.
- Bitumen BND 60/90 produced at the Kremenchug PJSC "Ukrtatnafta" (Ukraine) with the softening point 319 K, penetration $70 \cdot 10^{-4}$ m, ductility $63 \cdot 10^{-2}$ m, and adhesion to glass 47%.

### *5.2.2 SYNTHESIS PROCEDURE*

LFCT co-oligomerization was carried out in metal ampoules by the capacity of 100 ml. The ampoules were loaded by the corresponding mixture, blown with an inert gas, closed, and placed into a thermostat. After finishing the process the ampoules were cooled to the definite temperature and the matter was precipitated using petroleum ether. The precipitated product was dried in vacuum oven at 313 K.

#### *5.2.2.1 SYNTHESIS OF COUMARONE-INDENE RESINS (CIR) WITH CARBOXY GROUPS (CIRC)*

To synthesize CIRC the radical co-oligomerization of LFCT was carried out according to the procedure described in Section 2.2. 50 ml (48.5 g) of LFCT, 8 ml (7.2 g) of styrene, 2 ml (2.1 g) of MAA and 7.5 ml (0.25 g) of AMP were loaded. The co-oligomerization time was 6 h, co-oligomerization temperature was 353 K.

#### *5.2.2.2 SYNTHESIS OF COUMARONE-INDENE RESINS (CIR) WITH EPOXY GROUPS (CIRE)*

To synthesize CIRE the radical co-oligomerization of LFCT was carried out at 393 K for 6 h according to the procedure described in Section 2.2. The composition of the reaction mixture was: 50 ml (48.5 g) of LFCT, 8 ml (7.2 g) of styrene, 4 ml (4.2 g) of GMA and 2.5 g of PO.

#### *5.2.2.3 SYNTHESIS OF COUMARONE-INDENE RESINS (CIR) WITH METHACRYLIC FRAGMENTS (CIRM)*

To synthesize CIRM the radical co-oligomerization of LFCT was carried out at 353 K for 6 h according to the procedure described in Section 2.2. The composition of the reaction mixture was: 50 ml (48.5 g) of LFCT, 10 ml (9.35 g) of MMA, and 7.5 ml (0.25 g) of AMP.

### 5.2.3 CIR ANALYSIS

The yield relative to the initial reaction mixture ($X$) was calculated according to the formula (1).

$$X = \frac{m_r}{m_{in}} \cdot 100\% \tag{1}$$

where; $m_r$ is a weight of the resulted resin, g; $m_{in}$ is a weight of the initial reaction mixture, g.

The number-average molecular weight ($M_n$) of the synthesized CIR was determined using cryoscopy in dioxane. The content of carboxy groups was determined using a back titration method [21]. Content of epoxy groups (epoxy number (e.n.)) was determined according to the procedure described in [22]. The softening point of CIR was determined using ring-and-ball method [23].

### 5.2.4 SPECTRAL MEASUREMENTS

Infrared spectra of CIR were measures using Nicolet IR 200 (Thermo Electron Co., USA) with Golden Gate ATR diamond crystal. Every spectrum was recorded with 4 $cm^{-1}$ resolution. Samples were prepared as powders or were dissolved in acetone.

$^{1}H$ and $^{13}C$ NMR spectra were recorded at 400 MHz and 100 Hz, respectively, using Bruker Avance II 400 spectrometer (Poland) in deuterochloroform at room temperature.

### 5.2.5 THERMAL ANALYSES

Thermal studies were carried out to determine the decomposition temperature of PO relative to the peroxy group using Q-1500 D derivatograph (F. Paulik–J.

Paulik – L. Erdey). The samples were analyzed under dynamic mode at the heating rate of 2.5 K/min in air atmosphere. The sample was heated to a temperature of 523 K; its weight was 300 mg. The reference substance was alumina. TG sensitivity was 100 mg, DTA sensitivity 100 mV, DTG sensitivity 500 mV.

### 5.2.6 PROCEDURE FOR PREPARING BITUMEN-POLYMER MIXTURES (BPM)

Bitumen-polymer mixtures (BPM) were prepared as follows: the necessary amount of bitumen was heated till a definite temperature, then a modifier was added, and the mixture was stirred (Re = 1200) for a definite time.

### 5.2.7 ANALYSIS OF BPM

The softening point, penetration, ductility, adhesion to glass and the content of asphaltenes, resins, and oil were determined according to the methods described in Refs. [23–27].

## 5.3 RESULTS AND DISCUSSION

### 5.3.1 COUMARONE-INDENE RESINS WITH CARBOXY GROUPS (CIRC)

Under certain conditions styrene, coumarone, and indene present in LFCT are polymerized and able to form CIR. However, their amounts are insufficient. Therefore, to increase the yield of the resulting product we additionally introduce styrene and MAA into the reaction mixture. The initiator was 2,2'-azobis(2-methylpropionitrile) in the form of 0.2 M solution in toluene (AMP). To determine the optimum conditions for CIRC synthesis we studied the effect of AMP amount, temperature, and polymerization time on the yield and characteristics of the product.

#### 5.3.1.1 EFFECT OF VARIOUS PARAMETERS ON CIRC CHARACTERISTICS

The effect of AMP amount, temperature, and polymerization time on the yield and characteristics of CIRC are represented in Tables 5.2–5.4.

**TABLE 5.2** Effect of Initiator Amount on CIRC Characteristics

| AMP Amount, g | CIRC Characteristics | | | |
|---|---|---|---|---|
| | Softening Point, K | Content of COOH– Groups, % | $M_n$ | Yield Relative to the Initial Mixture, % |
| 0.00 | | Co-oligomerization Does Not Occur | | |
| 0.125 | 375 | 16.0 | – | 7.8 |
| 0.25 | 392 | 13.9 | 960 | 22.4 |
| 0.50 | 392 | 12.9 | 900 | 15.0 |
| 0.75 | 386 | 13.1 | – | 13.5 |
| 1.00 | 381 | 12.7 | – | 11.6 |

**Notes:** Co-oligomerization temperature 353 K, time 6 h. Amounts in the reaction mixture LFCT 48.5 g, styrene 7.2 g, MAA 2.1 g.

One can see from Table 5.2 that the highest yield is observed in the case of using AMP in the amount of 0.25 g per 48.5 g of the initial LFCT. Further increase in the AMP amount (0.5, 0.75, and 1.0 g) does not increase the CIRC yield but even decreases it. The reason is that apart from initiated co-oligomerization of unsaturated compounds the reaction resulting in the recombination of radicals formed by the AMP decomposition occurs (Scheme 5.1). The increase in AMP amount promotes this reaction. The extraction of formed free radicals from the system leads to the decrease in CIRC yield, the formation of which takes place only *via* initiated co-oligomerization.

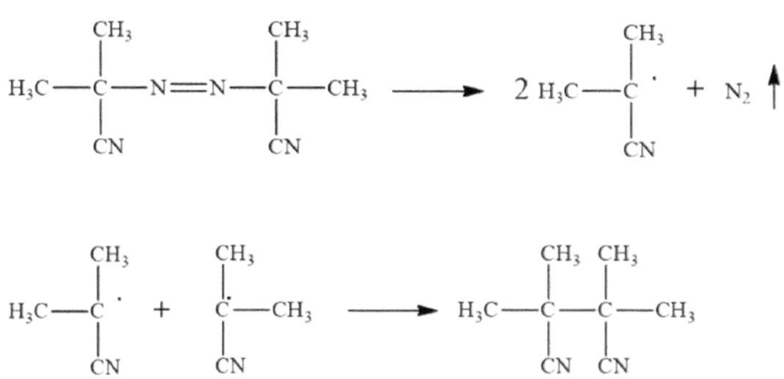


**SCHEME 5.1** Recombination of radicals formed by the AMP decomposition.

The results represented in Table 5.3 show that the maximum CIRC yield is observed at co-oligomerization time of 6 h. Further increase in a process time decreases the yield and content of carboxy groups in CIRC. It means that apart from co-oligomerization reaction, a complexation of

LFCT components takes place. The formation of additional complexes at the initial stage increases the CIRC yield. The heating for more than 6 h results in possible destruction of the complexes and further co-oligomerization providing the decrease in content of carboxy groups due to the entering of styrene, indene, and coumarone molecules into the structure.

**TABLE 5.3** Effect of Co-Oligomerization Time on CIRC Characteristics

| Time, h | CIRC Characteristics | | | |
|---|---|---|---|---|
| | Softening Point, K | Content of COOH– Groups, % | $M_n$ | Yield Relative to the Initial Mixture, % |
| 2 | 384 | 14.9 | – | 13.7 |
| 4 | 393 | 14.9 | – | 18.4 |
| 6 | 392 | 13.9 | 960 | 22.4 |
| 8 | 391 | 13.4 | – | 17.1 |
| 10 | 377 | 12.2 | – | 16.9 |

**Notes:** Co-oligomerization temperature 353 K. Amounts in the reaction mixture LFCT 48.5 g, styrene 7.2 g, MAA 2.1 g, AMP 0.25 g.

**TABLE 5.4** Effect of Co-Oligomerization Temperature on CIRC Characteristics

| *T*, K | CIRC Characteristics | | | |
|---|---|---|---|---|
| | Softening Point, K | Content of COOH– Groups, % | $M_n$ | Yield Relative to the Initial Mixture, % |
| 333 | Co-oligomerization Does Not Occur | | | |
| 343 | 391 | 13.7 | – | 11.7 |
| 353 | 392 | 13.9 | 960 | 22.4 |
| 363 | 394 | 14.3 | – | 12.3 |

**Notes:** Co-oligomerization time 6 h. Amounts in the reaction mixture LFCT 48.5 g, styrene 7.2 g, MAA 2.1 g, AMP 0.25 g.

Regarding the temperature effect on CIRC yield (Table 5.4) it was determined that CIRC are actually not formed at 333 K. At 363 K the CIRC yield is decreased (compared with that at 353 K) due to the rapid decomposition of initiator molecules and recombination of formed radicals according to Scheme 5.1. In addition, the co-oligomerization reaction in the presence of MAA is accelerated at 363 K. The increased content of the carboxy groups in CIRC confirms this fact.

Thus, the optimum conditions for obtaining CIRC were determined: co-oligomerization temperature 353 K; time 6 h; mixture consists of LFCT

48.5 g, styrene 7.2 g, MAA 2.1 g, and AMP 0.25 g. The resin obtained under optimum conditions is characterized by the molecular weight of 960 g/mol and carboxy group content of 13.9%. CIRC yield is 22.4% relative to the initial mixture.

### 5.3.1.2 IR-SPECTROSCOPIC INVESTIGATIONS OF CIRC

IR spectroscopy was used to confirm the CIRC structure synthesized under optimum conditions and the presence of free carboxy groups in its. IR-spectroscopic investigations were carried out according to the procedure described in Section 2.4. The obtained results are represented in Figure 5.1.

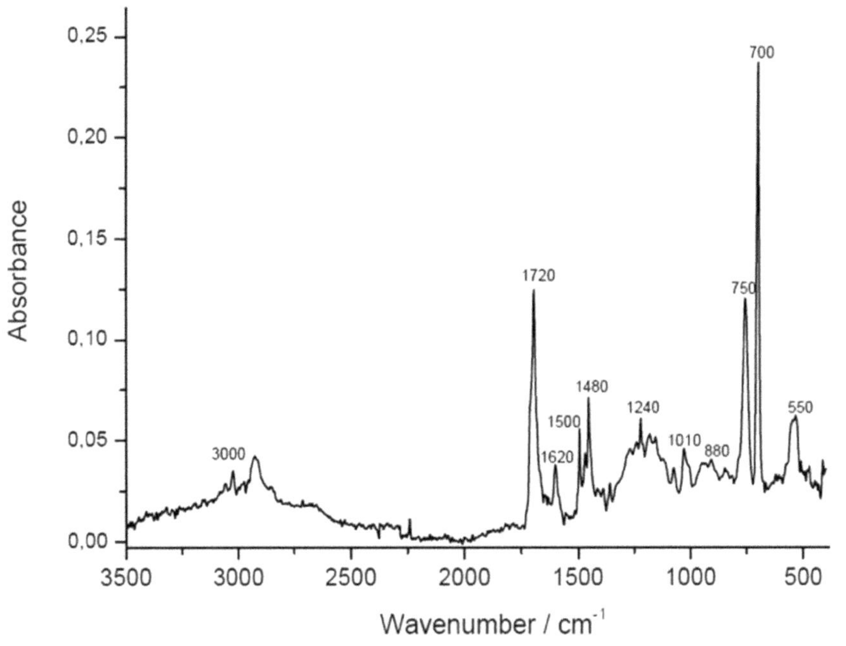


**FIGURE 5.1** IR-spectrum of CIRC.

The presence of coumarine fragments in the CIRC molecules is confirmed by the absorption bands at 1200–1100 $cm^{-1}$, typical of ether bond in alycyclic hydrocarbons. The fragments of aromatic compounds which characterize styrene, indene, coumarine, and hydroindene are confirmed by three absorption bands at 1620, 1500, and 1480 $cm^{-1}$. The presence of free carboxy groups is confirmed by the absorption band at 1720 cm$^{-1}$, corresponding to the stretching vibrations of carbonyl group in acids. The absorption bands at 3000 and 1010 cm$^{-1}$ indicate the presence of acid hydroxy group.

### *5.3.2 COUMARONE-INDENE RESINS (CIR) WITH EPOXY GROUPS (CIRE)*

Taking into account that content of unsaturated compounds in the initial LFCT is low (Table 5.1) we used additional styrene. GMA was used as a monomer to introduce epoxy groups into the structure of ICS. PO was used as the initiator of radical co-oligomerization of the reaction mixture.

In order to develop the CIRE synthesis procedure in the presence of PO, it was necessary to study the thermal stability of the initiator, as well as the effect of its amount, temperature, and reaction time on CIRE characteristics.

#### *5.3.2.1 PO THERMAL STABILITY*

In order to use PO as the initiator, it is necessary to know about its thermal stability, i.e., the temperature under which the –O–O– bonds are decomposed and free radicals are formed. The results of thermal studies are represented in Figure 5.2.

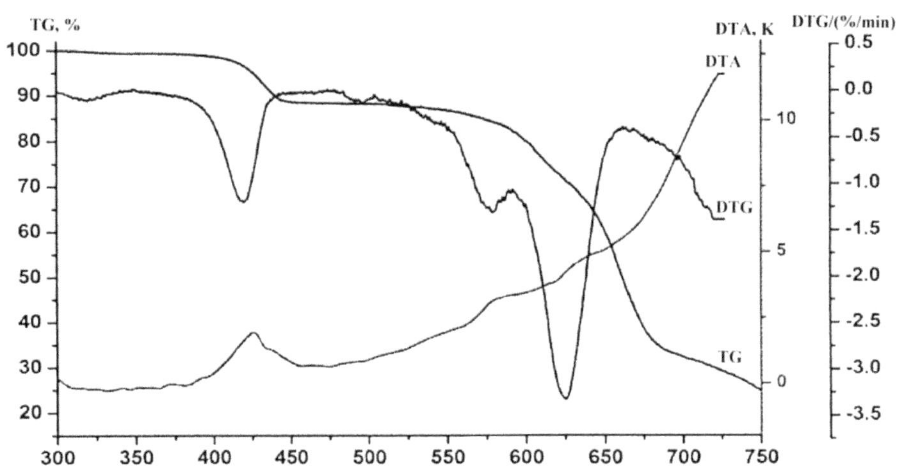


**FIGURE 5.2** TG, DTG, and DTA curves for PO.

Thermolysis kinetic parameters were calculated at the stage of peroxy groups using Mathcad 2011 Professional software on the basis of the modified kinetic equation and least-squares method, according to Eq. (2):

$$\ln \frac{dw}{w_k \cdot dt} = \ln \frac{Z}{q} \tag{2}$$

where; $w$ is a weight loss of PO at a definite temperature, mg; $w_k$ is a total weight loss of PO at the definite stage, mg; $Z$ is a preexponential factor, $s^{-1}$; $E$ is an activation energy, kJ/mol; $R$ is the universal gas constant ($R = 8.314$ J/mol·K); $q$ is a heating rate of PO, grad/min. The results obtained, according to Eq. (2), are represented in Table 5.5.

**TABLE 5.5** Kinetic Parameters of PO Thermal Stability

| Parameters | Values |
|---|---|
| Initial temperature of weight loss, K | 393 |
| Final temperature of weight loss, K | 475 |
| Maximum exoeffect, K | 440 |
| Reaction order | 1.4 |
| $Z$, $s^{-1}$ | $1.33 \cdot 10^{18}$ |
| $E$, kJ/mol | 168 |
| $k \cdot 10^2$, $s^{-1}$ | 0.89 |

The results presented in Figure 5.2 and Table 5.5 show that the –O–O– bond in PO is decomposed already at 393 K. Moreover, the most intense decomposition with the formation of free radicals is observed at 440 K. On the basis of obtained results and literature data [28] the formation of free radicals and chemistry of PO decomposition may be represented by the following equations:

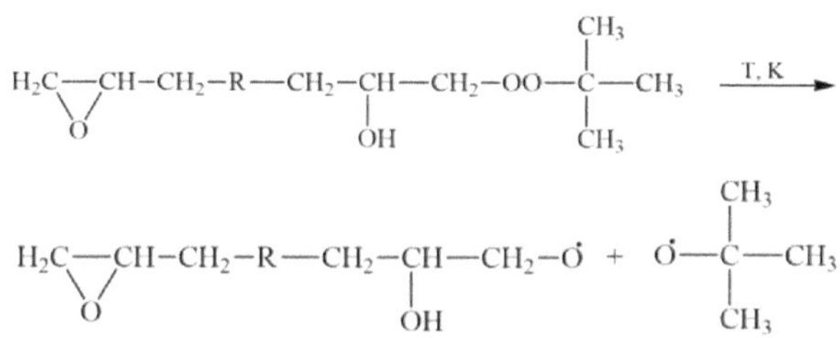


($\dot{R}_1$) ($\dot{R}_2$) (3)

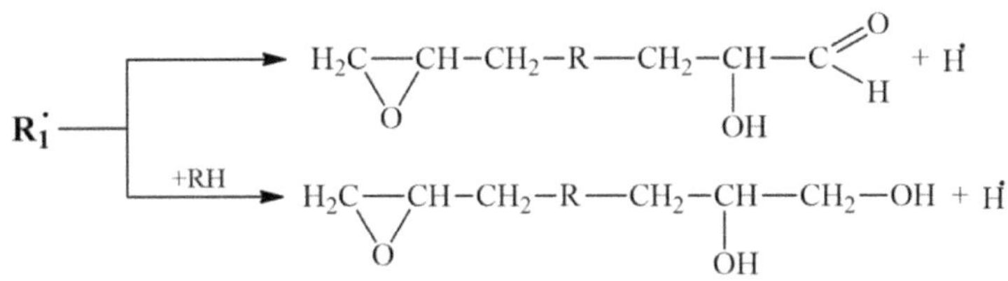

(4)

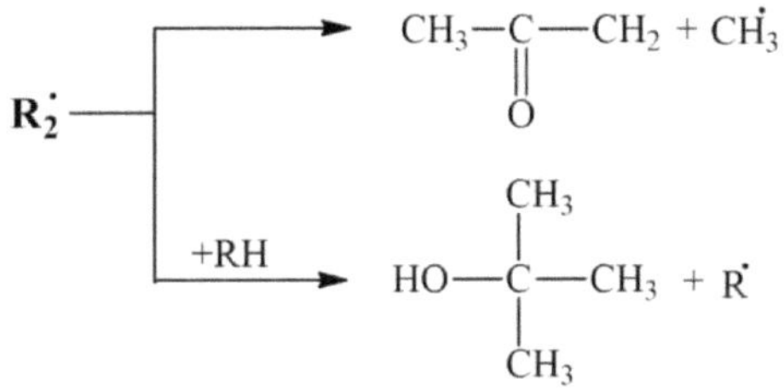

(5)

$$\dot{C}H_3 + \dot{H} \longrightarrow CH_4 \tag{6}$$

$$\dot{C}H_3 + \dot{R} \longrightarrow R{-}CH_3 \tag{7}$$

$$\dot{R} + \dot{R} \longrightarrow R{-}R \tag{8}$$

So, the weight loss of PO at low temperatures occurs due to the transformation of radicals and formation of volatile products such as methane, acetone, and *tert*-butyl alcohol. The formation of free radicals is observed at 393 K.

### 5.3.2.2 EFFECT OF VARIOUS PARAMETERS ON CIRE CHARACTERISTICS

The experimental regards regarding the effect of initiator amount, temperature, and process time on the yield and characteristics of CIRE are represented in Tables 5.6–5.8.

**TABLE 5.6** Effect of Initiator Amount of CIRE Characteristics

| Amount of PO, g | CIRE Characteristics | |
|---|---|---|
| | **Softening Point, K** | **Yield Relative to the Initial Mixture, %** |
| 0 | 349 | 6.6 |
| 1.25 | 337 | 10.0 |
| 2.50 | 336 | 14.9 |
| 5.00 | 376 | 20.4 |

**Notes:** Co-oligomerization temperature 393 K, time 6 h. Amounts in the reaction mixture: LFCT 48.5 g, styrene 7.2 g, GMA 4.2 g.

**TABLE 5.7** Effect of Co-Oligomerization Time on CIRE Characteristics

| *τ*, h | CIRE Characteristics | |
|---|---|---|
| | **Softening Point, K** | **Yield Relative to the Initial Mixture, %** |
| 2 | 308 | 10.9 |
| 4 | 325 | 18.6 |
| 6 | 336 | 14.9 |
| 8 | 315 | 12.0 |
| 10 | – | 5.1 |

**Notes:** Co-oligomerization temperature 393 K. Amounts in the reaction mixture: LFCT 48.5 g, styrene 7.2 g, GMA 4.2 g, PO 2.5 g.

**TABLE 5.8** Effect of Temperature on CIRE Characteristics

| *T*, K | CIRE Characteristics | |
|---|---|---|
| | **Softening Point, K** | **Yield Relative to the Initial Mixture, %** |
| 373 | 325 | 13.1 |
| 393 | 336 | 14.9 |
| 413 | 335 | 13.9 |

**Notes:** Co-oligomerization time 6 h. Amounts in the reaction mixture: LFCT 48.5 g; styrene 7.2 g; GMA 4.2 g, PO 2.5 g.

The results (Table 5.6) show that at 393 K the CIR are formed even in the absence of PO. It means that the monomers in the initial mixture are capable to form resins at this temperature according to the above-mentioned scheme. When using PO as the initiator, CIRE yield is increased. This fact confirms the assumption that apart from thermal polymerization of the mixture components, the initiation polymerization takes place as well due to the decomposition of PO molecules according to Eqns. (3–8) resulting in formation of free radicals. At the same time PO fragments are introduced into the structure of resulting resin. Thus, the highest yield was observed when using PO in the amount of 2.5 g relative to the initial mixture.

Table 5.7 demonstrates the effect of the process time on the CIRE yield. The highest yield is achieved at the process time of 4–6 h. Further increase in the process time decreases the yield. This fact is explained by destruction processes occurred at high temperature (393 K) and formation of low-molecular products soluble in petroleum ether and hence incapable of precipitation.

The reaction temperature also affects the CIRE yield (Table 5.8). The increase in temperature from 373 to 393 K increases the CIRE yield. The increase in temperature to 413 K decreases the yield of the target product. The reason is deactivation of free radicals at high temperature (*vide* Eqns. (3–8)) resulting in the reduction of the share of the reaction involved in the CIRE formation.

Thus, the following optimal conditions for CIRE synthesis are proposed: the reaction temperature 393 K, reaction time 6 h and PO amount (50% toluene solution) is 2.5 g relative to the initial mixture.

CIR with epoxy groups synthesized under optimal conditions are characterized by the yield of 14.9; softening temperature of 336 K; molecular weight of 1400 g/mol and epoxy number of 2.3%.

### *5.3.2.3 SPECTROSCOPIC INVESTIGATIONS*

IR-spectroscopy was used to confirm the structure of the synthesized resins. The investigations were carried out according to the procedure described in Subsection 2.4. The results are represented in Figure 5.3.

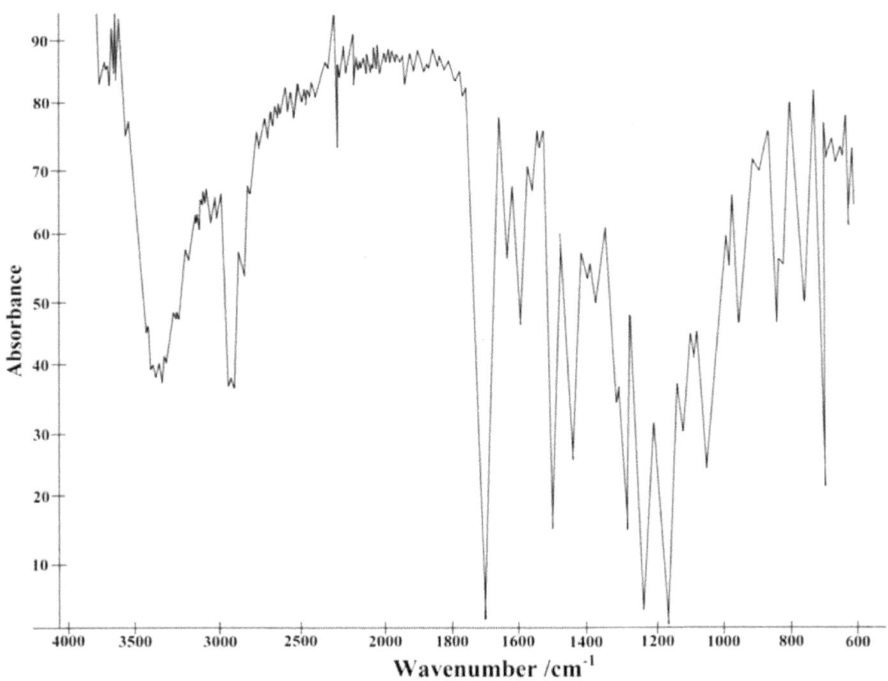


**FIGURE 5.3** IR-spectrum of CIRE.

The presence of epoxy groups in the synthesized resin is confirmed by the absorption band at 944 $cm^{-1}$. The presence of OH group in coumarone-indene resin formed due to the initiation of unsaturated compounds in the initial mixture by PO is confirmed by the absorption band at 3368 $cm^{-1}$. Moreover, the $H_3C-\overset{|}{\underset{|}{C}}-CH_3$ fragment in the resin structure is confirmed by gel-dimethyl vibrations at 1384 and 1356 $cm^{-1}$; the etheric bond at 1172 $cm^{-1}$; benzene rings at 1600 and 1512 $cm^{-1}$. The entering of GMA molecule with the esoteric bond into the resin structure is confirmed by absorption bands at 1720, 1244 and 1046 $cm^{-1}$.

### *5.3.2.4 COPOLYMERIZATION OF MODEL SYSTEMS*

The synthesized CIRE contains epoxy groups due to the presence of free epoxy and peroxy groups in the structure of initiator PO. At the polymerization temperature of 393 K, the decomposition occurs, and peroxy groups form two free radicals, according to Eq. (9):

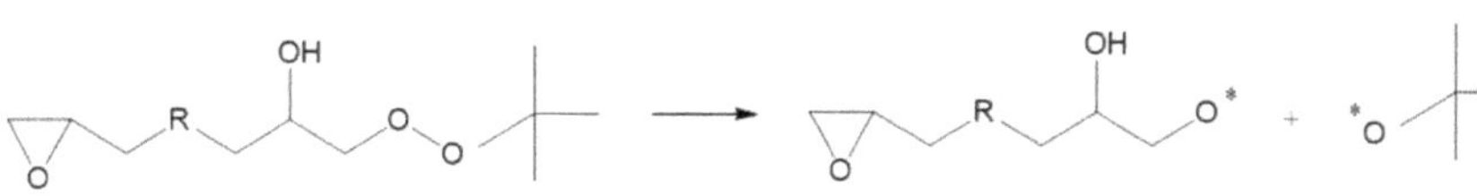


(A) (B) (9)

The radical (A) causes the copolymerization of resin-forming components of LFCT, namely indene, coumarone, styrene, as well as additionally introduced glycidyl methacrylate (GMA). Thus, we assume that the synthesized CIRE should have the following structure:

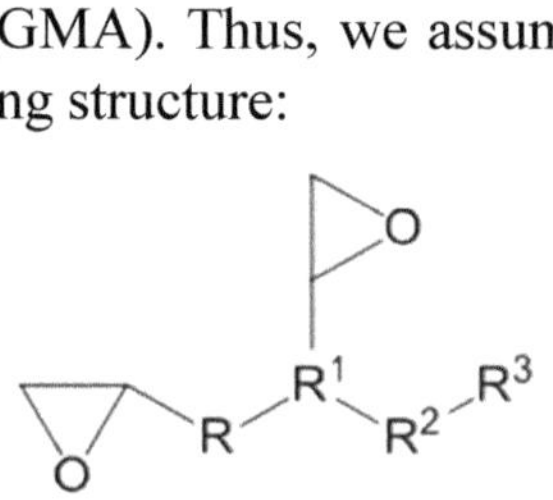


(10)

The epoxy groups in CIRE are at the ends of the molecule and in the side chains. The presence of epoxy groups in the latter ones is explained by GMA copolymerization with the molecules of styrene, indene, and coumarone. The epoxy group at the end of the molecule is introduced due to the initiation of components polymerization by the radical (A). The radical (B) actually does not participate in the copolymerization reactions but is converted into acetone and *tert*-butyl alcohol.

It was important to determine the capability of resin-forming components in LFCT to copolymerize and form the above-mentioned structure. For this purpose, we prepared the model systems consisting of the initiator PO and GMA of indene, or coumarone, or styrene, or the monomers mixture (Table 5.9).

**TABLE 5.9** Composition of the Model Systems and Characteristics of the Synthesized Resins

| Mixture Number | Initial Monomers/ Amount, g | Resins Characteristics | | |
|---|---|---|---|---|
| | | Yield, % | e.n., % | $M_n$, g/mol |
| I | Indene/50.0 | 74.9 | 15.6 | 550 |
| II | Coumarone/50.0 | 37.2 | 17.6 | 640 |
| III | Styrene/50.0 | 99.0 | 12.4 | not determined |
| IV | Indene/17.0<br>Coumarone/17.0<br>Styrene/17.0 | 84.8 | 14.4 | 790 |

**Notes:** The amount of GMA in the mixtures was 50.0 g, PO – 25.0 g. Toluene (400 g) was the reaction medium. Mixture II apart from the resin dissolved in toluene contained the residue in the amount of 15.7% with e.n. of 19.0%.

Therefore, IR spectra confirm that all resin-forming components in LFCT, as well as GMA, participate in the formation of resin molecule. The resin yield (*vide* Table 5.9) depends on component activity. GMA, styrene, and indene are more active, while coumarone has the lowest activity (Figure 5.4).

## 5.3.3 COUMARONE-INDENE RESINS WITH METHACRYLIC FRAGMENTS (CIRM)

The same, as in the previous cases, taking into account that the content of unsaturated compounds in the initial LFCT is low (Table 5.1), we used styrene in addition to MMA. AMP was used as the initiator. In order to develop the CIRM synthesis procedure, it was necessary to study the effect of the reaction mixture composition, temperature, and reaction time on CIRM yield and characteristics.

### 5.3.3.1 EFFECT OF VARIOUS PARAMETERS ON CIRM CHARACTERISTICS

Table 5.10 presents the results about the effect of initial mixture composition on CIRM characteristics. When adding only styrene to LFCT the CIRM yield

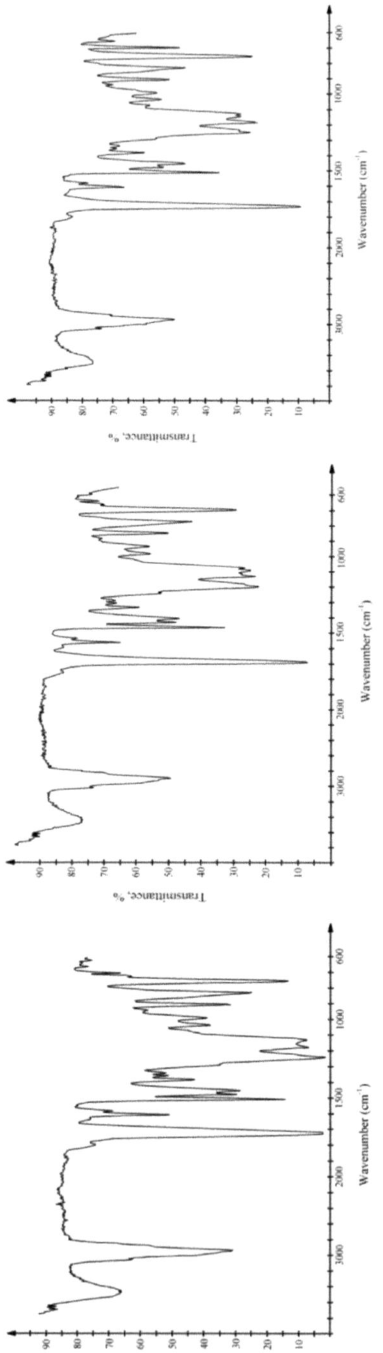


**FIGURE 5.4** IR spectra of the resins based on mixture I (a), II (b) and IV (c).

was found to be 5.1%. The decrease in styrene content and gradual increase in MMA amount to 9.35 g result in the increase of CIRM yield till 25.5%. At the same time the molecular weight and softening point of the resulting product decrease. Thus, we may assume that it is sufficient to introduce only MMA in the reaction mixture to increase the product yield.

**TABLE 5.10** Effect of the Reaction Mixture Composition on CIRM Characteristics

| Composition of the Initial Mixture, g | | | CIRM Characteristics | | | |
|---|---|---|---|---|---|---|
| LFCT | Styrene | MMA | Yield Relative to the Initial Mixture | | Softening Point, K | $M_n$, g/mol |
| | | | g | % | | |
| 48.5 | 9.09 | 0 | 2.8 | 5.1 | – | – |
| 48.5 | 7.27 | 1.87 | 3.8 | 7.0 | – | 840 |
| 48.5 | 3.64 | 5.61 | 4.3 | 13.4 | 382 | 700 |
| 48.5 | 1.82 | 7.48 | 9.5 | 17.4 | 375 | 670 |
| 48.5 | 0 | 9.35 | 13.9 | 25.5 | 379 | 600 |

**Notes:** Reaction temperature 353 K, time 6 h, AMP amount 0.25 g.

The next stage of investigations was to study the effect of initiator amount on CIRM characteristics. One can see from Table 5.11 that the increase in AMP amount to 0.5 g has a positive effect on the yield of the target product, which is 29.7%. The increase in AMP amount to 0.75 g practically does not affect the CIRM yield and its characteristics. Taking into account that AMP amount of 0.25 g provides a sufficiently high yield of the product, and the resulting resin has the highest molecular weight and softening temperature, such amount of initiator was chosen to study the effect of process time (Table 5.12) and temperature (Table 5.13) on the yield and characteristics of the resin.

It is seen from Table 5.12 that already in 4 hours the CIRM yield is 23.1%. The increase in reaction time to 6 hours allows obtaining the yield of 25.5%. The resin is characterized by a molecular weight of 600 g/mol and a softening point of 364 K. With a further increase in the reaction temperature to 8–10 hours, the resin yield and molecular weight are almost unchanged.

Analyzing the data of Table 5.13, we observe that the decrease in the reaction temperature to 343 K leads to the low resin yield and its molecular weight. The increase in the reaction temperature to 363 K slightly increases the resin yield and molecular weight, but decreases the softening point. This can be explained by the fact that at 363 K there is a rapid decomposition of the initiator and an increase in the recombination reaction rate. Under these conditions, not all monomers are capable to enter the oligomerization reaction resulting in the decrease of the resin yield. The optimum temperature at which the highest yield is achieved is the temperature of 353 K. Thus, the optimum synthesis conditions for CIRM synthesis are: reaction temperature 353 K; reaction time 6 h; AMP amount 0.25 g; MMA content in the reaction mixture 9.35 g. The resulting resin is characterized by the yield of 25.5%, softening point of 364 K and molecular weight of 600 g/mol.

**TABLE 5.11** Effect of AMP Concentration on CIRM Characteristics

| Composition of the Initial Mixture, g | | | CIRM Characteristics | | | |
|---|---|---|---|---|---|---|
| AMP | LFCT | MMA | Yield Relative to the Initial Mixture | | Softening Point, K | $M_n$, g/mol |
| | | | g | % | | |
| 0.1 | 48.5 | 9.35 | 5.1 | 9.3 | 363 | 540 |
| 0.25 | 48.5 | 9.35 | 13.9 | 25.5 | 364 | 600 |
| 0.5 | 48.5 | 9.35 | 16.2 | 29.7 | 337 | 510 |
| 0.75 | 48.5 | 9.35 | 16.0 | 29.3 | 341 | 490 |

**Notes:** Reaction temperature 353 K, time 6 h.

**TABLE 5.12** Effect of Reaction Time on CIRM Characteristics

| Time, h | CIRM Characteristics | | | |
|---|---|---|---|---|
| | Yield Relative to the Initial Mixture | | Softening Point, K | $M_n$, g/mol |
| | g | % | | |
| 2 | 10.1 | 18.5 | 318 | 470 |
| 4 | 12.6 | 23.1 | 328 | 500 |
| 6 | 13.9 | 25.5 | 364 | 600 |
| 8 | 13.9 | 25.5 | 321 | 590 |
| 10 | 13.8 | 25.3 | 325 | 650 |

**Notes:** Reaction temperature 353 K. Composition of the initial mixture: LFCT 48.5 g; MMA 9.35 g, AMP 0.25 g.

### *5.3.3.2 SPECTROSCOPIC INVESTIGATIONS*

IR- and NMR-spectroscopic investigations were carried out to confirm the structure of the synthesized resins. The results are represented in Figures 5.5–5.7.

**TABLE 5.13** Effect of Reaction Temperature on CIRM Characteristics

| Temperature, K | CIRM Characteristics | | | |
|---|---|---|---|---|
| | Yield Relative to the Initial Mixture | | Softening Point, K | $M_n$, g/mol |
| | g | % | | |
| 343 | 8.6 | 15.8 | 367 | 510 |
| 353 | 13.9 | 25.5 | 364 | 600 |
| 363 | 10.1 | 18.5 | 359 | 720 |

**Notes:** Reaction time 6 h. Composition of the initial mixture: LFCT 48.5 g; MMA 9.35 g, AMP 0.25 g.

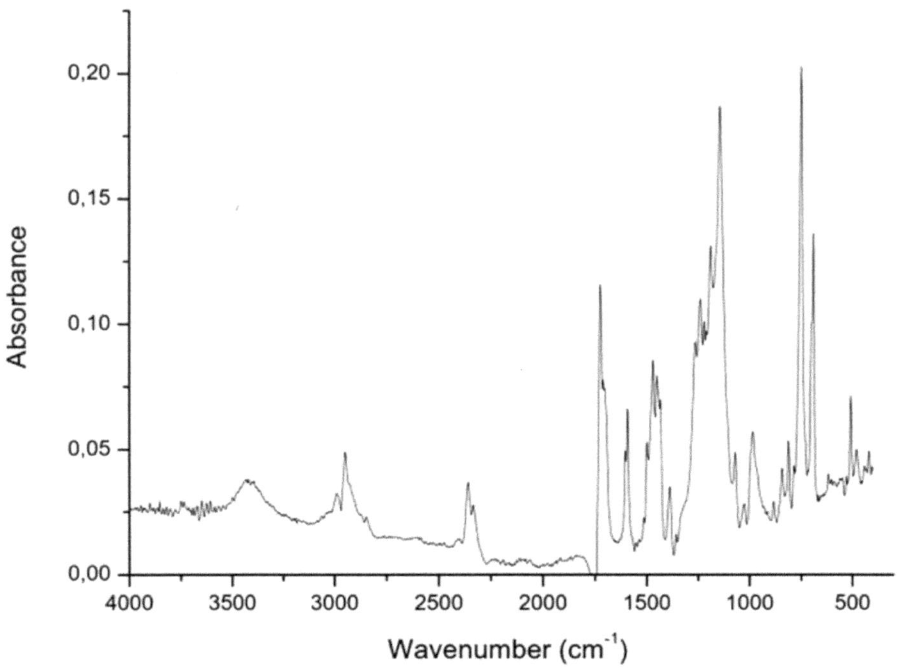


**FIGURE 5.5** IR-spectrum of CIRM.

The presence of methacrylic fragments in the synthesized resin was confirmed by stretching vibrations ($\nu_{C=O}$) of the carbonyl group at 1725 cm$^{-1}$ (Figure 5.5). Moreover, the vibrations at 3428 cm$^{-1}$ (Figure 5.3), which are characteristic of the carbonyl group in esters, also correspond to this group. The presence of $-CH_3$ group in $-C-O-CH_3$ is proved by symmetric deformation vibrations at 1386 cm$^{-1}$. The C–O–C group included in the methacrylic fragment is confirmed by asymmetric stretching vibrations at 1145 cm$^{-1}$. The presence of the indene fragment is proved by asymmetric valence vibrations at 2950 cm$^{-1}$ of the $CH_2$ group, which is directly linked to the benzene ring in the indene molecule. The fragments of coumarone, included to CIRM structure, are proved by stretching vibrations at 1190 cm$^{-1}$ of the C–O–C group contained in the coumarone molecule. Benzene rings, which can correspond to the molecules of styrene, coumarone, and indene, are proved by stretching vibrations in the range of 1592–1434 cm$^{-1}$, as well as by deformation vibrations at 752, 696, and 692 cm$^{-1}$ of CH group in the substituted benzene rings. The azoinitiator residues included in the structure of the obtained resins are proved by the stretching vibrations of –C≡N group at 2377 and 2358 cm$^{-1}$.

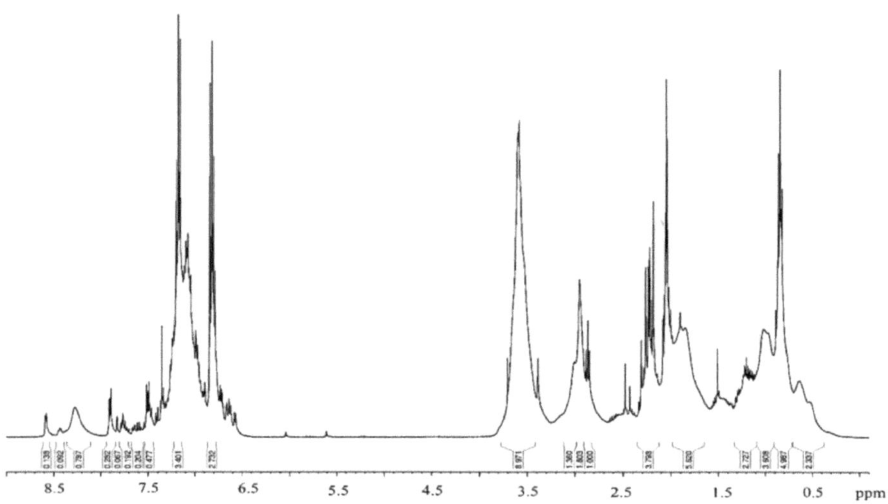


**FIGURE 5.6** $^{1}$H NMR spectrum of CIRM.

The fragment of methacrylate group $CH_3$–O in the $^{1}$H NMR spectrum (Figure 5.6) was confirmed by proton signals at 3.54–3.65 ppm, as well as by chemical shifts at 45.23 and 52.04 ppm in the $^{13}$C NMR spectrum (Figure 5.7). The $CH_3$ group, which is present in both the methacrylate fragment and the azoinitiator residues, is proved by proton signals in the region of 0.79–0.87 ppm (Figure 5.6) and chemical shifts at 29.04 and 33.33 ppm (Figure 5.7). The indene fragments corresponding to $CH_2$ group (Figure 5.6) are confirmed by proton signals in the region of 1.67–1.88 ppm. The presence of aromatic rings in the CIRM structure is proved by proton signals in the region of 6.6–8.0 ppm (Figure 5.6) and chemical shifts in the region of 116–132 ppm (Figure 5.7). The presence of –C≡N group was confirmed by a chemical shift at 120 ppm (Figure 5.7).

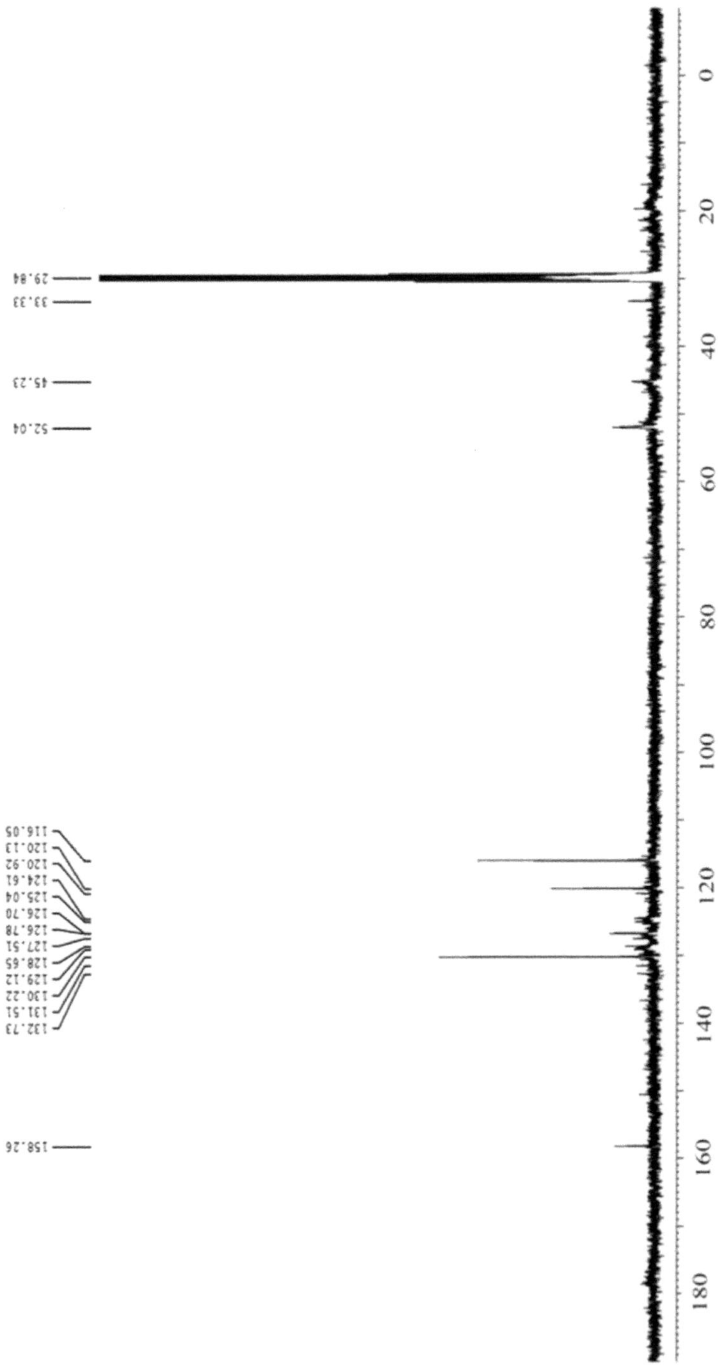


**FIGURE 5.7** $^{13}C$ NMR spectrum of CIRM.

The above results indicate that coumarone-indene resins (CIRMs), which are formed during LFCT co-oligomerization-using AMP, contain methacrylate fragments, apart from indene, coumarone, and styrene fragments.

## 5.4 APPLICATION OF CIR AS POLYMER ADDITIVE FOR THE PRODUCTION OF BITUMEN-POLYMER MIXTURES (BPM)

In recent years, a series of researches have been carried out at the Department of Oil and Gas Processing of Lviv Polytechnic National University (Ukraine) to obtain CIRs *via* co-oligomerization of unsaturated compounds from the CIF produced from "heavy benzene" of by-product coke plant. These CIRs are used as additives to bitumen in the amount of 7–8 wt.% relative to the final mixture [4, 19, 29, 30]. The introduction of different by nature functional groups into CIR structure should improve the performance characteristics of BPM with a significant reduction in CIR content in them. Therefore, the next stage of investigations was to study the possibility of using CIR with different functional groups as a polymeric additive to petroleum road bitumen.

### *5.4.1 BITUMEN-POLYMER MIXTURES (BPM) WITH CIRC*

To determine the possibility of using CIRC as a polymer additive to petroleum road bitumen, it was necessary to study the effect of polymer amount, temperature, and modification time on the quality characteristics of the original bitumen.

The modification was carried out at a temperature of 463K for 60 min. Such conditions were chosen on the basis of previous experiments [4, 19]. The results are shown in Table 5.14 and in Figure 5.8.

The increase in the modifier amount from 0 to 5 wt.% increases its softening point and adhesion but decreases the plastic properties of bitumen (penetration and ductility). The value of 5 wt.% may be considered as optimum one. This amount of modifier provides the required softening point and almost 100% adhesion. The amount of CIRC more than 5 wt.% leads to a significant deterioration of all characteristics due to incomplete dissolution of the modifier in bitumen. As a result, the heterogeneous sample is observed. The results of studies on the effect of modification temperature on the BPM properties are presented in Table 5.15 and Figure 5.9.

**TABLE 5.14** Effect of CIRC Amount on BPM Characteristics

| BPM Composition | | BPM Characteristics | | | | |
|---|---|---|---|---|---|---|
| **Bitumen Amount, wt.%** | **CIRC Amount, wt.%** | **Softening Point, K** | **Ductility, cm** | **Penetration, 0.1 mm** | **Adhesion, %** | **Homogeneity** |
| 100.0 | 0 | 46 | 63 | 70 | 33 | + |
| 99.0 | 1.0 | 49.5 | 62.5 | 59 | 58 | + |
| 97.0 | 3.0 | 51 | 60 | 57 | 91 | + |
| 95.0 | 5.0 | 53 | 60.5 | 55 | 97 | + |
| 93.0 | 7.0 | 50 | 31.5 | 30 | 69 | – |

**Notes:** Temperature is 463 K, time 60 min.

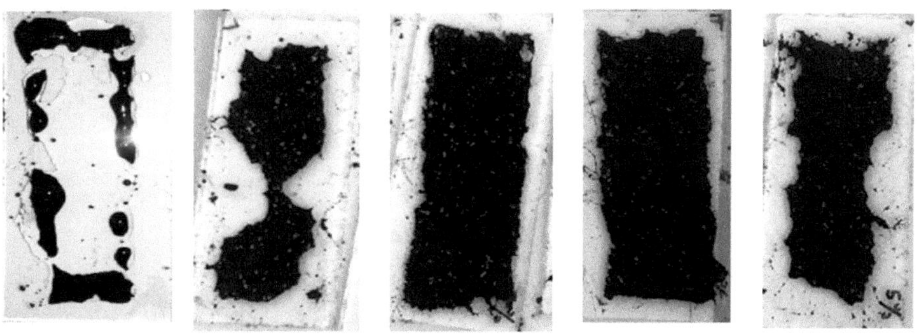

(a) (b) (c) (d) (e)

**FIGURE 5.8** BPM adhesion depending on CIRC amount: 0 wt.% (a), 1.0 wt.% (b), 3.0 wt.% (c), 5.0 wt.% (d) and 7.0 wt.% (e).

**TABLE 5.15** Effect of Temperature on BPM Characteristics

| Temperature, K | BPM Characteristics | | | | |
|---|---|---|---|---|---|
| | **Softening Point, K** | **Ductility, cm** | **Penetration, 0.1 mm** | **Adhesion, %** | **Homogeneity** |
| 423 | 319 | 71.5 | 60 | 32 | + |
| 443 | 321 | 69.5 | 57 | 48 | + |
| 463 | 326 | 63.5 | 55 | 97 | + |
| 483 | 322 | 62 | 55 | 78 | + |

**Note:** BPM consists of bitumen and CIRC (95 and 5 wt.%, respectively); time is 60 min.

The best results are observed at the temperature of 463 K. At this temperature it was possible to obtain BPM not only with excellent adhesive properties, but also with the highest softening point (Table 5.15). It should be noted that increasing the modification temperature above 463 K has a negative effect on ductility and penetration.

(a) (b) (c) (d)

**FIGURE 5.9** BPM adhesion depending on temperature: 423 K (a), 443 K (b), 463 K (c) and 483 (d).

All BPM samples were homogeneous, indicating complete dissolution of the modifier in the original bitumen.

The results of studies on the effect of modification time on the BPM properties are presented in Table 5.16 and Figure 5.10.

**TABLE 5.16** Effect of Time on BPM Characteristics

| Time, min | BPM Characteristics | | | | |
|---|---|---|---|---|---|
| | Softening Point, K | Ductility, cm | Penetration, 0.1 mm | Adhesion, % | Homogeneity |
| 30 | 319 | 79 | 70 | 87 | + |
| 60 | 326 | 63.5 | 55 | 97 | + |
| 90 | 326 | 66 | 57 | 60 | + |
| 120 | 327 | 44 | 54 | 81 | + |

**Note:** BPM consists of bitumen and CIRC (95 and 5 wt.%, respectively); temperature is 463 K.

It is seen from Table 5.16 that the increase in the modification time decreases both penetration and ductility, i.e., the plastic properties of the material deteriorate. At the same time, the softening point is increased.

The process time above 60 min does not have a significant influence on the softening point. The maximum adhesion is also achieved after 60 min of modification (Figure 5.10).

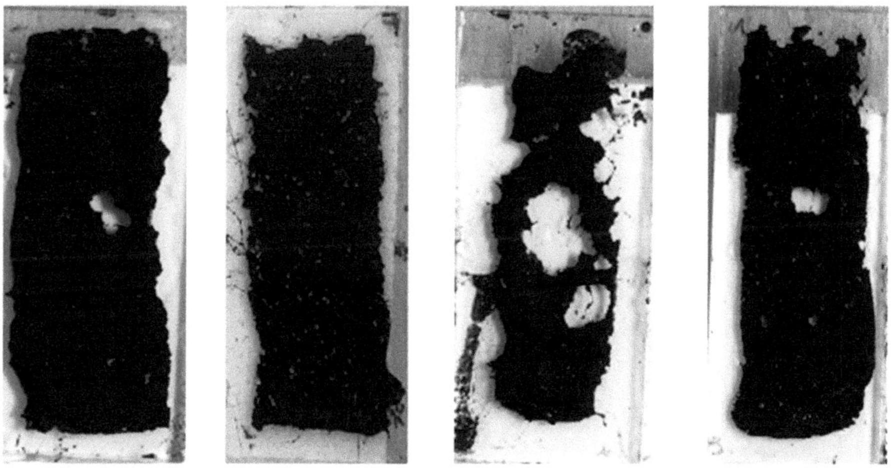

(a) (b) (c) (d)

**FIGURE 5.10** BPM adhesion depending on time: 30 min (a), 60 min (b), 90 min (c) and 120 min (d).

In view of all above, the optimal conditions for the preparation of BPM using coumarone-indene resin with carboxy groups are: the temperature of 463 K, time 60 min, amount of CIRC 5 wt.%. The introduction of CIRC in BND 60/90 bitumen improves its adhesion and softening point, but deteriorates its plastic properties-ductility and penetration.

### *5.4.2 BITUMEN-POLYMER MIXTURES (BPM) WITH CIRE*

To determine the effect of free epoxy groups on BPM properties, we examined the effect of CIRE amount, temperature, and modification time. The preparation conditions and characteristics of BPM are given in Tables 5.17–5.20 and Figures 5.11–5.13.

The introduction of CIRE increases the softening point and decreases penetration. The increase in CIRE amount increases ductility, but at CIRE content above 3 wt.% this value is reduced. All mixtures are homogeneous except that containing 7 wt.% of CIRE. In this case, the values of ductility and adhesion are essentially lower compared with other samples. The effect of CIRE amount on BPM adhesion is shown in Figure 5.11.

**TABLE 5.17** Effect of CIRE Content on BPM Characteristics

| BPM Composition | | BPM Characteristics | | | | |
|---|---|---|---|---|---|---|
| **Bitumen Amount, wt.%** | **CIRE Amount, wt.%** | **Softening Point, K** | **Ductility, cm** | **Penetration, 0.1 mm** | **Adhesion, %** | **Homogeneity** |
| 100.0 | 0 | 319 | 63 | 70 | 47 | + |
| 99.0 | 1.0 | 322 | 82 | 57 | 89 | + |
| 97.0 | 3.0 | 322 | 71 | 56 | 90 | + |
| 95.0 | 5.0 | 322 | 60 | 54 | 99 | + |
| 93.0 | 7.0 | 323 | 45 | 53 | 49 | – |

**Notes:** Temperature is 463 K, time 60 min.

The effect of temperature on BPM properties is represented in Table 5.18 and Figure 5.12. The increase/decrease in temperature decreases softening point and ductility but increases penetration, although this value is lower compared with that for bitumen without CIRE. Adhesion also depends on temperature and the highest value was observed at 463 K. Taking into account the obtained results, the effect of preparation time on BPM characteristics was studied at 463 K.

(a) (b) (c) (d) (e)

**FIGURE 5.11** BPM adhesion depending on CIRE content: 0 wt.% (a), 1.0 wt.% (b), 3.0 wt.% (c), 5.0 wt.% (d) and 7.0 wt.% (e).

**TABLE 5.18** Effect of Temperature on BPM Characteristics

| Temperature, K | BPM Characteristics | | | | |
|---|---|---|---|---|---|
| | Softening Point, K | Ductility, cm | Penetration, 0.1 mm | Adhesion, % | Homogeneity |
| Pure bitumen at 463 K | 319 | 63 | 70 | 47 | + |
| 443 | 320 | 75 | 60 | 62 | + |
| 463 | 322 | 82 | 57 | 89 | + |
| 483 | 319 | 68 | 65 | 80 | + |

**Note:** BPM consists of bitumen and CIRE (99 and 1 wt.%, respectively); time is 60 min.

(a) (b) (c)

**FIGURE 5.12** BPM adhesion depending on temperature: 443 K (a), 463 K (b), and 483 K (c).

As one can see from Table 5.19, the preparation time affects all values except softening point. The increase in time to 120 min decreases ductility but increases penetration. At the same time, adhesion is slightly reduced. The best adhesion was found for BPM prepared for 60 min (Figure 5.13).

**TABLE 5.19** Effect of Time on BPM Characteristics

| Time, min | BPM Characteristics | | | | |
|---|---|---|---|---|---|
| | Softening Point, K | Ductility, cm | Penetration, 0.1 mm | Adhesion, % | Homogeneity |
| Pure bitumen prepared for 60 min | 319 | 63 | 70 | 47 | + |
| 30 | 322 | 70 | 67 | 76 | + |
| 60 | 322 | 82 | 57 | 89 | + |
| 120 | 320 | 64 | 71 | 80 | + |

**Note:** BPM consists of bitumen and CIRE (99 and 1 wt.%, respectively); temperature is 463 K.

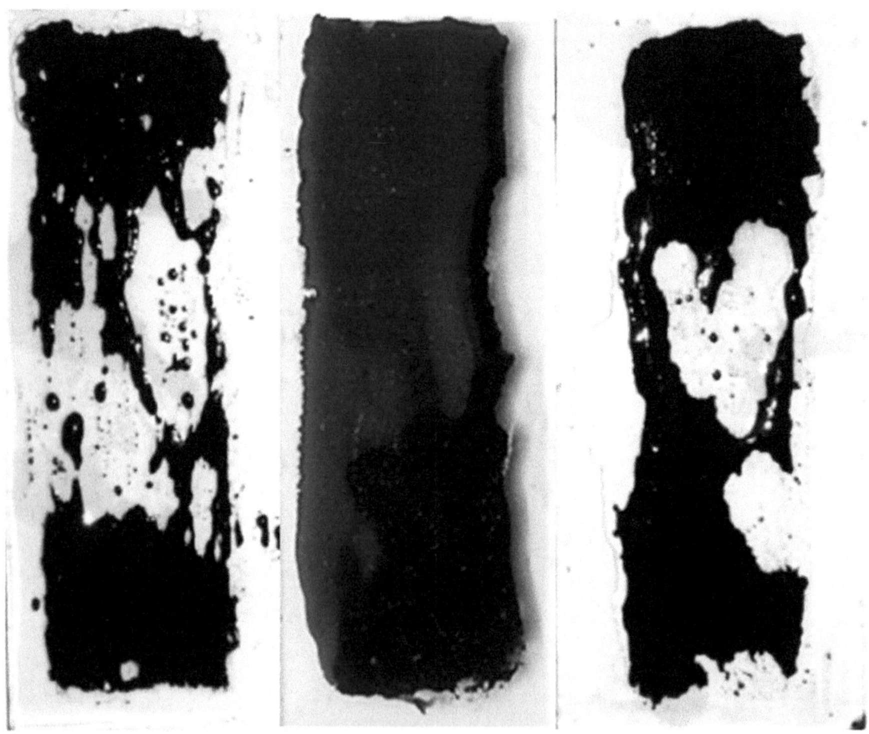

(a) (b) (c)

**FIGURE 5.13** BPM adhesion depending on time: 30 min (a), 60 min (b), and 120 min (c).

On the basis of obtained results, we can assert that CIRE as an additive for bitumen improves its characteristics, with the exception of penetration. Moreover, the adhesion of BPM is twice higher than adhesion of pure bitumen. The optimum conditions are: 99 wt.% of bitumen BND 60/90 + 1 wt.% of CIRE; temperature 463 K; preparation time 60 min.

To determine the reasons for properties improvement/deterioration we studied the group composition of BPM. One can see from Table 5.20 that introduction of CIRE increases the quantity of asphaltenes and resins but decreases the oils content. The reason is that CIRE provides the conversion of oils into resins and then into asphaltenes. Such conversions increase adhesion, softening point and ductility and decrease penetration.

**TABLE 5.20** Group Composition of BPM

| Composition | Content, wt.% | | | |
|---|---|---|---|---|
| | **Carbenes and Carboids** | **Asphaltenes** | **Resins** | **Oils** |
| Bitumen BND 60/90, 100% | absent | 20.7 | 35.0 | 44.3 |
| Bitumen BND 60/90 99% + CIRE 1% | absent | 22.7 | 36.9 | 40.4 |

### *5.4.2.1 SPECTROSCOPIC INVESTIGATIONS*

IR spectroscopy was used in order to determine the role of the components in the formation of BPM. Figure 5.13 represents spectra of CIRE (Figure 5.14a), pure bitumen (Figure 5.14b), and BPM consisting of 95 wt.% of bitumen and 5 wt.% of CIRE (Figure 5.14c).

The presence of epoxy and hydroxy groups in CIRE (Figure 5.14a) is confirmed by the absorption bands at 917 and 3400 $cm^{-1}$, respectively. Epoxy groups are introduced into the CIRE structure by PO molecules participated in the reaction and by GMA molecules as a result of co-oligomerization. Therefore, we suppose that the CIRE molecule contains epoxy groups at the ends of the molecule and in the side chains as well. The presence of indene molecule is confirmed by the absorption band at 755 $cm^{-1}$, corresponding to stretching vibrations of 1,2-disubstituted benzene. Coumarone fragment in CIRE is confirmed by absorption bands at 1050 and 755 $cm^{-1}$, corresponding to –C–O–C– bond and 1,2-disubstituted benzene, respectively.

In the spectrum of pure bitumen (Figure 5.14b), we did not observe the absorption bands typical of epoxy and hydroxy groups, as well as coumarone fragments. At the same time, we observed the absorption band at 750 $cm^{-1}$, indicating the presence of 1,2-disubstituted benzene rings. The absorption band at 1033 $cm^{-1}$ indicates the presence of naphthenic hydrocarbons in

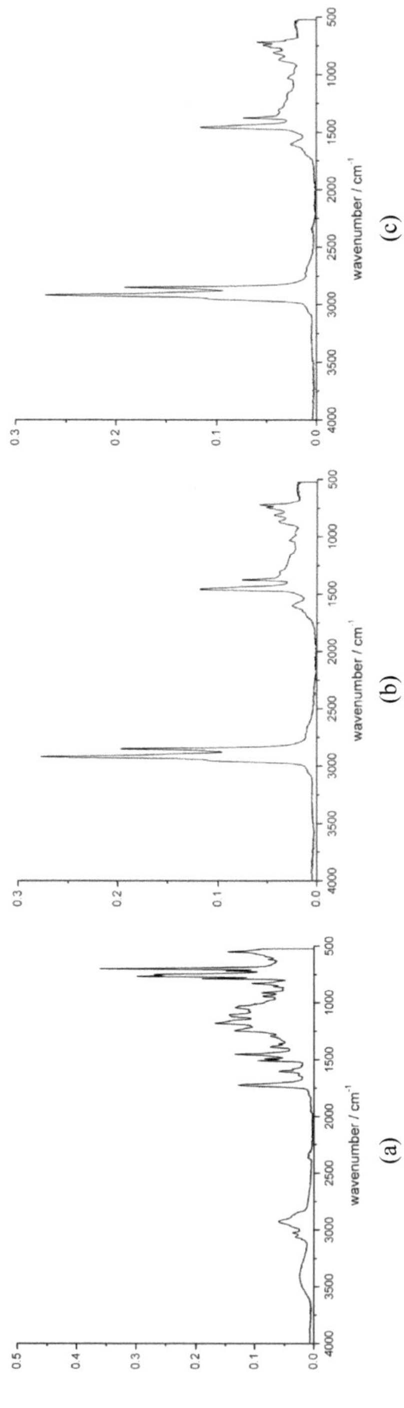


**FIGURE 5.14** IR spectra of CIRE (a), pure bitumen, (b) and BPM with content of bitumen 95 wt.% and CIRE 5 wt.% (c).

pure bitumen. Asymmetric stretching vibrations at 2933 cm$^{-1}$ and symmetric deformation vibrations at 1834 cm$^{-1}$ correspond to methyl groups in the benzene ring. The vibrations at 1603 cm$^{-1}$ also indicate the presence of compounds consisting of benzene rings. The presence of long aliphatic fragments is confirmed by symmetric deformation vibrations of $CH_2$ group at 2850 cm$^{-1}$ and asymmetric stretching vibrations of $CH_3$ group at 1460 cm$^{-1}$. The presence of substituted benzene rings is also confirmed by stretching vibrations of CH group at 867, 816, and 748 cm$^{-1}$.

The identity of BPM and pure bitumen spectra (*cf.* Figure 5.14c and 5.14b) is obvious. In the spectrum of BPM, the bands at 3400 and 917 cm$^{-1}$ corresponding to hydroxy and epoxy groups, respectively, are absent. This indicates that at sufficiently high temperature (463 K) of BPM formation the mentioned groups react with each other and perhaps with the bitumen components, improving the properties of BPM. The band at 1050 cm$^{-1}$ typical of coumarone is also not observed. At the same time, the band at 750 cm$^{-1}$ indicates the presence of 1,2-disubstituted fragments.

### 5.4.3 BITUMEN-POLYMER MIXTURES (BPM) WITH CIRM

It was necessary to determine the effect of polymer component amount, temperature, and time of BPM preparation on its characteristics. The results are shown in Tables 5.21–5.23 and Figures 5.15–5.17.

**TABLE 5.21** Effect of CIRM Amount on BPM Characteristics

| BPM Composition | | BPM Characteristics | | | | |
|---|---|---|---|---|---|---|
| **Bitumen Amount, wt.%** | **CIRM Amount, wt.%** | **Softening Point, K** | **Ductility, cm** | **Penetration, 0.1 mm** | **Adhesion, %** | **Homogeneity** |
| 100.0 | 0 | 319 | 70 | 63 | 33 | + |
| 99.5 | 0.5 | 319 | 70 | 69 | 40 | + |
| 99.0 | 1.0 | 319 | 85 | 67 | 45 | + |
| 97.0 | 3.0 | 320 | 80 | 65 | 94 | + |
| 95.0 | 5.0 | 321 | 75 | 60 | 85 | + |
| 93.0 | 7.0 | 321 | 65 | 55 | 68 | – |

**Notes:** Temperature is 463 K, time 60 min.

As can be seen from Table 5.21, the introduction of 0.5–7 wt.% of the resin into BPM practically does not change the softening point. At the same time, the penetration slightly decreases, and the ductility value passes through the maximum (when CIRM amount is 1 wt.%). The adhesion of the formed mixtures significantly increases. All investigated samples, with the exception of that with 7 wt.% of CIRM, are homogeneous. The reason of mixture non-homogeneity is the formation of large globules and insolubility of resin in bitumen.

(a) (b) (c) (d) (e) (f)

**FIGURE 5.15** BPM adhesion depending on CIRM content: 0 wt.% (a), 0.5 wt.% (b), 1.0 wt.% (c), 3.0 wt.% (d), 5.0 wt.% (e) and 7.0 wt.% (f).

The dependence of BPM adhesion on CIRM amount is demonstrated in Figure 5.15. The best adhesion (almost three times greater compared to bitumen without additive) are observed for CIRM amount of 3%.

Thus, the mixture consisted of 97 wt.% of bitumen and 3 wt.% of CIRM showed sufficient operational properties and was used for next experiments.

The effect of temperature on BPM characteristics are given in Table 5.22 and Figure 5.16.

**TABLE 5.22** Effect of Temperature on BPM Characteristics

| Temperature, K | BPM Characteristics | | | | |
|---|---|---|---|---|---|
| | Softening Point, K | Ductility, cm | Penetration, 0.1 mm | Adhesion, % | Homogeneity |
| 443 | 319 | 75 | 68 | 51 | + |
| 463 | 320 | 80 | 65 | 94 | + |
| 483 | 321 | 62 | 57 | 74 | + |

**Note:** BPM consists of bitumen and CIRM (97 and 3 wt.%, respectively); time is 60 min.

(a) (b) (c)

**FIGURE 5.16** BPM adhesion depending on temperature: 443 K (a), 463 K (b), and 483 K (c).

Analyzing the data presented in Table 5.22, it can be concluded that mixing the modifier and resin at 483 K slightly improves the softening point but degrades penetration and ductility. The best adhesion properties are observed at 463 K, which is illustrated in Figure 5.16. Therefore, the experiments regarding the effect of the preparation time were carried out at a temperature of 463 K.

The increase in preparation time (Table 5.23) decreases penetration and ductility, but slightly increases the softening point of BPM. The best values of adhesion were observed for 60 min (Figure 5.17).

**TABLE 5.23** Effect of Time on BPM Characteristics

| Time, min | BPM Characteristics | | | | |
|---|---|---|---|---|---|
| | Softening Point, K | Ductility, cm | Penetration, 0.1 mm | Adhesion, % | Homogeneity |
| 30 | 319 | 86 | 68 | 68 | + |
| 60 | 320 | 80 | 65 | 94 | + |
| 120 | 322 | 54 | 63 | 71 | + |

**Note:** BPM consists of bitumen and CIRM (97 and 3 wt.%, respectively); temperature is 463 K.

**FIGURE 5.17** BPM adhesion depending on time: 30 min (a), 60 min (b), and 120 min (c).

Thus, the optimum preparation conditions for CIRM-modified bitumen are the temperature of 463 K, time 60 min, CIRM amount of 3 wt.%. The obtained BPM has almost three times higher adhesion compared to pure bitumen. Therefore, coumarone-indene resin with methacrylic fragments can be used as an adhesive additive to petroleum road bitumen.

## 5.5 CONCLUSIONS

On the basis of LFCT, which is a product of PJSC "Zaporizhkoks" (Ukraine), the new indene-coumarone resins (CIR) with carboxy, epoxy, and methacrylic groups were obtained *via* initiated co-oligomerization with the addition of known monomers.

The effect of monomers nature, initiator amount, temperature, and reaction time on the yield and characteristics of the obtained CIR was established. On the basis of the conducted researches the optimum conditions for CIR obtaining were selected.

1. **CIR with Carboxy Groups:** Temperature 353 K; synthesis time 6 h; the mixture consists of LFCT (48.5 g), styrene (7.2 g), MAA (2.1 g) and AMP (0.25 g). The synthesized CIRC are characterized by a molecular weight of 960 g/mol and a content of carboxyl groups of 13.9%. CIRC yield is 22.4% relative to the initial mixture.

2. **CIR with Epoxy Groups:** Temperature 393 K; synthesis time 6 h; the mixture consists of LFCT (48.5 g), styrene (7.2 g), GMA (4.2 g), and PO (50% toluene solution, 2.5 g). The synthesized CIRE are characterized by the yield of 14.9%; softening point of 336 K; molecular weight of 1400 g/mol and content of epoxy groups 2.3%.
3. **CIR with Methacrylic Fragments:** Temperature 353 K; reaction time 6 h; mixture consists of LFCT (48.5 g), MMA (9.35 g), and AMP (0.25 g). The synthesized CIRM is characterized by the yield of 25.5%, softening point of 364 K and molecular weight of 600 g/mol.

The structure of the synthesized CIR was confirmed by IR spectroscopy. The presence of the carboxy group in CIR was confirmed by the absorption bands at 1750–1720 $cm^{-1}$, corresponding to the stretching vibrations of the carbonyl group and the band at 3000 $cm^{-1}$ corresponding to the stretching vibrations of the hydroxy group. The presence of epoxy groups was confirmed by the absorption band at 944 $cm^{-1}$. $^1H$ NMR spectroscopy demonstrated that the synthesized resins have epoxy groups at the ends of oligomeric chain and side branches.

The presence of methacrylic fragments in the synthesized resin was confirmed by stretching vibrations of the carbonyl group at 1725 $cm^{-1}$ corresponding to the stretching vibrations of C=O group in esters; the $CH_3$-O group in $^1H$ NMR spectrum (proton signals at 3.54–3.65 ppm); as well as by chemical shifts at 45.23 and 52.04 ppm ($^{13}C$ NMR spectrum).

The possibility of using CIR as a polymer additive to BPM was investigated. The effect of CIR amount, temperature, and time on the operational characteristics of mixtures was established. It was found that the introduction of CIR with different functional groups in the road petroleum bitumen of the BND 60/90 brand allows to significantly improving its adhesive properties.

## KEYWORDS

- **bitumen-polymer mixtures**
- **co-oligomerization**
- **coumarone-indene**
- **diglycidyl ether**
- **spectroscopic investigations**
- **thermal analyses**

## REFERENCES


1. McNally, T., (2011). *Polymer Modified Bitumen: Properties and Characterization.* Woodhead Publ.
2. Airey, G., (2004). *J. Mater. Sci., 39*, 951. https://doi.org/10.1023/B:JMSC.0000012927.00747.83.
3. Fang, C., Li, T., Zhang, Z., & Jing, D., (2008). *Polym. Composite, 29*, 500. https://doi.org/10.1002/pc.20390.
4. Pyshyev, S., Gunka, V., Grytsenko, Y., & Bratychak, M., (2016). *Chem. Chem. Technol., 10*, 631. https://doi.org/10.23939/chcht10.04si.631.
5. Bahl, J., Atheya, N., Singh, H., et al., (1993). *Erdoel Kohle, Erdgas, Petrochem., 46*, 22.
6. Bratychak, M., Chervinskyy, T., Astakhova, O., et al., (2010). *Chem. Chem. Technol., 4*, 325.
7. Bratychak, M., Grynyshyn, O., Astakhova, O., et al., (2010). *Ecolog. Chem. Eng. S., 17*, 309.
8. Grynyshyn, O., Astakhova, O., & Chervinskyy, T., (2010). *Chem. Chem. Technol., 4*, 241.
9. Strap, G., Astakhova, O., Lazorko, O., et al., (2013). *Chem. Chem. Technol., 7*, 279. https://doi.org/10.23939/chcht07.03.279.
10. Bratychak, M., Iatsyshyn, O., Shyshchak, O., et al., (2017). *Chem. Chem. Technol., 11*, 49. https://doi.org/10.23939/chcht11.01.049.
11. Bratychak, M., Gagin, M., Shyshchak, O., & Waclawek, W., (2004). *Ecological Chemistry and Engineering, 11*, 15, 21.
12. Gagin, M., Bratychak, M., Shyshchak, O., & Waclawek, W., (2004). *Ecological Chemistry and Engineering, 11*, 27.
13. Chervinskyy, T., Bratychak, M., Gagin, M., & Waclawek, W., (2004). *Ecological Chemistry and Engineering, 11*, 1225.
14. Skibitskiy, V., Grynyshyn, O., Bratychak, M., & Waclawek, W., (2004). *Ecological Chemistry and Engineering, 11*, 41.
15. Bratychak, M., Grynyshyn, O., Shyshchak, O., et al., (2007). *Ecolog. Chem. Eng., 14*, 225.
16. Grynyshyn, O., Bratychak, M., Krynytskiy, V., & Donchak, V., (2008). *Chem. Chem. Technol., 2*, 47.
17. Wu, Y., Li, J., Xu, J., et al., (2010). *J. Shanghai Univ. Engl. Ed., 14*, 313. https://doi.org/10.1007/s11741-010-0651-3.
18. Sokolov, V., (1978). Coumarone-Indene Resins. Moscow.
19. Pyshyev, S., Grytsenko, Y., Solodkyy, S., et al., (2015). *Chem. Chem. Technol., 9*, 359. https://doi.org/10.23939/chcht09.03.359.
20. Bazyliak, L., Bratychak, M., & Brostow, W., (1999). *Mater. Res. Innovat., 3*, 132. https://doi.org/10.1007/s100190050138.
21. Bratychak, M., Zubal, O., Bashta, B., et al., (2017). *Chem. Chem. Technol., 11*, 180. https://doi.org/10.23939/chcht11.02.180.
22. Braun, D., Cherdron, H., Rehahn, M., et al., (2013). *Polymer Synthesis: Theory and Practice: Fundamentals, Methods, Experiments* (5th edn.) Springer-Verlag, Berlin Heidelberg. https:/doi.org/10.1007/978-3-642-28980-4_1.
23. ASTM- D36/D36M-14e1. *Standard Test Method for Softening Point of Bitumen (Ring-and-Ball Apparatus)*. https://www.astm.org/Standards/D36.htm (accessed on 19 October 2020).
24. BS EN 1426:2000. *European Standard.* Bitumen and bituminous binders. Methods of tests for petroleum and its products. Determination of needle penetration.

25. BS EN 13589:2008. *European Standard.* Bitumen and bituminous binders. Determination of the tensile properties of modified bitumen by the force ductility method.
26. BS EN 13614:2011. *European Standard.* Bitumen and bituminous binders. Determination of adhesivity of bituminous emulsions by water immersion test.
27. Gagin, M., Bratychak, M., Shyshchak, O., & Waclawek, W., (2004). *Ecological Chemistry and Engineering, 11*, 27.
28. Wilson, G., Henderson, J., Caruso, M., et al., (2010). *J. Polym. Sci. A, 48*, 2698. https://doi.org/10.1002/pola.24053.
29. Pyshyev, S., Grytsenko, Y., Danyliv, N., et al., (2015). *Petrol. Coal, 57*, 303.
30. Pyshyev, S., Grytsenko, Y., Nikulyshyn, I., & Gnativ, Z., (2014). *J. Coal Chem., 5, 6*, 41.

## CHAPTER 6

# Oxidation and Exfoliation of Powdered Graphite Foil and Its Wastes: Preparation of Graphene and Its Oxides

T. DUNDUA,[1] V. UGREKHELIDZE,[1] L. NADARAIA,[1] N. NONIKASHVILI,[2] V. GABUNIA,[2] M. JAPARIDZE,[2] N. BARBAKADZE,[2] and R. CHEDIA[2]

[1]*Department of Agricultural Sciences and Biosystems Engineering, Georgian Technical University, 17, D. Guramishvili Str. Tbilisi–0192, Georgia*

[2]*Laboratory of Problems of Chemical Ecology, Petre Melikishvili Institute of Physical and Organic Chemistry, Iv. Javakhishvili Tbilisi State University, 31, Ana Politkovskaia Str., Tbilisi – 0186, Georgia; E-mail: chemicalnatia@yahoo.de (N. Barbakadze)*

## ABSTRACT

Graphene oxide (GO) and reduced graphene oxide (rGO) obtained from graphitic precursors such are flake graphite, graphite intercalated compounds, expanded, and synthetic graphite. In this study, by oxidation of graphite foil and its wastes with strong oxidizing systems, GO was obtained. Graphite foils are used in many fields; therefore, a large amount of wastes are accumulated. Obtaining of graphene from these wastes was implemented according to the following stages: graphite foil wastes grinding → powdered graphite foil wastes (pGFW) oxidation (II) → graphite oxide isolation → graphene oxide reduction → rGO thermal treatment → graphene. Each step requires some physical-chemical processes: I. Waste grinding and removal of impurities from powders using chemical reagents; II. Oxidation of pGFW and isolation of graphite oxide from the reaction mixture by decantation and filtration; III. Exfoliation of graphite oxide by sonication to obtain GO; its isolation from the solution; IV. Reduction GO to rGO by chemical reagents or physical methods; V. Thermal treatment of rGO in order to obtain graphene with defective structure. From obtained GO and rGO membranes are prepared and reinforced methods of matrix ceramic materials with graphene are developed. The membranes were obtained by the methods of vacuum-filtration, evaporation of solvents from homogeneous suspensions, and thermal treatment GO foils in the air or inert atmosphere. Membranes with a diameter of 3–15 cm were obtained.

## 6.1 INTRODUCTION

Graphene oxide (GO) and reduced graphene oxide (rGO) are obtained from graphitic precursors such are flake graphite, graphite intercalated compounds, expanded, and synthetic graphite. One of the most widely used methods is the oxidation of graphite precursors by strong oxidizing systems ($HNO_3$, $HClO_4$, $H_2SO_4$, $H_3PO_4$, $KClO_3$, $NaNO_3$, $KMnO_4$, etc.). Chemical oxidation of graphite is based on methods of Brodie, Staudenmaier, Hofmann, and Hummers [1–3]. Oxidation degree, hydrophilicity, and microstructure of GO depend both on the nature of used graphite, also on the synthesis methods. In the scientific literature, there are provided several structural models of GO (Hofmann, Ruess, Scholz-Boehm, Nakajima-Matsuo, Lerf-Klinowski, Szabo, etc.). GO structure is still under discussion. This introduction compares solution-processed methods of GO synthesis and reviews all the noteworthy and original methods from 1859 to the present [4–6] (Figure 6.1).

In laboratory practice, several versions of the Hummers method are most often used for the oxidation of graphite. The composition of the oxidizing systems is as follows: $NaNO_3$-$H_2SO_4$-$KMnO_4$ [3]; $H_2SO_4$-$H_3PO_4$-$KMnO_4$ ($H_2SO_4$/$H_3PO_4$ = 9:1) [8]; $H_2SO_4$-$KMnO_4$ [9]; Pretreatment of graphite powder or intercalation by $H_2SO_4$-$P_2O_5$-$K_2S_2O_8$ or $KMnO_4$-$HNO_3$-$H_2SO_4$ system and further oxidation of pre-oxidized graphite [10]; $NaNO_3$-free Hummers methods $K_2FeO_4$-$H_2SO_4$ [11]; $K_2FeO_4$-$H_2SO_4$-$KMnO_4$ [12]. Synthesis is mainly carried out at low temperature (0–20°C) by the addition of $KMnO_4$ to the reaction mixture in small portions so that the temperature does not exceed 20°C, as this can lead to an uncontrolled exothermic reaction and the instantaneous eruption of the reaction mixture from the reactor [4]. In the case of using the improved Hummers method and optimized improved Hummers method, $KMnO_4$ is added at 50°C [8, 9]. Reagent mixing sequence is modified in works [13, 14]. $H_2SO_4$ (98%) is added to the expanded graphite (<30 μm)-$KMnO_4$ mixture at low temperature, reaction mixture increases in volume several times, after dissolution with water, a yellow suspension of GO is obtained. There are many industrial and laboratory methods of obtaining graphene, GO, and rGO. This issue is discussed in many scientific papers and is also protected by patents. Areas of their application are growing rapidly [15–22].

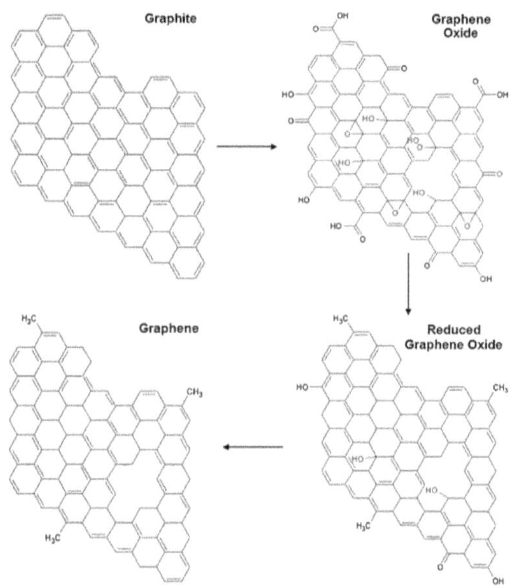


**FIGURE 6.1** Chemical structures of graphite, graphene oxide, reduced graphene oxide, and graphene [7].

Graphene with defective structure obtained from GO by chemical reduction or thermal treatment. The transformation of GO to rGO can be achieved through chemical or physical reduction. Through the reduction process, the oxygenous functional groups in the GO are eliminated to form rGO with carbon to oxygen (C/O) ratio of 8:1–246:1 [22]. During the thermal treatment of GO under atmosphere air, it begins to transform to rGO from 200°C. It is confirmed, that by rapidly heating confined GO films on a hot plate set to 400C under an atmosphere of air, GO films were readily converted to intact, electrically conductive, reduced thin films [23]. In this way, rGO membranes can be obtained. With rapid heating in air or vacuum, an explosion occurs as rapid heating leads to a sudden release of intercalated $H_2O$, and evolution of CO and $CO_2$ that can rapidly expand the layers, similar to the 'thermal-shock' exfoliation of graphite oxide. By exploiting the internal and the applied pressure differential, vacuum has been used to increase the mechanical driving force and promote the thermal exfoliation of graphite oxide [24–26].

One of the methods of obtaining graphene is direct liquid-phase exfoliation of graphite using various polar organic solvents. For example, there is obtained the high concentration exfoliation of graphene nanosheets in dimethyl formamide (DMF) as a solvent from graphite powder [27]. A viable liquid-phase exfoliation route is carried out in an aqueous solution through suitable surfactants. To reduce the damage to the graphene sheets by ultrasonic processing, the maximum sonication time was 3 h [28]. The performances of different kinds of surfactants for dispersing hydrophobic graphene in aqueous solutions were studied. These surfactants included cationic, non-ionic, anionic, and polymer types [29].

Modern technological processes require improvement of membrane properties, and application of new materials with improved properties for their manufacturing. Among these requirements are: improved mechanical properties, continuity, and simplicity of the separation process, ecological safety, less labor energy costs end, etc. One of the perspective ways in solving this problem, together with the traditional methods is use of nanotechnological methods, in particular, use of 2D and 3D carbon substances for making new types of membranes. From this point of view, very interesting results are obtained of application membranes in various fields of technology that are obtained on the basis of graphene oxides. There are various technologies of making membranes from graphene oxide (GO): vacuum filtration, pressure filtration, spinning-casting/coating, drop-casting, dip-casting, dip-coating, doctor blade-casting, layer-by-layer assembly, hybrid approach, evaporation-assembled method, templating method, Langmuir-Blodgett (LB) assembly, shear-alignment method [30–32].

Our earlier studies confirmed, that it is possible to obtain GO as well as rGO, by the oxidation of graphite foils and powders of its wastes [33]. Their oxidation was carried out using versions of the Hummers method ($NaNO_3$-$H_2SO_4$-$KMnO_4$) and optimized improved Hummers method ($H_2SO_4$-$KMnO_4$) The purpose of the present work is to study the possibility of pGFW oxidation by other methods, in particular the improved Hummers method ($H_2SO_4$-$H_3PO_4$-$KMnO_4$) ($H_2SO_4/H_3PO_4$ = 9:1) and changing the order of addition the reagents to the reaction mixture. We also consider it is promising to study the liquid-phase exfoliation of GO obtained from pGFW and its application areas.

## 6.2 EXPERIMENTAL

### 6.2.1 *MATERIALS AND METHODS*

Graphite flake, natural, 325 mesh–99.98% (metals basis), sodium cholate hydrate (99%) was purchased from Alfa Aesar. $KMnO_4$, $NaNO_3$, $H_2SO_4$ (98%), HCl (37%), HI (57%), DMF and ascorbic acid were purchased from Sigma Aldrich. The morphology of the samples was studied by optical and scanning electron microscopes (Nikon ECLIPSE LV 150, LEITZ WETZLAR, and JEOL JSM-6510 LV-SEM). Samples x-ray diffraction (XRD) patterns were obtained with a DRON-3M diffractometer (Cu–$K\alpha$, Ni filter, 2°/min). FTIR spectra were recorded on PerkinElmer Spectrum Version 10.4.2 using KBr disks (450–4000 $cm^{-1}$). After purification, graphite foil powders were analyzed using EDX. The initial powders were obtained by grinding of graphite foil and its wastes in grain lab mill. Removal of inorganic impurities from the powders was carried out with chemical reagents and their subsequent sifting. Fractions with particles <140 μm were collected. Powdered graphite foil (PGF) and powdered graphite foil wastes (pGFW) are obtained. Thermal treatment of GO and rGO (<1500°C) was implemented in high temperature vacuum furnace (Kejia furnace). For the ultrasound treatment and homogenization of suspensions was used Ultrasonic cleaner (40 KHz) and JY92-IIDN Touch Screen Ultrasonic Homogenizer (20–25 KHz, 900 W). GO and rGO particle sizes were determined by Winner 802 DLS Photon correlation nanoparticle sizer analyzer.

### 6.2.2 *SYNTHESIS OF GO*

1. **Graphene Oxide (GO):** Its synthesis was carried out with one of the versions of Hummers method, as described previously [33].
2. **Oxidation pGFW with Optimized Improved Hummers Method [9]:** We have used a simplified oxidation system-$KMnO_4$-$H_2SO_4$. In more detail, 40 ml of 98% sulfuric acid is added to 1 g pGFW (<140 μm) in a glass reactor. The mixture was stirred at 35–40°C for half an hour, and 3 g of potassium permanganate was added at 40÷45°C under 1 h (the temperature can reach 50°C). The mixture was stirred for 3 h. A gray viscous mass was obtained, which was cooled to 10°C and 100 ml of ice-water was added to the glass reactor. After adding 3–4 drops of hydrogen peroxide (30%), the reaction mixture

becomes sharp yellow. The mixture was diluted to 500 ml. During the washing of the sediment its color gradually changed to dark brown. 20 min later solution was removed by decantation. This operation was repeated twice. For the rapid precipitation of graphite oxide from the suspension, a 5% solution of hydrochloric acid (500 ml) was added. Decantation was repeated 3 times in 10 min intervals. An aqueous gel-like mass was obtained, centrifugation, and washing of this precipitate was continued until the pH of the solution 5÷6. The sediment was washed with acetone 3 times and is dried in vacuum at 70°C for 4 h.

3. **Method Adding Sulfuric Acid in the $KMnO_4$-pGFW Mixture:** 0.5 g of pGFW and 1.5 g of $KMnO_4$ is placed into 250 ml three-neck flask. ($KMnO_4$ is finely grinded and then mixed with graphite powder). Mixture is cooled to 0 ÷ +3°C using ice under stirring. Then 10 ml of 98% $H_2SO_4$ is added once. The temperature of the reaction mixture is raised to 5–8°C. Mixture is stirred during 1 hour and then ice bath is removed and continue stirring at 30–35°C for 1 h. Reaction mixture has increased in volume only for 1.5–2 times, while in case of use of commercial expanded graphite, its volume increases several times [13, 14]. Mixture is cooled and 50 g of ice is added. It is stirred again and being heated to 90–98°C. Yellow suspension should form (if pink suspension is formed 1–2 drops of $H_2O_2$ is added). Stirring lasts 1 hour. The separation of GO from suspension is carried out like previously described methods.
4. **Direct Liquid-Phase Exfoliation pGFW in DMF:** to 400 mg of pGFW with particle size <50 μm is added 200 ml DMF and sonicated for 3 h. using JY92-IIDN Touch Screen Ultrasonic Homogenizer (20–25 KHz, 900 W). In this process suspension temperature changes within 20–60°C. The obtained dark dispersion was left to stand overnight for the sufficient sedimentation of large particles. The precipitate is removed from black dispersion by decantation. Centrifugation was carried out at 3000 rpm for 20 minutes to remove large aggregates. The homogeneous colloidal suspension of graphene sheets in the DMF is obtained, which is stable for 3 months. Graphene isolation was carried out by evaporation of DMF in vacuum at 75°C.
5. **Direct Liquid-Phase Exfoliation of pGFW in Sodium Cholate Hydrate Water Solution:** 200 mg of pGFW was places in 250 ml beaker and 100 ml of 1 mg/ml sodium cholate hydrate solution was added. Sonication was performed using JY92-IIDN Touch Screen Ultrasonic Homogenizer (20–25 KHz, 900 W). The sonication lasted

8 h and resulted in black suspension was left to stand overnight and after was centrifuged (2500 rpm, 60 min.) in order to remove non-exfoliated agglomerates of pGFW. The supernatant is separated from the precipitate using a pipette. The resulting colloidal suspension does not contain agglomerates.

## 6.3 RESULTS AND DISCUSSIONS

The main steps in manufacturing of graphite foil are: intercalation of natural graphite, heat treatment of resulting product and production of expanded graphite, that are self-bonded by a rolling operation, where the expanded graphite particles form a continuous strip, without the addition of a binder. It has various applications in many fields, as it is characterized by excellent performance properties, such as excellent resistance to chemical agents; No material alteration over time; A very wide temperature range without property alteration: −196°C to 2500°C in an inert atmosphere, and up to 450°C/550°C in air; very high compressibility; The material's elastic recovery, and particularly low creep within its working temperature range, guarantee a proper seal; It is easy to cut, allowing it to be transformed into flat seals; areas of its application are as follows:

- It is optimum for components in semiconductor or nuclear energy industry application;
- Thermal protection in high-temperature furnaces owing to its reflecting power;
- Protection of rigid insulation from erosion and corrosion caused by vapors;
- Heat distribution and dissipation of hot spots, notably during brazing operations;
- Protection of molds during casting operations [34].

Graphite foil is used in many technological processes and therefore is collected in large quantities as waste. One of the ways to use these wastes and make innovative materials out of them is to make graphene and its oxides, which are in great demand in many areas of modern science and technology. Figure 6.2 shows the products of one of the companies.

Obtaining of graphene from these wastes was implemented according to the following stages: Graphite foil wastes grindig → pGFW oxidation (II) → Graphite oxide isolation → Graphene oxide reduction → rGO thermal treatment → Graphene. Each step requires some physical-chemical processes:

I. Waste grinding and removal of impurities from powders using chemical reagents; II. Oxidation of pGFW and isolation of graphite oxide from reaction mixture by decantation and filtration; III. Exfoliation of graphite oxide by sonication to obtain GO; its isolation from the solution; IV. Reduction GO to rGO by chemical reagents or physical methods; V. Thermal treatment of rGO in order to obtain graphene with defective structure.

I II

**FIGURE 6.2** Industrial graphite sheets (perma-foil); http://www.toyotanso.com/index.html (I); image: powders that were obtained by grinding of graphite foil and its wastes in grain lab mil (II).

Figure 6.3 shows the comparison of the XRD patterns of flake graphite and graphite foil powder. The XRD pattern of the natural flake graphite exhibited peaks located at 26.5°, 44.3°, and 54.5°, corresponding to planes (002), (101), and (004), respectively. Their diffractive maximums coincide with each other. In the case of graphite foil powder, no peak (101) was observed at all, and a peak of high intensity (004) appeared. This is caused by the structure of the expanded graphite particles (layers) in the rolling operation process. pGFW used in our works contains various impurities. In wastes that are exfoliated at high temperatures, the oxygen content is within the range of 1.8–10.4% (mass), whereas commercial products are characterized by high carbon content (99%–99.99%). It can be concluded that pGFW is partially oxidized, although this do not interfere its subsequent oxidation to graphite oxide.

The morphology of flake graphite and pGFW particles is different (Figure 6.4). Non-powdered flake graphite powder particles are flat and have a thickness of 10–20 μm and a width of 50–250 μm. pGFW is obtained by expanded graphite particles rolling operation process: it is pressed into layers that have good elasticity, so pGFW particles get certain non-angular rolled shapes during the grinding process.

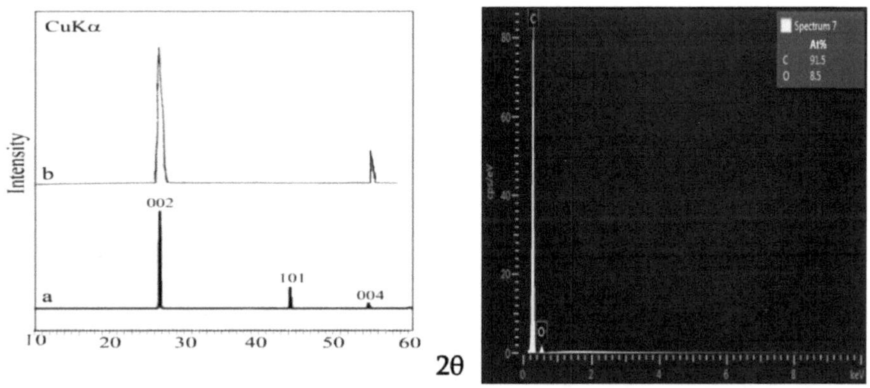


I II

**FIGURE 6.3** XRD patterns of flake graphite (Ia, PDF#41-1487) and powdered graphite foil wastes (pGFW) used in our experiments (Ib); EDX spectrum pGFW (II).

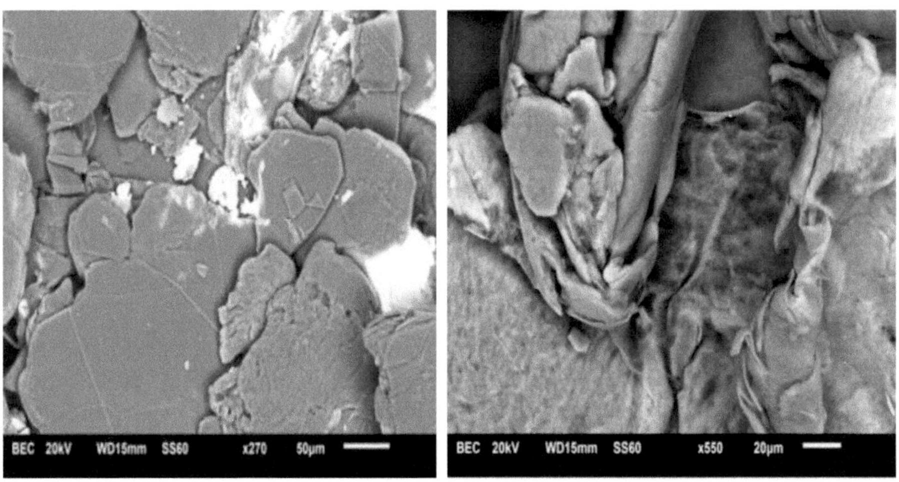


I II

**FIGURE 6.4** SEM images of simples of flake graphite (I) and purified pGFW (II).

pGFW with a particle size of <140 μm was used in the experiments. Their oxidation was conducted using the methods described in the experimental section. Based on our previous studies, it is confirmed that the synthesis of

GO from pGFW was successfully carried out with one of the versions of Hummers method at 0°C [33] and optimized improved Hummers method $KMnO_4$-$H_2SO_4$ at 40–50°C [9]. At this stage, an improved Hummers method ($H_2SO_4$-$H_3PO_4$) at 50°C for pGFW oxidation was tested, and also the method, where concentrated $H_2SO_4$ is added to the $KMnO_4$-graphite mixture. It is confirmed that both methods ensure obtaining of GO (2θ = 10–12.5°), which can easily be reduced to rGO (2θ = 20.03–26.16°) using organic and inorganic reducing agents.

FTIR data showed that the obtained GO contains OH, C = O, CO, C-OH, C = C functional groups. Broad peaks between 3200–3700 $cm^{-1}$ and at 1463 $cm^{-1}$ due to the stretching and bending vibration of OH groups, indicating the presence of adsorbed water molecules and OH functional group. The peaks at 2924 $cm^{-1}$ and 2854 $cm^{-1}$ represent the symmetric and anti-symmetric stretching vibrations of $CH_2$ group. The peak at 1641 $cm^{-1}$ indicated the formation of an aromatic ring (C = C stretching). Peaks at 1744 $cm^{-1}$ and at 1213 $cm^{-1}$ correspond to stretching vibration C = O and epoxy C–O groups (Figure 6.5).

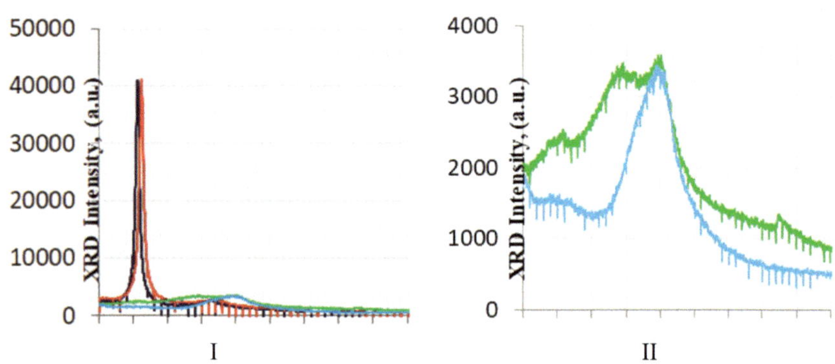


**FIGURE 6.5** XRD patterns of GO dried at 70°C (I, 2θ = 10.61°) and GO dried at 120°C (I, 2θ = 11.16°). XRD patterns of rGO obtained by reducing GO with tannin (II, blue, 2θ = 24.41°) and rGO obtained by reducing GO with extracts of Alder bark (*Alnus*) at 100°C (II, green, 2θ = 24.66°). GO was prepared with adding concentrated $H_2SO_4$ to the $KMnO_4$-graphite mixture. Simples are dried in air 3 h at 120°C.

By increasing the drying temperature of the samples, the diffraction maximums of CO peak gradually shifts to the right: the 2θ value increases, and from 180°C, the rGO begins to form. During the heat treatment of thin films of GO at 200°C for 5 min on air, give rGO phase. One of the stages of GO synthesis is diluting the reaction mixture with water, during which the temperature of the resulting suspension rises rapidly. Depending on the amount of added water, the temperature can reach 135–145°C. In this case, GO is partially reduces and transform to rGO. Thermal reduction of GO begins from 175–180°C, and its formation at a relatively low temperature (135–145°C) is due to the presence of sulfuric acid in the suspension (Figure 6.6).

As mentioned above, the aim of our study is the synthesis of GO and rGO from pGFW and the production of graphene by further heat treatment, as well as the identification of the possibility of liquid-phase exfoliation of pGFW.

Currently, graphene is obtained by direct liquid-phase exfoliation of graphite using various polar organic solvents and surfactants aqueous solutions and ultrasonication. As already mentioned, graphite foil is a product of expanded graphite, and its structure is different from graphite. The decomposition of pGFW was performed by ultrasonication in an aqueous solution of sodium cholate and dimethylformamide. 1 mg/ml sodium cholate hydrate water solution and pGFW (50 μm) suspension with concentration- 2 mg/ml were used. Ultrasonic treatment of this suspension leads to the formation of a black suspension. After removal of non-exfoliated pGFW agglomerates, a stable colloidal solution is obtained, from which no precipitation occurs for 3 months. XRD phase analysis revealed that 13–17% of the initial pGFW undergoes exfoliation and produces few-layered graphene. The size of the main particles decreases to 32–45 nm, which corresponds to 96–135 graphene layers. Particle size was determined by the Scherrer equation. Different results are obtained when particle sizes were determined by the Winner802 DLS Photon correlation nanoparticle size analyzer (Figure 6.7). The average size of exfoliated particles is 270 nm, indicating their agglomeration in suspension. Similarly was carried out direct liquid-phase exfoliation in DMF. In this case, the average size of exfoliated particles reaches 1207 nm, indicating sodium cholate efficacy in the liquid-phase exfoliation process of pGFW compared with DMF. The morphology of non-exfoliated particles is well illustrated in Figure 6.8. As we can see, pGFW particle layers are separated and morphology is violated (see Figure 6.3(II)), but it is not transformed to few-layered graphene. An interesting fact is that the initial pGFW contained 8.5% (mas) of oxygen, the amount of which decreases sharply during the liquid phase exfoliation process (2.8%). This effect can be used in the purification process of used graphite foil wastes.

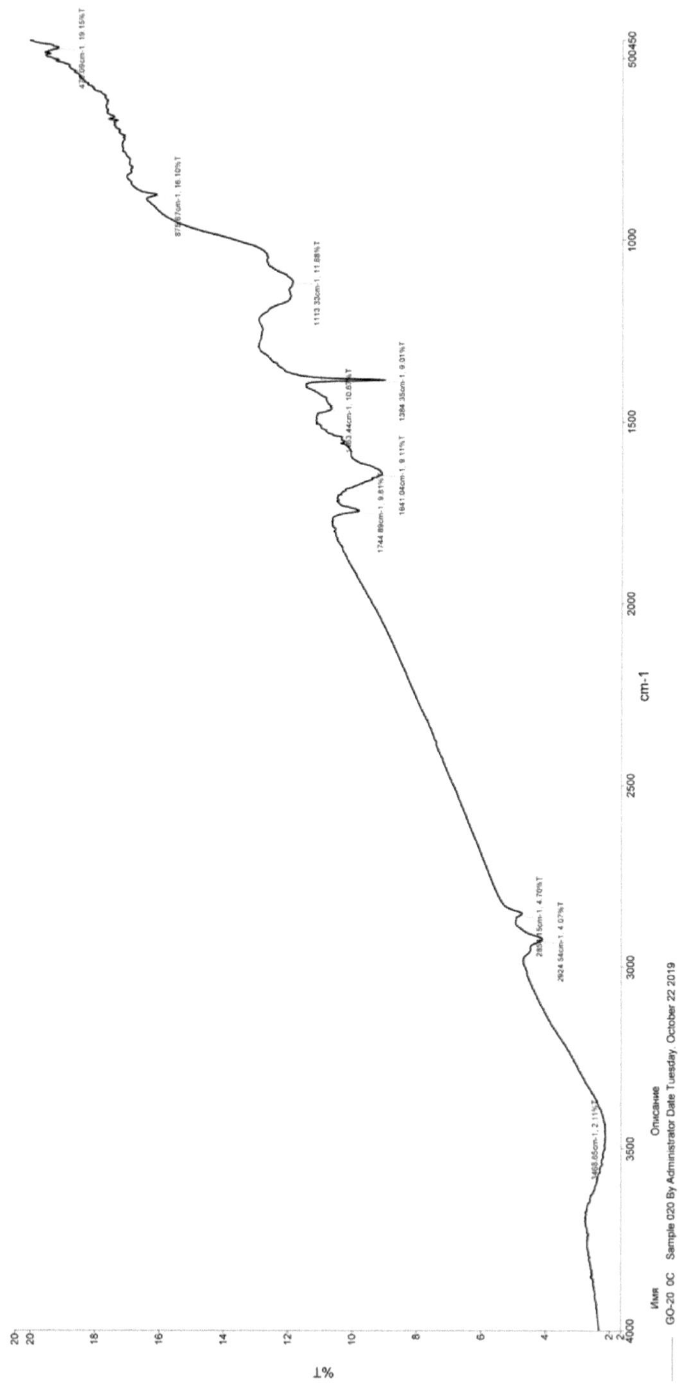


**FIGURE 6.6** FTIR spectra of GO were obtained by adding $H_2SO_4$ (98%) to the $KMnO_4$-graphite mixture.

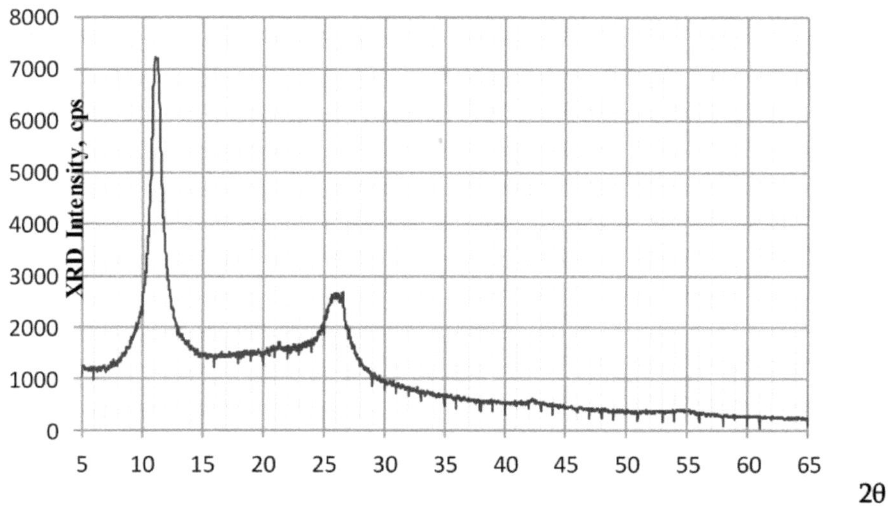


**FIGURE 6.7** XRD patterns of GO (2θ = 10.03°) and rGO (2θ = 25.61°) mixture, obtained by optimized improved hummers method and diluting the mixture with water, obtained by the pGFW oxidation (135–145°C).

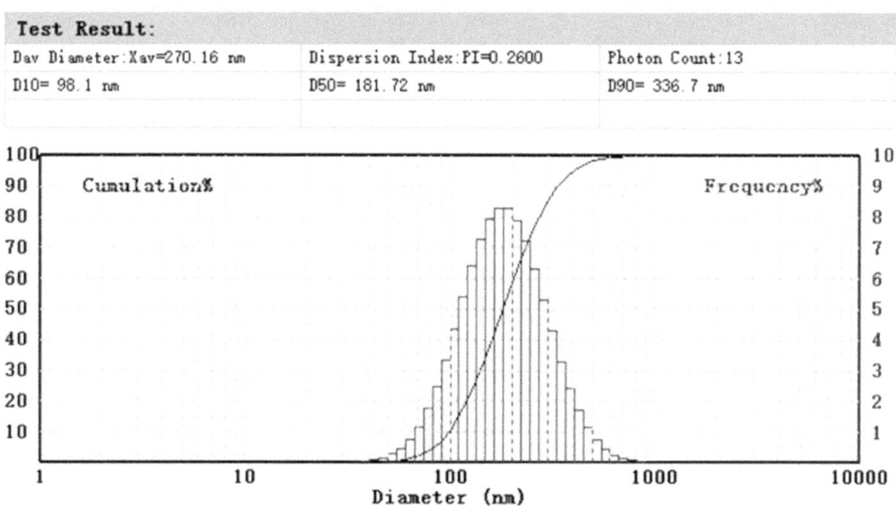


**FIGURE 6.8** Particle size distribution few-layered graphene obtained with ultrasonication of pGFW water suspension in the presence of sodium cholate.

Industrial graphite foils and their wastes are to be appropriate precursors for eco-friendly synthesis of GO and rGO. Thermally reduced graphene was obtained by GO, using the above-mentioned methods; membranes are obtained also; Doping methods of powdered ceramic composites with graphene are developed as well.

Graphene was obtained by thermal treatment of GO and rGO films or powders in hydrogen or inert gas atmosphere at 1000°C. According to EDX spectrum the graphene obtained under these conditions does not contain any metal impurities and consists only of carbon and oxygen in the ratio of 20: 1 (Figure 6.9). This result is fully in line with the literature [22], showing that the C: O ratio varies 8: 1–246:1 depending on thermal treatment modes.

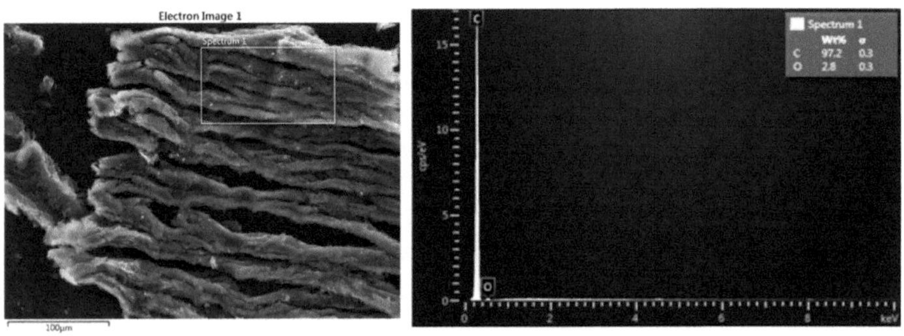


I II

**FIGURE 6.9** SEM images of non-exfoliated particle of pGFW and its EDX spectrum (II).

Modern achievements in nanotechnology have caused use of 2D and 3D carbon substances for making new types of membranes. From this point of view, very interesting results are obtained of application membranes in various fields of technology that are obtained on the basis of GOs. The membranes were obtained by the methods of vacuum-filtration, evaporation of solvents from homogeneous suspensions and thermal treatment GO foils in air or inert atmosphere. Samples of membranes based on GO-obtained from pGFW are shown in Figure 6.10.

Adding graphene and its oxides to organic, inorganic, or organic-inorganic hybrid materials sharply change the physicochemical and mechanical properties of products made from them and, therefore, increases their operational characteristics. They are used to reinforce oxide and non-oxide ceramic composites. It is known that the addition of graphene in small amount increases crack resistance and plasticity of materials.

We have developed a doping method for ceramic composite powders with GO-based on GO and rGO thermal exfoliation in vacuum. Based on this method, $Al_2O_3$-MgO-rGO (0.5%), $Al_2O_3$-$ZrO_2$-$Y_2O_3$-rGO (0.5%) ceramic composite powders are obtained, which were rapidly consolidated using spark-plasma sintering technology (SPS) at 1400–1600°C in 5–10 min. Unlike the traditional method where the powders take a long time to dry (7–10 hours), the SPS method is fast; therefore, the graphene structures are maintained in the resulting product, and the mechanical properties of the ceramic materials are increased (Figure 6.11).

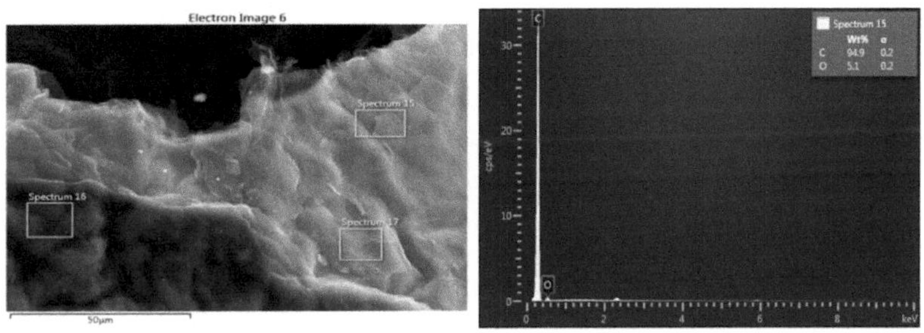


I II

**FIGURE 6.10** SEM images of graphene obtained from rGO by thermal treatment at 1000°C in argon flow and its EDX spectrum (II).

I II III

**FIGURE 6.11** Images of obtained rGO membrane simples by vacuum-filtration (I), evaporation of solvent from homogeneous suspensions rGO (II), and thermal treatment GO foils in Ar flow at 200°C for 5 min (III).

## 6.4 CONCLUSIONS

Powdered Industrial graphite foils and its wastes are new graphite precursors for the synthesis of graphene and its oxides. pGFW with particle size <140 µ easily are oxidized with various methods, including one of the versions of Hummers method ($KMnO_4$-$NaNO_3$-$H_2SO_4$), the improved Hummers method oxidation ($H_2SO_4$-$H_3PO_4$); Optimized improved Hummers method $KMnO_4$-$H_2SO_4$ and by adding concentrated $H_2SO_4$ to the $KMnO_4$-graphite mixture. As a result of a reaction, GO (2θ = 10–12.5°) is obtained, which can easily be reduced to rGO (2θ = 20.03–26.16°) using organic and inorganic reducing agents. Industrial wastes of graphite foils and its wastes are to be appropriate precursors for eco-friendly synthesis of GO and rGO, which can be transformed into graphene by thermal treatment at high temperature of 1000°C. Based on GO and rGO were obtained membranes by the methods of vacuum-filtration, evaporation of solvents from homogeneous suspensions, and thermal treatment GO foils in the air or inert atmosphere. A doping method of ceramic composite powders with GO is also developed.

## ACKNOWLEDGMENTS

This study was partially funded by the Shota Rustaveli National Science Foundation of Georgia (SRNSFG) in the scope of project #YS-18–061. We would like to thank Dr. Archil Mikeladze (Ferdinand Tavadze Institute of Metallurgy and Materials Science, Tbilisi, Georgia), who provided us with graphite foil and its waste samples.

## KEYWORDS

- **exfoliation**
- **graphene**
- **graphene oxides**
- **graphite foils waste**
- **membranes**
- **oxidation**

## REFERENCES

1. Brodie, B. C., (1859). On the atomic weight of graphite. *Philos. Trans. R. Soc., 14,* 249–259.
2. Staudenmaier, L., (1898). Process for the preparation of graphite acid. *Ber. Deut. Chem. Ges., 31,* 1481–1487.
3. Hummers, W. S., & Offeman, R. E., (1958). Preparation of graphitic oxide. *J. Am. Chem. Soc., 80*, 1339.

4. Sun, L., (2019). Structure and synthesis of graphene oxide, *Chinese Journal of Chemical Engineering*, Article in press. https://doi.org/10.1016/j.cjche.2019.05.003.
5. McCoy, T. M., Turpin, G., Teo, B. M., & Tabor, R. F., (2019). Graphene oxide: A surfactant or particle? *Current Opinion in Colloid and Interface Science, 39*, 98–109.
6. Al-Gaashani, R., Najjar, A., Zakaria, Y., Mansour, S., & Atieh, M. A., (2019). XPS and structural studies of high-quality graphene oxide and reduced graphene oxide prepared by different chemical oxidation methods. *Ceramics International, 45*(11), 14439–14448.
7. Mohan, V. B., Jayaraman, R. B. K., & Bhattacharyya, D., (2015). Characterization of reduced graphene oxide: Effects of reduction variables on electrical conductivity. *Materials Science and Engineering, B, 193,* 49–60.
8. Marcano, D. C., Kosynkin, D. V., Berlin, J. M., Sinitskii, A., Sun, Z., Slesarev, A., Alemany, L. B., et al., (2010). Improved synthesis of graphene oxide. *ACS Nano, 4,* 4806–4814.
9. Kovtyukhova, N. I., Ollivier, P. J., Martin, B. R., Mallouk, T. E., Chizhik, S. A., Buzaneva, E. V., & Gorchinskiy, A. D., (1999). Layer-by-layer assembly of ultrathin composite films from micron-sized graphite oxide sheets and polycations. *Chem. Mater., 11,* 771–778.
10. Del, P. L. L. M., Valverde, P. J. L., Sanchez, S. M. L., & Romero, I. A., (2016). Recent advances in graphene research. *Intech.,* 122–133.
11. Peng, L., Xu, Z., Liu, Z., Wei, Y., Sun, H., Li, Z., Zhao, X., & Gao, C., (2015). An iron-based green approach to 1-h production of single-layer graphene oxide. *Nature Comm., 6*(5716), 9. doi: 10.1038/ncomms6716.
12. Yu, H., Zhang, B., Bulin, C., Li, R., & Xing, R., (2016). High-efficient synthesis of graphene oxide-based on improved Hummers method. *Scientific Reports, 6*(36143), 7. doi: 10.1038/srep36143.
13. Sun, L., & Fugetsu, B., (2013). Mass production of graphene oxide from expanded graphite. *Mater. Lett., 109,* 207–210.
14. Yang, H., Li, H., Zhai, J., Sun, L., & Yu, H., (2014). Simple synthesis of graphene oxide using ultrasonic cleaner from expanded graphite. *Ind. Eng. Chem. Res., 53*(46), 17878–17883.
15. Geim, A. K., & Novoselov, K. S., (2007). The rise of graphene. *Nature Materials, 6,* 183–191.
16. Rem, W., & Cheng, H. M., (2014). The global growth of graphene. *Nature Nanotechnol., 9,* 726–730.
17. Dimiev, A. M., & Eigler, S., (2016). *Graphene Oxide: Fundamentals and Applications* (p. 432). Chichester, Wiley.
18. Muzyka, R., Kwoka, M., Smedowski, L., Diez, N., & Gryglewicz, G., (2017). Characterization of graphite oxide and reduced graphene oxide obtained from different graphite precursors and oxidized by different methods using Raman spectroscopy. *New Carbon Mater., 32*(1), 5–20.
19. Liu, J., Wang, H., Li, X., Jia, W., Zhao, Y., & Ren, S., (2017). Recyclable magnetic graphene oxide for rapid and efficient demulsification of crude oil-in-water emulsion. *Fuel, 189*, 79–87.
20. Keyte, J., Pancholi, K., & Njuguna, J., (2019). Recent developments in graphene oxide/epoxy carbon fiber-reinforced composites. *Front. Mater., 6.* doi: 10.3389/fmats.2019.00224.

21. Hiew, B. Y. Z., Lee, L. Y., Lee, X. J., Thangalazhy, G. S., Gan, S., Lim, S. S., Pan, G. T., et al., (2018). Review on synthesis of 3D graphene-based configurations and their adsorption performance for hazardous water pollutants. *Process Saf. Environ., 116*, 262–286.
22. Hiew, J., Yan, B., Lai, Z., Chiew, K., Lai, L., Suyin, G., Thangalazhy-Gopakumar, S., & Rigby, S., (2019). Review on graphene and its derivatives: Synthesis methods and potential industrial implementation. *Journal of the Taiwan Institute of Chemical Engineers, 98*, 163–180.
23. Chen, X., Meng, D., Wang, B., Li, B. W., Li, W., Bielawski, C. W., & Ruoff, R. S., (2016). Rapid thermal decomposition of confined graphene oxide films in air. *Carbon, 101*, 71–76.
24. McAllister, M. J., Li, J. L., Adamson, D. H., Schniepp, H. C., Abdala, A. A., & Liu, J., (2007). Single sheet functionalized graphene by oxidation and thermal expansion of graphite. *Chem. Mater., 19*(18), 4396–4404.
25. Zhang, H. B., Wang, J. W., Yan, Q., Zheng, W. G., Chen, C., & Yu, Z. Z., (2011). Vacuum-assisted synthesis of graphene from thermal exfoliation and reduction of graphite oxide. *J. Mater. Chem., 21*(14), 5392–5397.
26. Lv, W., Tang, D. M., He, Y. B., You, C. H., Shi, Z. Q., & Chen, X. C., (2009). Low-temperature exfoliated graphenes: Vacuum-promoted exfoliation and electrochemical energy storage. *ACS Nano, 3*(11), 3730–3736.
27. Durgea, R., Kshirsagara, V. R., & Tambeb, P., (2014). Effect of sonication energy on the yield of graphene nanosheets by liquid-phase exfoliation of graphite. *Procedia Engineering, 97*, 1457–1465.
28. Gomez, C. V., Tene, T., Guevara, M., Usca, G. T., Colcha, D., Brito, H., Molina, R., et al., (2019). Preparation of few-layer graphene dispersions from hydrothermally expanded graphite. *Appl. Sci., 9*, 25–39.
29. Pu, N. W., Wang, C. A., Liu, Y. M., Sung, Y., Wang, D. S., & Ger, M. D., (2012). Dispersion of graphene in aqueous solutions with different types of surfactants and the production of graphene films by spray or drop coating. *Journal of the Taiwan Institute of Chemical Engineers, 43,* 140–146.
30. Zhu, Y., Murali, S., Cai, W., Li, X., Suk, J. W., Potts, J. R., & Ruoff, R. S., (2010). Graphene and graphene oxide: Synthesis, properties, and applications. *Advanced Materials, 22*, 3906–3924.
31. Huang, H., Mao, Y., Ying, Y., Liu, Y., Sun, L., & Peng, X., (2013). Salt concentration, pH and pressure-controlled separation of small molecules through lamellar graphene oxide membranes. *Chemical Communications, 49*, 5963–5965.
32. Chong, J. Y., Aba, N. F. D., Wang, B., Mattevi, C., & Li, K., (2015). UV-enhanced sacrificial layer stabilized graphene oxide hollow fiber membranes for nanofiltration. *Scientific Reports, 5*(15799). doi: 10.1038/srep15799.
33. Barbakadze, N. G., Tsitsishvili, V. G., Korkia, T. V., Amiridze, Z. G., Jalabadze, N. V., & Chedia, R. V., (2018). Synthesis of graphene oxide and reduced graphene oxide from industrial graphite foil wastes. *Eur. Chem. Bull., 7,* 329–333.
34. www.mersen.com; http://www.toyotanso.com/index.html; http://www.geegraphite.com/; https://www.canadacarbon.com/; https://sealwiz.com/; https://www.graflex.ru/contacts/ (accessed on 19 October 2020).

## CHAPTER 7

# Liquid Crystal Microspheres Based Light and pH Controlled Smart Drug Delivery Systems

K. CHUBINIDZE,[1,2] D. DZIDZIGURI,[1] O. MUKBANIANI,[1] M. CHUBINIDZE,[3] A. PETRIASHVILI,[3] G. PETRIASHVILI,[2] M. P. DE SANTO,[4] M. D. LUIGI BRUNO,[4] and R. BARBERI[4]

[1]*Faculty of Exact and Natural Sciences, Iv. Javakhishvili Tbilisi State University, I. Chavchavadze 1, Tbilisi, Georgia, E-mail: chubinidzeketino@yahoo.com (K. Chubinidze)*

[2]*Institute of Cybernetics of Georgian Technical University, Tbilisi, Georgia*

[3]*Tbilisi State Medical University, Tbilisi, Georgia*

[4]*CNR-IPCF, UOS Cosenza, Physics Department, University of Calabria, Rende (Cs), Italy*

## ABSTRACT

The development of a smart microencapsulation system programmed to actively respond to environmental stimuli is gaining increasing importance compared to the traditional forms of drug administration. In this study, we introduce two new concepts of the drug delivery system, based on the light and pH-sensitive liquid crystal (LC) microspheres. The proposed system represents an emulsion formed by the immiscibility between the LC microspheres and a water environment.

## 7.1 INTRODUCTION

Delivering drugs locally, in the targeted placement and at the controlled portions, is a common way to decrease side effects due to the drug toxicity and, consequently, maximally reduce the undesirable side effects. The drug can either be adsorbed, dissolved, or dispersed throughout the nanoparticle complex or, alternatively, it can be covalently attached to the surface [1].

Assorted types of remotely-triggerable drug delivery systems have been developed, which rely on applying an external stimulus to release the drug load [2] (Figure 7.1). Ideally, such systems could determine the timing, duration, dosage, and even location of drug release, and could allow remote, noninvasive, repeatable, and reliable switching of therapeutic agent flux [3]. As an example, the thermosensitive polymers consisting of drugs and gold nanoparticles have been shown to effectively release the embedded drug upon local heating of the gold nanoparticles via near-infrared light (NIR) irradiation [4, 5]. In other variants of such a system, magnetic nanoparticles have been used in combination with an alternating magnetic field to impose local heating [6, 7]. Recently, a wide range of electromagnetic waves has been proposed to control drug release [8, 9]. Light-activated drug delivery offers distinct advantages over other stimuli because it can release a drug at a desired time and place, so that they hit only targeted cells and not surrounding healthy tissues [10]. At present, different mechanisms are elaborated and proposed that can generally be distinguished based on illumination mode, which may range from femtosecond pulses to continuous-wave illumination [11]. In this study, we introduce a new concept of the drug delivery system, based on the spiropyran emulsion formed by the immiscibility between the NLC microspheres and a water environment (Figure 7.2).

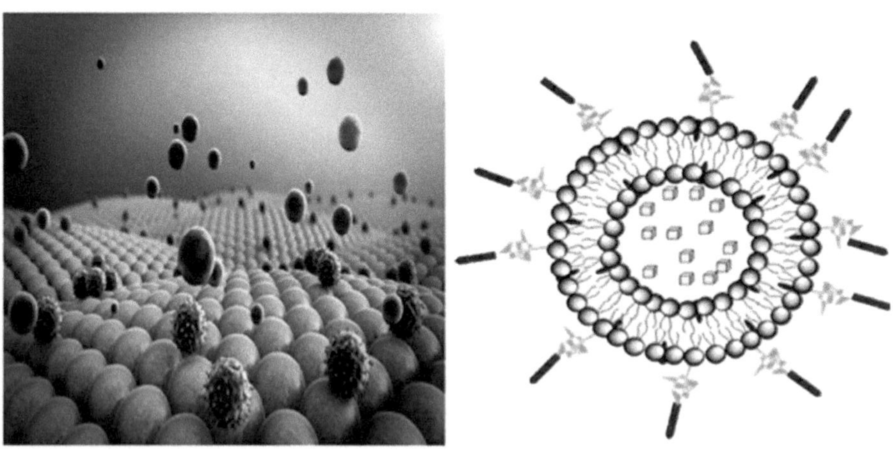

**FIGURE 7.1** Smart microencapsulation for the drug delivery systems.

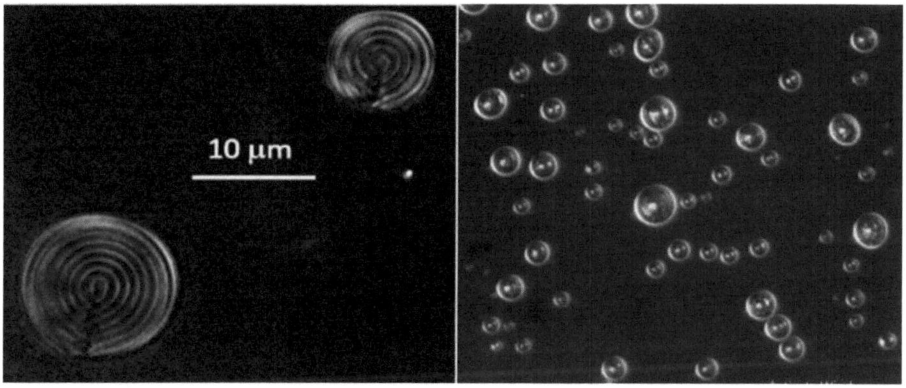


**FIGURE 7.2** An emulsion formed by the immiscibility between the LC microspheres and the water environment.

The stable state of SP is a non-colored closed molecular form that can be transformed into their merocyanine (MC) state, upon UV irradiation (Figure 7.3). The MC conformation spontaneously tends to come back to the SP state, and this transition can be stimulated by irradiation with visible light or by heating. More specifically, the carbon oxide bond of the SP molecules is cleaved when it is transformed into the polar-colored MC form [12]. Since the interconversion between the closed SP form and the open MC one, involves a large molecular rearrangement, some compounds of this class do not exhibit photo-isomerization in solid-state conditions [13]. To efficiently switch SP molecules, require conformational freedom, which usually is not available within the densely packed arrays of molecules in the solid-state. Due to the unique properties of liquid crystalline materials, SP doped NLCs are assumed to have many advanced optical characteristics, quite different from those of SP doped isotropic liquids, semiconductors, and polymer systems. Further, SP doped NLCs present additional advantageous features: an extremely high solubility of SP in the NLC host, which can vary in the range 3–4% (by weight) without destroying the liquid crystalline phase, a high orientational order parameter for SP molecules provided by the spatial orientation of the NLC host molecules. The widespread utility of the SP switch lies in the fact that SP and MC isomers have vastly different physicochemical properties. The charge separation in MC gives rise to a large electric dipole moment, particularly in comparison with the SP isomer. Density functional theory calculations, as well as electrical interferometry and electro-optical absorption measurements, have shown that while the dipole moment of the SP form is in the range of ~4–6 D, this changes drastically to ~14–18 D for the MC form [12]. Therefore, upon the exposure to UV light, a hydrophobic form of SP molecules undergoes photoisomerization into the hydrophilic MC form, which can potentially destabilize the lipid membrane of the NLC-water interface so that the MC molecules can escape inside the aqueous medium.

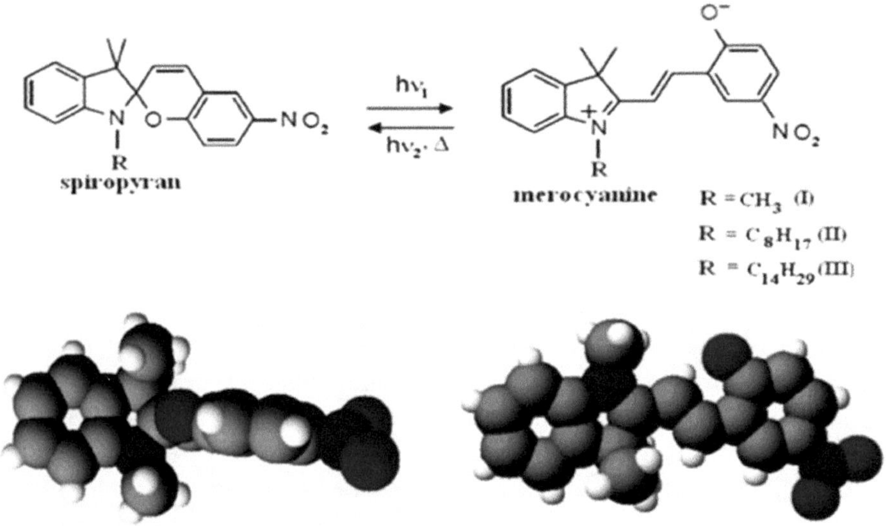


**FIGURE 7.3** Structure of spiropiran and merocyanin.

## 7.2 MATERIALS AND METHODS

### *7.2.1 SP-NLC MIXTURE PREPARATION*

To prepare the SP-NLC mixture, we used the next commercially available compounds: BL-038, as a nematic host (from Merck), 1,' 3′, 3′-Trimethyl-6-nitro-1,' 3′-dihydrospiro [chromene-2, 2'-indole] as a photochromic material (from Sigma-Aldrich). It should be noted that the SP, we used in our experiments, is essentially insoluble in water, whereas the solubility in NLC is very high-3–4 wt.%. The photo-switching behavior and absorption spectra of the mixtures were investigated with a UV/VIS Spectrometer (AvaSpec 2048, Avantes) at room temperature. For a light-induced generation of the polar-colored MC form, we irradiated samples using a 100 W mercury lamp (HG 100 AS, Jelosil) equipped with a 320–410 nm bandpass filter, and the diode-pumped solid-state laser (Shangai Dream Lasers Technology Co. Ltd.), with a maximum power 50 mW at wavelength $\lambda$ = 405 nm. For the vigorous agitation of SP-NLC-water emulsion, we used a Vortex test tube mixer, with controlled mixing speed in the range of 0–2800 rpm. To enclose the emulsions, small laboratory glass bottles were used. The images of SP-NLC emulsified microspheres were obtained with the use of a polarizing microscope coupled to a digital camera (from Polam), and a fluorescence microscope (from Nikon). At first, an SP doped NLC mixture was prepared by mixing the nematic host and SP material with the following concentration ratio: 96 wt.% BL-038 + 4 wt.% SP. Before preparing the SP doped NLC microspheres emulsified in water, to demonstrate a light stimulated photoisomerization, a macro size drop (3 mm) of the SP doped NLC mixture was slowly immersed in the bottle filled with deionized water. Then a laser beam was forwarded to a sphere. During 4 seconds of laser irradiation, the color of the SP doped NLC sphere was dramatically changed from red to dark purple one (Figure 7.4).

a b

**FIGURE 7.4** SP-NLC macro sphere in water, before the laser irradiation (a), and after 4 s of laser irradiation (b).

### *7.2.2 SP-NLC-WATER EMULSION PREPARATION*

To form an SP-NLC-water emulsion, a mixture consisting of SP-NLC and water was prepared with the following percentage in weight: 95%wt. (90%wt.

water + 10%wt. glycerol) + 5%wt. (96% wt. BL-038 + 4 wt.% SP). Glycerol was added to the aqueous phase to obtain a homogeneous distribution of SP-NLC microspheres in the water-glycerol matrix and for the emulsion stabilization [14]. The prepared mixture was agitated at 600 rpm, at room temperature for 10 min. As a result, a homogeneous distribution of SP-NLC microspheres suspended in an aqueous matrix was obtained. The diameters of SP-NLC microspheres were ranging from 10 to 15 μm.

## 7.3 OPTICAL MEASUREMENTS

After the preparation, an SP-NLC-water emulsion was sandwiched between two cover glasses and was assembled in an optical cell, whose gap was set to 100 μm. The prepared optical cell was placed under the polarizing microscope. Figure 7.5(a), from the top view, shows a distribution of SP-NLC microspheres inside the water environment. The colors of the microdroplets are produced by the interference between the ordinary and extraordinary rays of transmitting light, which is caused by the radial configuration of the NLC molecules inside the sphere, Figure 7.5(b). The location of the SP in the NLC matrix causes the alignment of SP molecules parallel to the host NLC molecules so that the orientation of SP molecules inside the NLC microsphere follows the radial configuration of the NLC molecules. To visualize the alignment of SP-MC molecules, we irradiated an optical cell by UV/vis mercury lamp, equipped with a 350–410 nm bandpass filter. The distance from the lamp to the sample was adjusted to 25 cm. The light intensity at the sample, measured by optical power/energy meter, was 0.30 mW/cm$^2$. Exposure time was 5 s. During this period, a large quantity of SP form was photoisomerized to the MC form. Figure 7.5(c) shows an image of SP doped NLC microsphere between crossed polarizers, when irradiated by a mercury lamp. Red circles correspond to the regions in which the MC molecules are aligned either horizontally or vertically in the field of view, confirming a radial orientation of MC molecules, similar to the orientation of NLC molecules. Figure 7.5(d) demonstrates the same microsphere, but the image was taken in the reflectance mode.

Chemically, NLC consist of amphiphilic molecules. Therefore, they have two distinctly different characteristics, polar, and nonpolar in different parts of the same molecule that induces a strong homeotropic anchoring at the NLC-aqueous interfaces. The competition between elasticity and interface tension formed around the NLC microspheres can be controlled by the difference between the polarities of the two liquids. Photoisomerizable molecules that experience a conformational change upon light illumination are promising candidates for the controlling of elastic/tension forces between two liquid mediums. The change of the anchoring condition has a striking effect on the nematic order parameter, which is presented at the interface of the NLC-water phases. In our case, upon UV illumination, photoisomerization of SP molecules into MC ones, can potentially disrupt the NLC-water barrier and allow for the release of MC molecules. To prove that SP-MC photoisomerization is linked to the destabilization of the NLC-water barrier and MC molecules can pass through this phase boundary, we carried out the next measurements. For the experimental manipulation, we used five glass bottles filled with just prepared [95% wt. (90%wt. water + 10% wt. glycerol) + 5% wt. (96wt.%. BL-038 + 4% wt. SP)] emulsion. Then each bottle, except one, was irradiated by mercury lamp, equipped with a 350–410 nm bandpass filter. The light intensity, measured at the samples, was about 0.50 mW/cm$^2$. The exposure time for each bottle was different so that, the first one was irradiated for 4 s, the second one-for 8 s, the third one-for 12 s and the fourth one-for 16 s. Immediately, after the light irradiation, each bottle, including non-irradiated one, was agitated using a test tube shaker with shaking orbital speed equal 80 rpm. A time of agitation for each bottle was equal to 10 s. After this procedure, a non-irradiated bottle appears as a non-colored milk-like substance, whereas the color of each irradiated bottle gradually changes from the pale red-orange to the deep red-violet one (Figure 7.6).

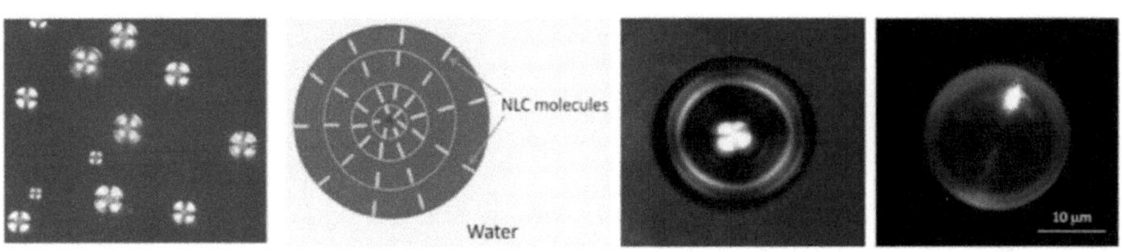


a b c d

**FIGURE 7.5** Polarizing microscope observation of SP-NLC microspheres emulsified in an aqueous medium (a), a schematic representation of the spatial distribution of the molecules inside NLC microsphere (b), which results in a radial director profile with a point defect in the center of the microsphere. Images of MC-NLC microsphere upon irradiation of UV light in the transmittance (c), reflectance, and (d) modes, respectively.

Using a volumetric pipette, from each bottle we extracted a small quantity of the emulsions, which then were injected into the optical cells. A gap between two glass plates was fixed 100 µm. Under the optical microscope, for each optical cell, we selected and marked the areas, free of SP-NLC and MC-NLC microspheres, to minimize a light scattering caused by the microspheres. By using a UV/Vis spectrometer, we recorded the light intensities passed through the selected areas. Figure 7.7(a) shows the time-dependent increase of the absorption of five optical cells. The first cell was filled with non-irradiated SP-NLC-water emulsion, and the other four samples were filled with MC-NLC-water emulsions, after being irradiated by UV lamp with the different exposure times. According to this dependence, light absorption is gradually increasing with respect to the increase of exposure time. It is known, that unlike from SP form, an MC form is characterized by fluorescent properties. Figure 7.7(b) demonstrates a light emission of the MC-NLC-water emulsion, produced by MC molecules.

**FIGURE 7.6** Bottles with SP-NLC-water emulsions: (a) exposure time 0 s. shaking time 10 s. (b) exposure time 4 s. shaking time 10 s. (c) exposure time 8 s. shaking time 10 s. (d) exposure time 12 s. shaking time 10 s. (e) exposure time 16 s. shaking time 10 s.

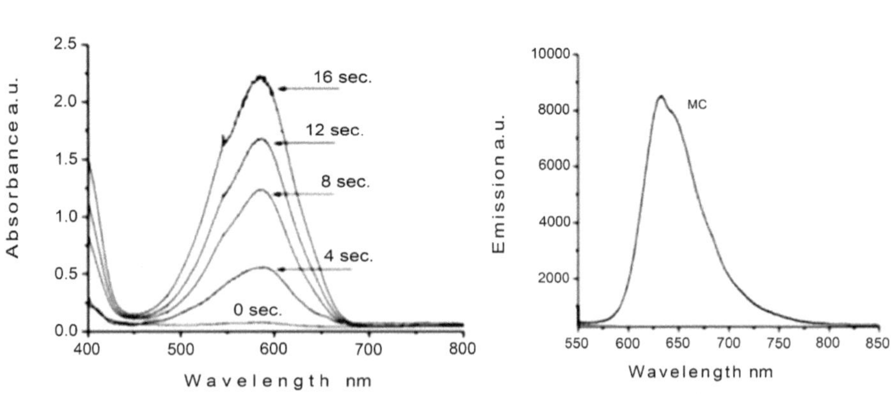


a b

**FIGURE 7.7** Light absorption as a function of the exposure time for the five optical cells filled with SP-NLC-water and MC-NLC-water emulsions (a). Fluorescence emission from MC-NLC microspheres (b).

A rate of SP-MC photo-transformation and the quantity of MC molecules that penetrate through the NLC-water barrier may increase dramatically using laser irradiation. An optical cell filled with SP-NLC-water emulsion was irradiated by the diode laser beam with λ = 405 nm, and 5 s of exposure time. Figure 7.8 shows 150 × 150 μm size area of the sample, before (a), and after (b), irradiation by a laser beam. Due to light absorption, caused by the MC molecules, an exposed area looks much darker, than that of a non-exposed one. We note that not all MC molecules are transferred into the water environment, and some amount of them is permanently located inside the MC-NLC microspheres. As mentioned above, upon UV irradiation, an emulsion of MC-NLC-water microspheres emits a bright red light. This fluorescence property of MC molecules is an attractive way for *in vivo* imaging and tracking of SP-NLC micro container-based drug delivery systems. By using a fluorescent microscope, we captured the images of MC-NLC micro-droplets emulsified in water (Figures 7.8(c) and 7.8(d)).

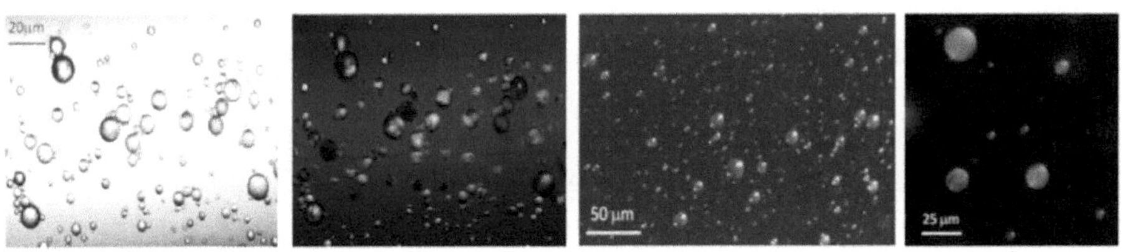


a b c d

**FIGURE 7.8** Images of SP-NLC-water emulsion, before (a), and after (b), exposure to the laser beam. Distribution of MC-NLC microspheres inside the optical cell, observed under the fluorescent microscope (c, d).

According to our concept, initially, SP molecules are bound to the specific therapeutic drug molecules, and this assemblage physically is entrapped inside the NLC micro containers. Upon exposure to UV light, which results in SP-MC photoisomerization, the combination of MC-drug molecules translocate across the NLC-water barrier and disseminates evenly throughout in an extracellular environment (Figure 7.9).

A major drawback is that the UV irradiation that stimulates SP-MC photoisomerization, has a limited tissue penetration because the light in this range of the optical spectrum is highly absorbed and scattered by the different kinds of absorbing chromophores like blood, melanin, fat, yellow pigment, etc. One way to overcome this obstacle is to introduce a two-photon excitation. This method can be highly advantageous for drug delivery systems because the reduced scattering and absorption of NIR irradiation (750–900 nm), results in deep penetration of light in the biological tissues.

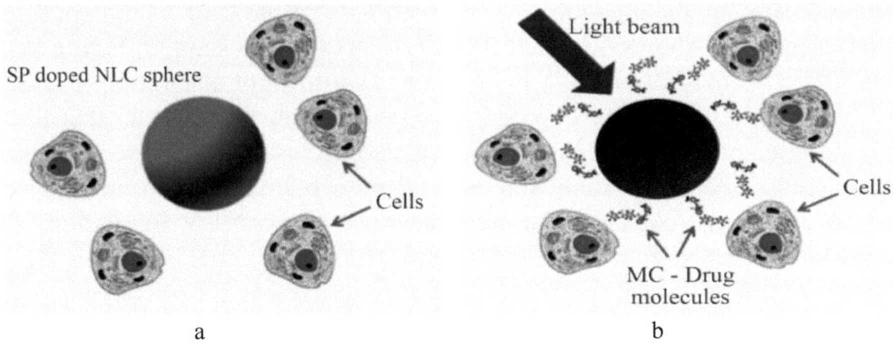


**FIGURE 7.9** Schematic of light-controlled drug release. SP-NLC microsphere before (a) and after (b) UV activation.

## 7.4 pH CONTROLLED DRUG DELIVERY MICRO CONTAINERS

Acidity/alkalinity (pH) controlled drug delivery systems are gaining importance as these systems, because different organs, tissues, and cellular compartments have different pH values, which make the pH value a suitable stimulus for controlled drug release. For example, pH decrease is observed in most solid tumors, drug-delivery systems responsive to the slightly acidic extracellular. Due to this reason, as the second concept, we utilize the same LC/water microemulsion, but instead of SP molecules inside the LC microspheres are entrapped pH-sensitive fluorescent dye (FD) molecule that intelligently responds to the environmental pH change, resulting in immediate translocation across the LC-water barrier and disseminates evenly throughout in an aqueous environment. To prepare a FD-LC mixture, we used the next commercially available compounds: BL-038, as an LC (from Merck), FD as a photochromic material (from Sigma-Aldrich). To prepare an LC microdroplet/glycerol emulsion, first, the LC mixture was dispersed in the water, with a 1:99 weight ratio of LC mixture and water, respectively. Preparation of LC microdroplets/water emulsion was obtained with a magnetic stirrer at a rotation speed of 350 rpm and a rotation time of 10 minutes. After that, a small concentration of nitric acid (2:98) was doped in water/LC emulsion (Figure 7.10).

Cholesteric liquid crystals (CLCs) are a class of soft materials, which presents peculiar optical properties [14, 15] due to their ability to self-organize in periodical helical structures. CLCs may be obtained adding

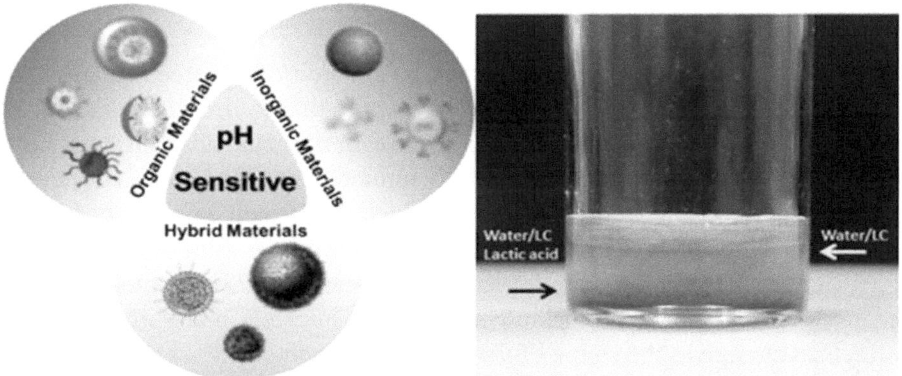


**FIGURE 7.10** LC/water emulsion with nitric acid (blue color) and without nitric acid (red color).

chiral molecules in a suitable quantity into a nematic matrix. Since liquid crystal molecules are centrosymmetric, the periodicity of these structures is equal to half the pitch p of the helix. CLCs, then, maybe considered as an example of self-assembled one-dimensional photonic materials, as they present a one-dimensional periodic variation of the refractive index. Light propagating along the helical axis will experience Bragg reflection at a wavelength $l_0 = np$, where n is the average of the refractive indices defined as $n = (n_e + n_o)/2$, where ne and no are the extraordinary and ordinary indices of refraction, respectively. Because the LCs is birefringent materials, there will be a whole interval of wavelengths, known as selective reflection band (SRB) that will not propagate inside the material. CLCs can selectively reflect the 100% of circularly polarized light. The full width at half-maximum Dl of the SRB equals to $Dl = P \times Dn$, where $Dn = n_e - n_o$ is the birefringence of a nematic layer perpendicular to the helical axis. Since the helical pitch may be tuned varying the concentration of the chiral dopant in the nematic matrix, the SRB can be shifted along the whole visible range. Further, the selective reflection wavelength can also be tuned changing external or internal factors (electric and electromagnetic fields, temperature, local order, etc.) [16, 17]. Recently, CLC microdroplets have attracted much attention because the geometric confinement of CLCs in micrometer-sized spheres leads to a variety of configurations. CLC microdroplets can be easily formed when the hydrophobic liquid crystal is dispersed in an immiscible fluid as, for example, glycerol. Stirring, shaking the blend, or using microfluidic devices generates LC droplets emulsions. Due to the LC hydrophobicity, the droplets acquire a spherical shape and, in case of the planar orientation of the LC molecules at the interface and of droplet size larger than the cholesteric pitch, the helical axes assume a radial configuration. Such droplets may be considered as three-dimensional micro-resonators. It is known from the literature that environmental changes such as pH variation, protein binding, enzymatic catalysis can cause the functional groups on CLC droplets to induce a change in the anchoring condition at the interface [18–21]. Besides, it is well known that some species of organic dyes change their optical properties following exposure to different chemical and biological substances [22, 23]. In the proposed case, the infiltration of acid molecules inside microdroplets produces two effects: a change in the shape of the photonic bandgap and a shift in the dye absorption and fluorescence band spectral position.

## 7.5 MATERIALS AND METHODS

A dye-doped cholesteric CLC was prepared using the following commercially available compounds: BL036 as a nematic host, MLC-6247 as chiral dopant (both from Merck) and 9-diethylamino-5-benzo[a] phenoxazinone, known as Nile Red (from Sigma Aldrich) as a FD. The mixture was prepared in the following percentages in weight: 99.6% [77% BL036þ23% MLC-6247] + 0.4% Nile Red. Nitric acid was used (Fluka) to lower the pH of the mixture.

The emulsion was created dispersing the chiral mixture in glycerol and stirring the blend at 20 Hz for 40 sec in a laboratory vortex mixer. Due to the immiscibility of LCs in glycerol, LC droplets with a spherical shape and a diameter of few tenths of microns were formed (Figure 7.11). Glycerol imposes parallel anchoring conditions on the CLC molecules at the interface promoting the formation of an internal radial configuration of the cholesteric helices. The optical properties of the materials were analyzed using an optical fiber spectrometer (AVASPEC-2048, Avantes) with a spectral resolution of 1 nm, confining the mixture between two glass plates separated by a 36 mm spacer. Chiral microdroplets were studied using a polarizing optical microscope confining the emulsion between two glass plates separated by a 200 mm spacer.

## 7.6 RESULTS AND DISCUSSIONS

It is known from the literature that Nile Red is a highly solvatochromic dye, i.e., the absorbance and fluorescence band spectral positions are sensitive to the polarity of the solvent [22–24]. The dye fluorescence was investigated dispersing 0.4% of Nile Red in a glycerol matrix (Figure 7.12, red curve). The maximum of fluorescence was found at around 610 nm. Then 5% in volume of nitric acid was added to the glycerol matrix and a spectral shift of about 25 nm towards longer wavelengths was observed (Figure 7.12, blue curve).

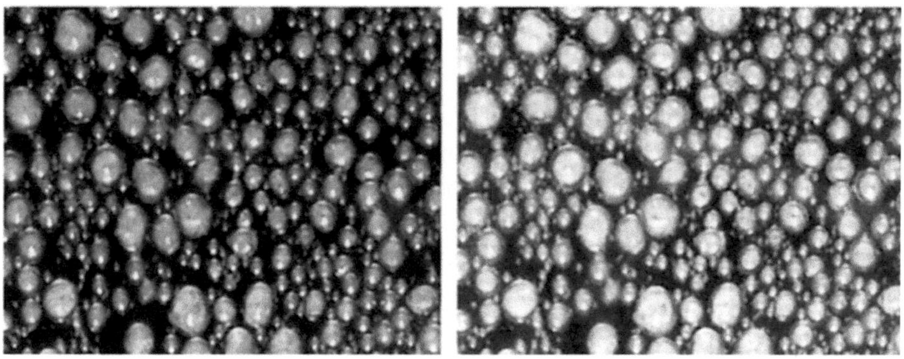

a b

**FIGURE 7.11** LC microdroplets under an optical microscope: (a) no nitric acid, (b) with nitric acid.

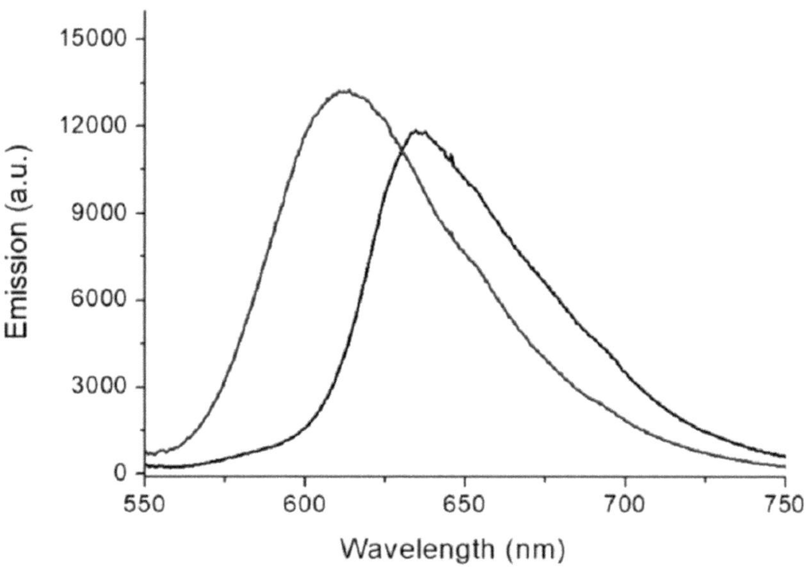


**FIGURE 7.12** Dye fluorescence from a solution containing glycerol doped with 0.4% of Nile red (red curve) and after the addition of about 5% nitric acid (blue curve).

To evaluate the effect of the presence of nitric acid on the optical properties of the mixture, i.e., on the photonic bandgap shape and position, an optical cell was filled with the dye-doped chiral mixture. Then, 20 mL of acid was placed at one of the cell openings, allowing its diffusion inside the cell. As expected, the optical properties of the cholesteric mixture were influenced by the presence of acid molecules. Figure 7.2 shows a polarized light optical microscopic image of the optical cell, and the acid was infiltrated from the right. The red color on the left is the one shown by the cell before the nitric acid infiltration. As soon as the acid starts to spread inside the cell, it causes a variation in the SRB. Transmission spectra were recorded in three different conditions of the cell. In Figure 7.14, the blue line spectrum was recorded before the acid infiltration, i.e., using the pristine LC mixture, the red line spectrum was obtained in an area close to the infiltration point, where the concentration of nitric acid was higher than in the central area of the cell, where the green line spectrum was registered These observations are qualitative of the photonic response of the LC cholesteric mixture in the presence of nitric acid as it is impossible to evaluate the quantity of nitric acid present in the different areas of the cell. In fact, the spreading of the nitric acid inside the cell is fast, and it could lead, close to the infiltration point, to a transition of the liquid crystal mixture to its isotropic phase. This experimental procedure was chosen to induce a gradient of acid inside the cell, allowing an easy way to obtain a measurable modification of the local optical properties. The acquired spectra show that the presence of acid molecules has a negligible effect in the position of the middle point of the SRB, which is almost unchanged, while the shape of the optical band is affected, being narrower and narrower, increasing the acid concentration. The data of Figures 7.12 and 7.13 show that the concentration of the chiral dopant is optimal so that at least one wavelength edge of the cholesteric SRBs always overlaps the fluorescence wavelength region of the dye with or without acid doping. Therefore, these materials can be all used for stimulated laser emission by pumping the emulsion with the second harmonic of a Q-switched Nd: YAG laser at 532 nm. The stimulated laser emission is obtained by using LC microdroplets dispersed in glycerol. By adding different amounts of nitric acid in the cuvette containing nitric acid from the right side. The concentration of acid molecules increases from left to right.

The presence of acid molecules inside the liquid crystal mixture may lead to both a variation of the microdroplets interface conditions and a change of free ions density.

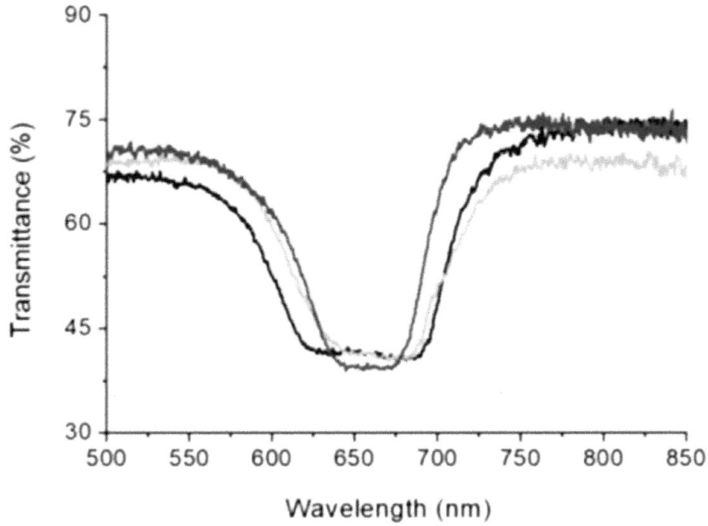


**FIGURE 7.13** Selective reflection band of the pristine chiral mixture (blue line), in the presence of a high concentration of acid molecules (red line) and an intermediate situation (green line).

In this case, electrostatic effects could shield van der Waals forces that align the liquid crystal molecules, varying the liquid crystal order parameter and hence the birefringence properties, as shown in Figures 7.14 and 7.15. Moreover, the acid molecules are responsible for the shift of the fluorescence: this effect, known as solvatochromism, is due mainly to the dielectric constant and hydrogen bonding of the solvent. In this case, the increasing of the acid concentration inside the emulsion implies a polarity growth, and this is reflected by the absorption and emission spectra of the dye towards longer wavelengths.

## 7.7 CONCLUSIONS

In summary, we have demonstrated a novel, light, and pH-controlled drug delivery systems, based on SP and FD doped LC microspheres. Experimental results have shown that upon exposure to UV light and pH change, the photochromic and/or pH-responsive molecules located inside the LC microspheres; experience an interconversion from the hydrophobic, oil-soluble, nonpolar state to the hydrophilic, water-soluble, highly polar state. Light and/or pH-induced isomerization destabilizes the LC water interface, stimulates its translocation across the LC-water barrier, and results in their homogeneous distribution throughout in an aqueous environment. The proposed strategy can be considered as a new platform for the photostimulated/pH-sensitive drug delivery systems that offer the possibilities of the controlled delivery and release of a wide variety of drugs into the body, at the suitable time and desired site, to fight different kinds of diseases including cancer diseases.

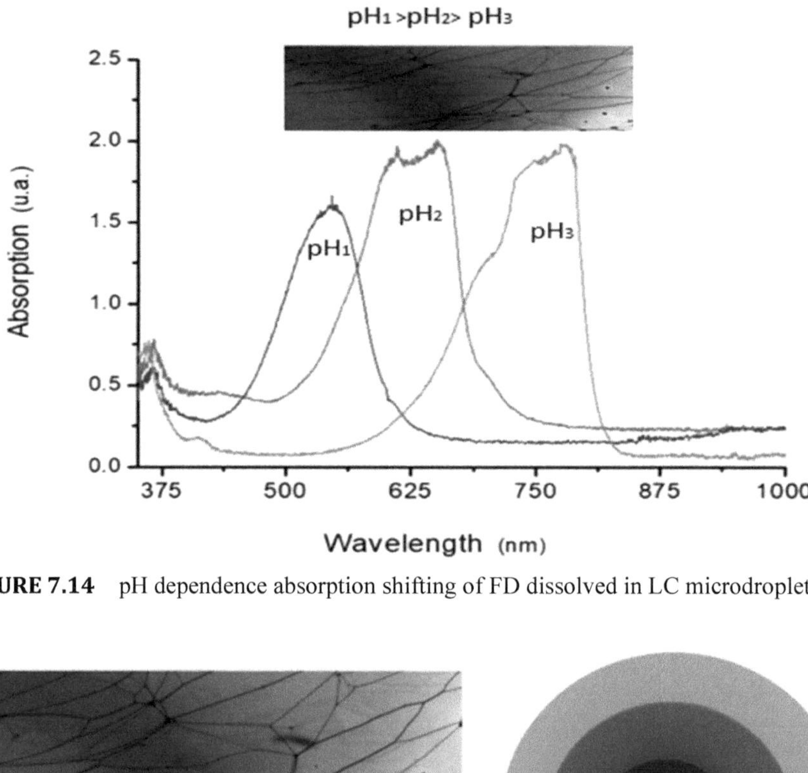


**FIGURE 7.14** pH dependence absorption shifting of FD dissolved in LC microdroplets.

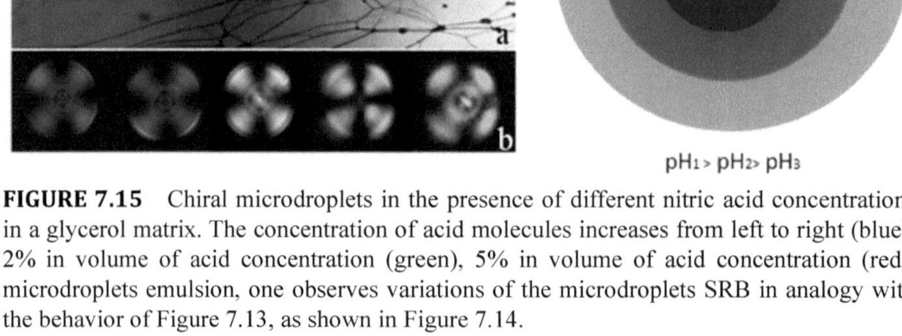


**FIGURE 7.15** Chiral microdroplets in the presence of different nitric acid concentrations in a glycerol matrix. The concentration of acid molecules increases from left to right (blue), 2% in volume of acid concentration (green), 5% in volume of acid concentration (red), microdroplets emulsion, one observes variations of the microdroplets SRB in analogy with the behavior of Figure 7.13, as shown in Figure 7.14.

## KEYWORDS


- **cholesteric liquid crystals**
- **liquid crystals**
- **luminescent dye**
- **photo-isomerization**
- **pH-sensitive**
- **spiropyran**


## REFERENCES


1. Thakor, A. S., & Gambhir, S. S., (2013). Nanooncology: The future of cancer diagnosis and therapy. *CA Cancer J. Clin., 63*(6), 395–418. 10.3322/caac.21199.
2. Shah, S., Sasmal, P. K., & Lee, K. B., (2014). Photo-triggerable hydrogel-nanoparticle hybrid scaffolds for remotely controlled drug delivery. *J. Mater. Chem. B Mater. Biol. Med., 2*(44), 7685–7693. 10.1039/C4TB01436G.
3. Timko, B. P., Dvir, T., & Kohane, D. S., (2010). Remotely triggerable drug delivery systems. *Adv. Mater., 22*(44), 4925–4943. 10.1002/adma.201002072.
4. Timko, B. P., Arruebo, M., Shankarappa, S. A., McAlvin, J. B., Okonkwo, O. S., Mizrahi, B., Stefanescu, C. F., et al., (2014). Near-infrared-actuated devices for remotely controlled drug delivery. *Proc. Natl. Acad. Sci. U.S.A., 111*(4), 1349–1354. 10.1073/pnas.1322651111.
5. Hribar, K. C., Lee, M. H., Lee, D., & Burdick, J. A., (2011). Enhanced release of small molecules from near-infrared light responsive polymer-nanorod composites. *ACS Nano, 5*(4), 2948–2956. 10.1021/nn103575a.
6. Hoare, T., Timko, B. P., Santamaria, J., Goya, G. F., Irusta, S., Lau, S., Stefanescu, C. F., et al., (2011). Magnetically triggered nanocomposite membranes: A versatile platform for triggered drug release. *Nano Lett., 11*(3), 1395–1400. 10.1021/nl200494t.
7. Kumar, C. S., & Mohammad, F., (2011). Magnetic nanomaterials for hyperthermia-based therapy and controlled drug delivery. *Adv. Drug Deliv. Rev., 63*(9), 789–808. 10.1016/j.addr.2011.03.008.
8. Hernot, S., & Klibanov, A. L., (2008). Microbubbles in ultrasound-triggered drug and gene delivery. *Adv. Drug Deliv. Rev., 60*(10), 1153–1166. 10.1016/j.addr.2008.03.005.
9. Derfus, A. M., Von, M. G., Harris, T. J., Duza, T., Vecchio, K. S., Ruoslahti, E., & Bhatia, S. N., (2007). Remotely triggered release from magnetic nanoparticles. *Adv. Mater., 19*(22), 3932–3936. 10.1002/adma.200700091.
10. Fan, N. C., Cheng, F. Y., Ho, J. A., & Yeh, C. S., (2012). Photo controlled targeted drug delivery: Photocaged biologically active folic acid as a light-responsive tumor-targeting molecule. *Angew. Chem. Int. Ed. Engl., 51*(35), 8806–8810. 10.1002/anie.201203339.
11. Leung, S. J., & Romanowski, M., (2012). Light-activated content release from liposomes. *Theranostics, 2*(10), 1020–1036. 10.7150/thno.

12. Klajn, R., (2014). Spiropyran-based dynamic materials. *Chem. Soc. Rev., 43*(1), 148–184. 10.1039/C3CS60181A.
13. Lyergar, S., & Biewer, M. C., (2005). Solid-state interactions in photonic host-guest inclusion complexes. *Cryst. Growth Des., 5*(6), 2043–2045. 10.1021/cg050313b.
14. Chilaya, G., (2001). *Chirality in Liquid Crystals*. Springer: New York, USA.
15. Chilaya, G., (1981). *Rev. Phys. Appl., 16*(5), 193–208.
16. Chilaya, G., Chanishvili, A., Petriashvili, G., Barberi, R., Bartolino, R., De Santo, M. P., Matranga, M. A., & Collings, P., (2006). *Mol. Cryst. Liq. Cryst., 453*(1), 123–140.
17. White, T. J., Bricker, R. L., Natarajan, L. V., Tondiglia, V. P., Green, L., Li, Q., & Bunning, T. J., (2010). *Opt. Express, 18*(1), 173–178.
18. Wang, Y., & Li, Q., (2012). *Adv. Mater., 24*(15), 1926–1945.
19. Khan, M., & Park, S. Y., (2014). *Sensor Actuat. B, 202*, 516–522.
20. Khan, W., Choi, J. H., Kim, G. M., & Park, S. Y., (2017). *Lab Chip, 11*(20), 3493–3498.
21. Jang, J. H., & Park, S. Y., (2017). *Sensor Actuat. B, 241*, 636–643.
22. Lee, H. G., Munir, S., & Park, S. Y., (2016). *ACS Appl. Mater. Inter., 8*, 26407–26417.
23. Wolfbeis, O. S., (1997). In: Dakin, J., & Culshaw, B., (eds.), *Optical Fiber Sensors* (p. 53). Chapter 8, Hartec House, Boston-London.
24. Briggs, M. S., Burns, D. D., Cooper, M. E., & Gregory, S. J., (2000). *Chem. Commun., 2323–2324015), Opt. Express, 23*, 22922.

# CHAPTER 8

# Coating of Cordierite Monolith Substrate by Washcoat and Hybrid Nanocomposite

N. MAKHALDIANI, M. DONADZE, and M. GABRICHIDZE

*Faculty of Chemistry and Metallurgy, Georgian Technical University, Kostava Ave., 69, 0171, Tbilisi, Georgia,*
*E-mail: makhaldianinino@gmail.com (N. Makhaldiani)*

## ABSTRACT

The washcoat is usually applied of metal or ceramic honeycomb monolith by impregnation of the support with slurry. The main goal is to develop and test ceramic filters modified by nanomaterials (nanocomposite) obtained via innovative technology. Ceramic filters containing metal nanoparticles will be used in the model water purification process. As nanofiller substrate used the cellular-structure cordierite with a washcoat, on which a nanocomposite material will be layered. Cellular-structure cordierite substrate, due to its small specific surface area, has to be layered by porous aluminum oxide ($gAl_2O_3$) with a large specific surface or by boehmite (AlOOH), in order to load a sufficient amount of hybrid nanocomposite into cordierite.

Attempts to optimize the primary layer of g-$Al_2O_3$ (washcoat) coating of a monolith honeycomb substrate, as well as service characteristics of the wash coat, are described. The optimum technology for filtering material has been developed. Cordierite substrate covered with washcoat is analyzed with SEM and EDS that shows the coating is stability and homogeneity.

## 8.1 INTRODUCTION

Safe drinking water is the biggest problem in the world. Environmental risk increases every year. Old methods of water purification are already withdrawn from circulation in the leading countries of the world. Modern water treatment technologies that do not imply disinfection with chlorine are used. The researchers in the past few decades have revealed that the use for disinfection purposes of traditional formation of harmful including carcinogenic disinfectants (chlorine, chloramines, and ozone) is related to the formation of compounds that are extremely hazardous for human health, including carcinogenic substances.

Natural sources of drinking water (underground reservoirs, arterial wells) are depleted by bacteria, persistent organic pollutants, and heavy metals. Water filtration by nanomaterials has a number of advantages compared to the conventional, membrane filtration (MF) and treatment by ultraviolet light (UVT) technologies. In contradistinction from conventional disinfection, nanomaterials do not create harmful disinfection byproducts. Integration of nanomaterials into water treatment systems in many cases successfully substitutes harmful chemical disinfectors.

The stability of the filter substrate and nanocomposite material loaded into it is a main problem encountered in the nanofiltration process. As a filtering material there, we are use textiles, perlite, pyrolusite, and other materials, with the use of which the water purification from bacteria and heavy metals was recorded, though the washout of a main active phase was registered in the purified water, too. Due to this fact selection of a substrate still remains the biggest challenge.

Nanoporous membranes for water purification can generally be divided into three types based on their material composition: inorganic, organic, and inorganic-organic hybrid membranes. Inorganic membranes are mainly made on the basis of $Al_2O_3$, $TiO_2$, $ZrO_2$, $SiO_2$, $TiO_2$-$SiO_2$, $TiO_2$-$ZrO_2$ [1]. As nanofilter substrate, we use the cellular-structure cordierite with a washcoat, on which a nanocomposite material will be layered. Cellular-structure cordierite substrate due to its small specific surface has to be layered by a porous aluminum oxide ($gAl_2O_3$) with a large specific surface or by boehmite (AlOOH), in order to load sufficient amount of hybrid nanocomposite into cordierite [2].

## 8.2 EXPERIMENTAL

### 8.2.1 EXPERIMENT SETUP I-PREPARATION OF WASHCOAT

The carrier is usually applied by the impregnation of the honeycomb in slurry of finely ground alumina powder (dip-coating or wash-coating). For the washcoat material was used high purity white powder of AlOOH (Pural SB

Sasol UK) with a surface area of 230 $m^2/g$. It is distinguished by its inertness and thermal stability, porous, and is characterized by high catalytic activity; polyvinyl alcohol (PVA) (Mowiol, Sigma-Aldrich) with an average molar weight of 67,000 g $mol^{-1}$ was used as a binder, nitric acid is purchased from Merck; for preparation, the slurry was used deionized water. As the substrate was used, the cellular structure cordierite (Redgan, China) composition-magnesium, silicon, and aluminum-molar ratio 2:5:2; temperature stability 1465°C, who characterized by a small surface area (0.7 $m^2/g$). Commercial monolith cell density 400 cpsi (cells per square $inch^2$).

For the characterization of slurry and wash-coating layers were used dynamic light scattering (DLS Malvern), Glass Capillary Viscometers, and SEM (JSM-6510LV) images. Many methods are available to deposit catalytic powders on complex geometrical substrates. Usually, a two-phase process is performed. It implies first a pretreatment of the support in order to promote surface interactions between the substrates and the washcoat and then the deposition of the washcoat via a proper coating technique [2, 3]. The washcoat can be either a bare morphological support or the final catalyst already. Among the different deposition methods, dip-coating from a sol or a slurry liquid phase is one of the most applied one, being a good compromise between cost, complexity, and final product effectiveness [4].

To obtain a porous layer of-g$Al_2O_3$ with a large surface area, a suspension was prepared. Distilled water and nitric acid ($HNO_3$) were added with 2% PVA, with constant stirring was heated to 85°C, and powder AlOOH (20 wt.%), the solution was stirred for 24 hours to obtain a homogeneous suspension, the acidity was monitored using a pH meter.

The colloidal system consists of two parts: micelles (a colloidal particle surrounded by a double electric layer) and micellar fluid (electrolyte, non-electrolyte, surfactants)-a dispersion medium that prevents the adhesion of micelles. On the surface of a micellar colloidal particle is formed a double electric layer, consisting of an adsorption and diffusion layer. Electro-kinetic potential-the zeta potential depends on the thickness of the diffusion part of the double layer. The thicker a diffusion layer, the greater the force of electrostatic attraction between colloidal particles and the higher the stability of the colloidal system. When ions move from the diffusion layer to the adsorption layer, the thickness of the diffusion layer decreases, the electro-kinetic potential decreases, and the solution loses its stability. In some cases, the potential becomes equal to zero (the diffusion layer disappears); the colloidal system is in the isoelectric state, when the colloidal particles are stuck together to form large aggregates; that is, the system becomes unstable. The value of the zeta potential depends on the pH of the solution.

In one case, the slurry was prepared without a binder. Studies have shown that the suspension is most stable at pH = 4, since the value of the zeta potential, in this case, is maximum and is 42 mv. When pH = 8.4, the zeta potential is zero, the system is in an isoelectric state, and the system becomes unstable. Acidity pH = 4 is optimal, because relative viscosity (1.05) and particle size determined by dynamic light scattering (DLS Malvern) is also minimal (200 nm) (Figure 8.1(a, b)). In other cases, slurry as binder contains PVA (1–5%), acidity is 4 (pH = 4). Studies have shown that the suspension is most stable at PVA = 2%, relative viscosity (1.8), and particle size determined by DLS is also minimal (200 nm) (Figure 8.2(a, b)).

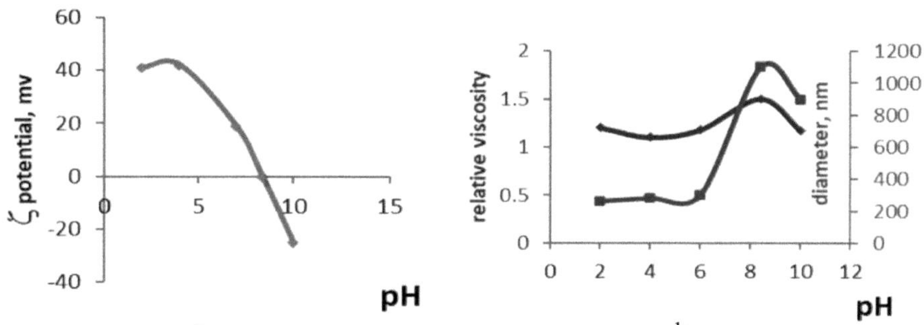


a b

**FIGURE 8.1** Dependence of zeta potential (a), particle size, and viscosity (b) on pH slurry.

Based on the presented graphic data (Figures 8.1 and 8.2), the optimal conditions for the preparation of the suspension were chosen.

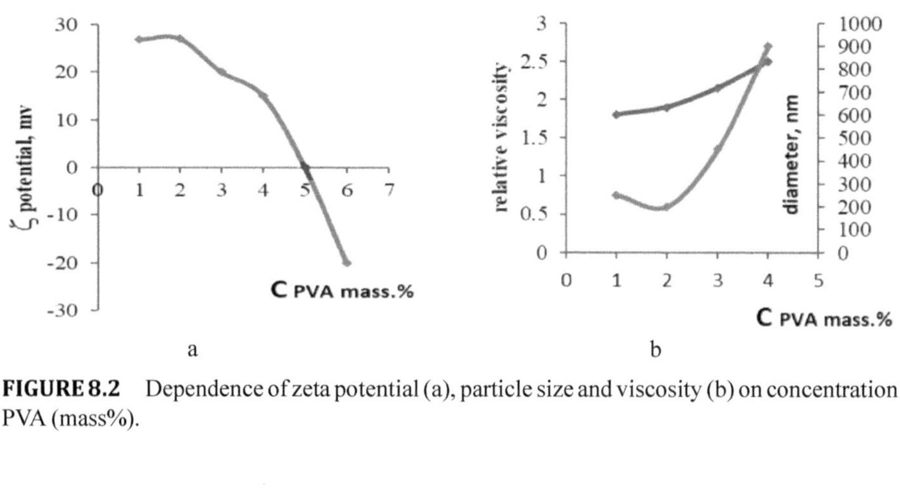


a b

**FIGURE 8.2** Dependence of zeta potential (a), particle size and viscosity (b) on concentration PVA (mass%).

### *8.2.2 COATING OF HONEYCOMB STRUCTURE CORDIERITE WITH WASHCOAT*

The cordierite of the cellular structure to increase the specific surface area is coated with a layer of $gAl_2O_3$ or AlOOH. Before covering with washcoat the cordierite monolith is degreased with the mixture solution-deionized water ($H_2O$): $H_3PO_4$: $CH_3CO_2H_5$ (5: 1: 1: 1). The cordierite is cleaned for 20 minutes in an ultrasonic bath (Power 100 Watt, Frequency 50 Hz), then washes with distilled water until the acid is neutralized. It dried in the oven at 150°C for 5 h and weighed after cooling.

To preparation of slurry, in one case, distilled water was added 20 mass% AlOOH (Pural SB). In the second case, distilled water was added 20 mass% AlOOH (Pural SB) and 2 mass% PVP (binder). The acidity is maintained through pH = 4–5 with nitric acid. In both cases, the suspension was prepared for 24 hours under constant stirring.

For coating cordierite with a wash coat and nanocomposite for circulation, a solution was made a stand (Figure 8.3). The solution can be supplied in both directions (from top to bottom and top to bottom) via compressed nitrogen (1). When the valve (4) is opened using compressed nitrogen, the slurry (Washcott or Hybrid Catalyst) from the vessel (5) is supplied from the bottom to the cordierite (6), the remaining liquid is collected in the vessel (7), the residual gas flows from the valve (8). If valve (8) is open and valve (4) is closed, cordierite is impregnated from top to bottom. The device provides primary drying without removing cordierite. During drying, nitrogen passes through a coil, which is located in the thermostat, and dries the wash coat or catalyst layer.

After bilateral circulation, the cordierite was dried at 110°C for 2 hours. After triple impregnation and drying, they were calcined in the oven for 4 hours at 650°C. After cooling, the weight gain of cordierite was fixed.

With scanning electron microscopy (SEM) was studied the structure of the cordierite without washcoat (Figure 8.4); wash coat layer and the degree of distribution on the surface of cordierite (Figures 8.5 and 8.6).

As shown in the SEM images, the wash coat obtained by the binder (Figure 8.6) is much better in quantity and distribution. In the case of a sample without a binder (Figure 8.5) layer thickness is 1–50 microns, the addition of mass 77 mg/g and in the sample with binder layer thickness of wash coat is 15–20 microns and weight gain is 120 mg/g.

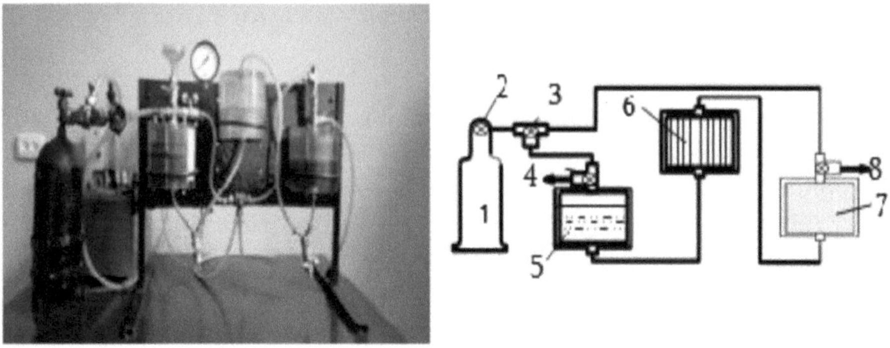


**FIGURE 8.3** Impregnation stand for coating cordierite with washcoat and catalyst layer: 1-compressed nitrogen, 2-reducer, 3-switcher, 4,8-valve, 5,7-vessel to suspension, 6-vessel to cordierite.

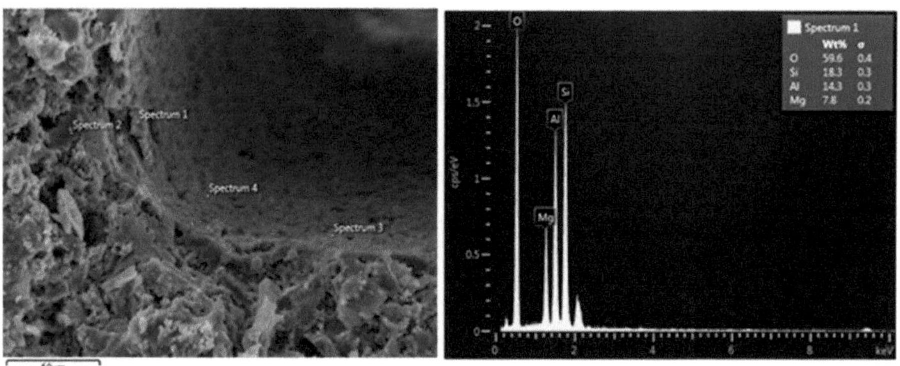


a b

**FIGURE 8.4** Cordierite without washcoat: (a) SEM image; (b) elemental analysis.

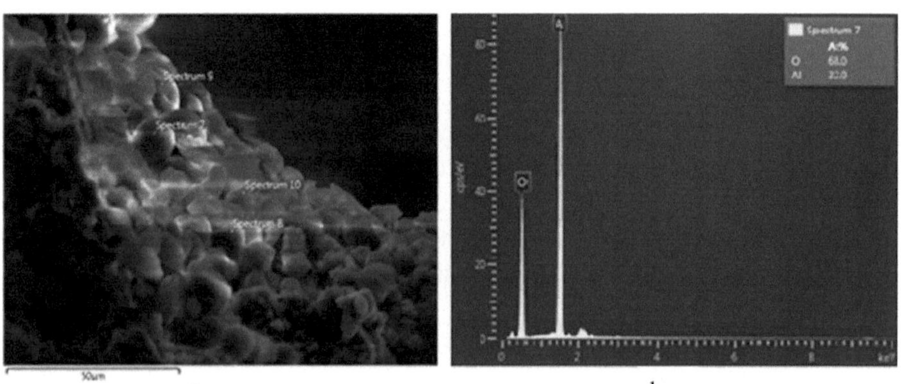


a b

**FIGURE 8.5** Washcoat without binder: (a) SEM image; (b) and (c) elemental analysis.

a b

**FIGURE 8.6** Washcoat with binder (PVA): (a) SEM image; (b) elemental analysis.

### *8.2.3 EXPERIMENT SETUP II-PREPARATION HYBRID NANOCOMPOSITE*

As hybrid nanocomposite material was used Ag@MnOx nanoparticles. Silver nanoparticles have been used extensively as antimicrobial agents in health industry, food storage, textile coatings, and a number of environmental applications. Nano-crystalline silver particles have been found tremendous applications in the fields of antimicrobials, therapeutics, implants, catalysis, microelectronics, for water and air purification. Manganese dioxide, which is characterized by improved catalytic and sorption properties, provides effective water purification from iron, manganese, arsenic, other heavy metals, persistent organic pollutants, and radionuclide.

The sols of silver NPs in a hexane were synthesized using an electrochemical reactor consisting of a sacrificial silver anode (99.9% purity), and aluminum (99.9%) ring cathode, which upon rotation crosses immiscible layers of an aqueous (0.05 M $AgNO_3$, doubled distilled water) and an organic (hexane, 1% OA) solvents. The experimental set-up allows silver ions that are formed at the anode to discharge at the cathode surface poisoned by a surfactant (OA), which adsorbs at sites favorable for silver atoms and inhibits the growth of silver nanoclusters. The latter being weakly adsorbed at the surface and strongly bonded to amphiphile OA molecules, they are easily washed out from the cathode upon rotation, forming the stable sols of Ag@OA core-shell NPs in hexane. In a previous study [5, 6], the authors demonstrated the ability to tune a particle size by variation in residence time $\tau_r$, during which a metal cluster formed at a ring cathode in an aqueous electrolyte is allowed to adsorb amphiphile molecules of a surfactant, from an organic phase In the present study, electrosynthesis has been carried out at the experimental conditions (cathode current density: 7500 A $cm^{-1}$; ring cathode rotation rate: 960 rev $min^{-1}$; electrolyte temperature: 20°C; $\tau_r$ = 36 s), which leads to the formation of an Ag@OA sol containing 0.54 g $L^{-1}$ NPs with an average particle size of 10–15 nm. With oleic acid stabilized Ag NPs characterized with TEM, SEM, EDS, and DLS techniques, is shown in Figure 8.7.

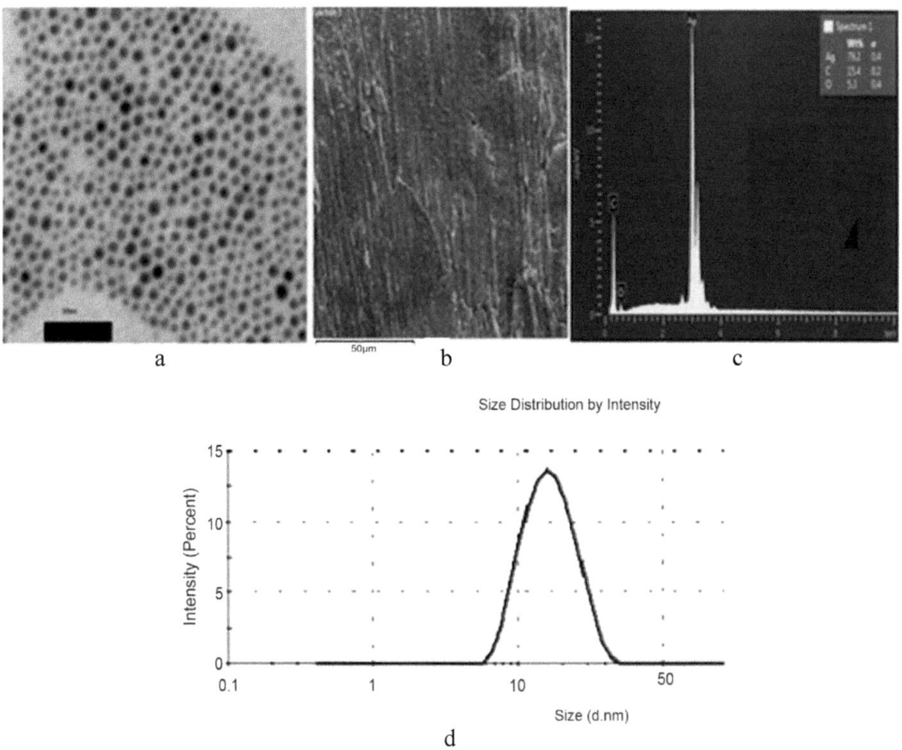


**FIGURE 8.7** Images by oleic acid stabilized Ag NPs: (a) TEM, (b) SEM, (c) EDS, and (d) DLS.

For the preparation of a hybrid material 100 ml sole of Ag NPs in a hexane was mixed with 40 ml 0, 2 M $KMnO_4$ aqueous solution under vigorous shaking during 1 hour (until discoloring of permanganate solution) at ambient temperature. The chemical step involves the partial substitution of the oleic acid shell by a $MnO_2$ formed via interfacial reduction of precursor-potassium permanganate accompanied by broken of the double bond in the oleic acid molecule and formation of pelargonic and azelaic acids according to the following reaction [6–9]:

$$3CH_3\text{-}(CH_2)_7\text{-}CH{=}CH\text{-}(CH_2)_7\text{-}COOH + 4KMnO_4 + 2H_2O® $$
$$3CH_3\text{-}(CH_2)_7\text{-}COOH + 3HOOC\text{-}(CH_2)_7\text{-}COOH + 4KOH \quad (1)$$
$$(\textit{Pelargonic Acid}) \qquad (\textit{Azelaic Acid})$$

After the separation of organic and aqueous phases by centrifugation (0, 5 hours, 8,000 rev min$^{-1}$), a sole of spherical Ag@MnO$_2$ particles with an average size of 20–35 nm was obtained (Figure 8.8).

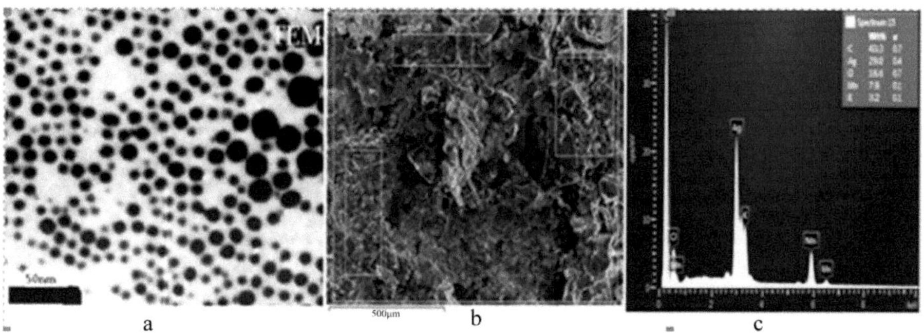

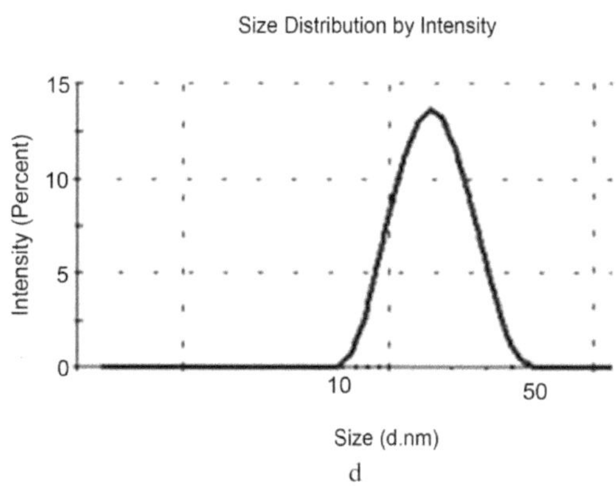


**FIGURE 8.8** Images of Ag-MnO$_2$ particles: (a) TEM, (b) SEM, (c) EDS, and (d) DLS.

### 8.2.4 *XRD SPECTROMETRY*

The structural properties of Ag NPs prior to and post oxidation of an OA shell as well as of a synthesized MnO$_2$ were characterized by the XRD spectra (Figure 8.3 (a, b)). The XRD patterns of as-prepared silver sol demonstrate the amorphous character of the material. The size of the silver particles calculated according to the Debye Sheerer equation is about 18 nm (Figure 8.3(a)). After calcinations of Ag@MnOx at 400°C, the X-ray patterns show clear crystalline structure of a composite $Ag_2Mn_8O_{16}$ (Figure 8.9(b)).

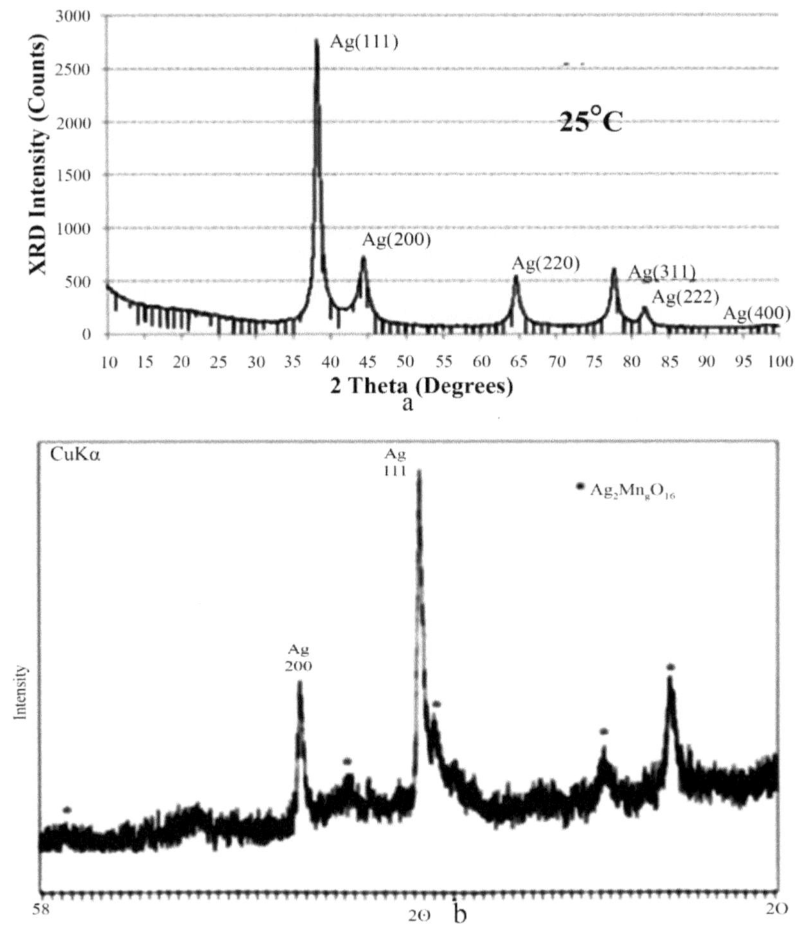


**FIGURE 8.9** XRD patterns of: (a) As-prepared Ag@OA NPs, (b) $Ag_2Mn_8O_{16}$ NPs after calcination at 400°C for 2 h.

## 8.2.5 IMPREGNATION WASH COAT WITH $AG_2MN_8O_{16}$ NANOPARTICLES

Calcined cordierite (with washcoat) is impregnated with hybrid nano-composite-$Ag_2Mn_8O_{16}$. To circulate the solution is used pressure nitrogen

(Figure 8.3). The impregnated cordierite is dried with warm nitrogen. Previously dried cordierite was dried in an oven at 110°C for 1 hour. After triple impregnation and drying, they were calcined in the oven for 3 hours at 450°C. After cooling in the desiccators, the active mass of AgMnOx is fixed. The adhesion test was performed in an ultrasonic bath (50 Hz), where cordierite was vibrated at 40°C for 1 h. The total weight loss is 0.1%, which indicates the stability of the layer.

## 8.3 CONCLUSION

The technological scheme of coating cordierite honeycomb structure with boehmite has been developed. The results showed that a suspension containing a binder is better in terms of both load mass and degree of distribution. Stand allows cover washcoat homogeneous layer hybrid nanocomposite. Ultrasonic bath testing showed slight weight loss, which is a good prerequisite for catalytic testing in harsh conditions.

Ag@$MnO_2$ hybrid nanoparticles were synthesized following the novel bottom up two-step strategy, which involves the electrochemical formation of oleic acid caped silver nanoparticles and partial substitution of the oleic acid shell by the $MnO_2$ via interfacial chemical reduction of a precursor the potassium permanganate by the oleic acid shell. At elevated temperatures amorphous Ag@$MnO_2$ nanocomposite loses organic components and acquires crystallinity [5–7] (Figure 8.10).

Cordierite with a honeycomb structure, coated with a primer and nano-hybrid catalysts, can be used both in air, water, POU and in exhaust gas treatment systems.

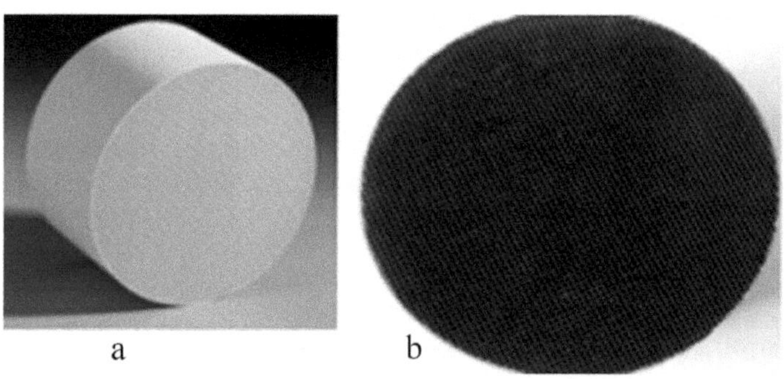

a b

**FIGURE 8.10** Honeycomb structure cordierite: (a) before impregnation with Ag@MnOx, (b) after impregnation with nano Ag@MnOx.

## KEYWORDS


- **binder**
- **cordierite**
- **dynamic light scattering**
- **nanoparticles**
- **substrate**
- **washcoat**


## REFERENCES


1. Wang, Z., Wu, A., Ciacchi, L. C., & Wei, G., (2018). Recent advances in nanoporous membranes for water purification. *Nanomaterials, 65*(8), 1–19.
2. Agrafiotis, C., & Tsetsekou, A., (2002). Deposition of meso-porous g-alumina coatings on ceramic honeycombs by sol-gel methods. *J. of the European Ceramic Society, 22*, 423–434.
3. Balzarotti, R., Ciurlia, M., Cristiani, C., Paparella, F., & Pelosato, R., (2015). Low surface area cerium oxide thin film deposition on ceramic honeycomb monoliths. *J. Chem. Eng. Transactions,* (43), 1747–1752.
4. Cristiani, C., Valentini, M., Merazzi, M., Neglia, S., & Forzatti, P., (2005). Effect of ageing time on chemical and rheological evolution in gamma-$Al_2O_3$ slurries for dip-coating. *J. Catalysis Today,* (105), 492–498.
5. Agladze, T., Donadze, M., Gabrichidze, M., Toidze, P., Shengelia, J., Boshkov, N., & Tsvetkova, N., (2013). Synthesis and size tuning of metal nanoparticles. *Z. Phys. Chem., 227*, 1187–1198.
6. Toidze, P., Machaladze, T., Donadze, M., Gabrichidze, M., Latsuzbaia, R., & Agladze, T., (2018). Examination of silver nanoparticle core-oleic acid shell bonding characteristics. *Bulletin of the Georgian National Academy of Sciences, 12*(3), 70–75.
7. Donadze, M., & Agladze, T., (2017). Strategy for nanohybridezed synthesis of MaMbOx systems. In: Mukbaniani, O., (ed.), *Production of Functional and Flexible Materials.* New Jersey, USA, Apple Academic Press, Chapter 13.
8. Donadze, M., Gabrichidze, M., Agladze, T., & Calvache, S., (2016). Novel method for fabrication of hybrid metal(I)/Metal(II) oxides nanoparticles. *Transactions of the IMF: The International Journal of Surface Engineering and Coatings, 94*(1), 16–23.
9. Donadze, M., Gabrichidze, M., Toidze, P., & Agladze, T., (2016). Multifunctional inorganic core-shell hybrid nanopartcles; synthesis and applications. *Bulletin of the Georgian National Academy of Sciences, 42*, 263–271.
10. Khutsishvili, S., Toidze, P., Donadze, M., Gabrichidze, M., Agladze, T., & Makhaldiani, N., (2019). Structural and magnetic properties of silver oleic acid multifunctional nanohybrids. *Annals of Agrarian Science, 17*, 242–250.

# CHAPTER 9

# Negative Photoresists on the Basis of Copolymers of 2-Chloromethyl-1-(p-Vinyl Phenyl)Cyclopropane with Glycidyl Methacrylate

K. G. GULIYEV,[1] A. I. SADYGOVA,[2] Ts. D. GULVERDASHVILI,[2] and D. B. TAGIYEV[2]

[1]*Institute of Polymer Materials of Azerbaijan National Academy of Sciences, Az5004, Sumgait, S. Vurgun Str.124, Azerbaijan, E-mail: guliyev.kazim_pm@mail.ru*

[2]*Azerbaijan Medical University, Biophysical, and Bioorganic Chemistry (Sub)Department, Az1022, Baku, Bakikhanov Str., 23, Azerbaijan*

## ABSTRACT

The radical copolymerization of 2-chloromethyl-1-(p-vinyl phenyl)cyclopropane (CMCP) with glycidyl methacrylate (GMA) has been carried out. The composition and structure of the obtained copolymer have been established. The constant values of relative activity of monomers have been determined and Q-e parameters on Alfrey-Price have been calculated. The photosensitivity of new cyclopropane and epoxy-containing photosensitive copolymers has been studied. The photochemical structuring has been investigated and it has been established that the synthesized polymer has photosensitivity (56 $cm^2/J$) and can be used for the creation of photosensitive material.

## 9.1 INTRODUCTION

The synthesis and application of the technical polymers with photosensitive groups attract great attention in recent decades due to their wide spectrum of applications in new fields such as advanced microelectronics [1], photolithography [2], holographic [3], photo-cured coatings [4], etc.

The choice of photoresist and also the conditions of its application are determined by the intended purpose. In this regard, there is a tendency to produce the photoresists of narrow purpose, but this requires the creation of a wide assortment of photosensitive materials. Now, the negative photoresists do not sufficiently satisfy all technological requirements of integrated scheme production. Therefore, there is a relentless interest of the specialists in the preparation of new types of negative photoresists meeting certain requirements. Due to this, there is a great interest of the researchers in the preparation of new types of photosensitive polymers for microelectronics [5–8]. We have also tried to solve this problem by means of the synthesis and polymerization of functionally substituted cyclopropyl styrene [9–11].

The polymers containing reactive fragments in basic or in the position of the suspension chain, in UV-irradiation are subjected to structuring processes forming negative photoresists with high sensitivity [12]. As a result of copolymerization of the functional cyclopropane-containing vinyl compounds being one of the perspective reactive monomers there have been synthesized the polymers containing cyclopropane groups, regularly located in the side appendages or in macrochain [13, 14].

This paper has been devoted to the investigation of regularity of the copolymerization of 2-chloromethyl-1-(p-vinyl phenyl)cyclopropane (CMCP) with glycidyl methacrylate (GMA) and to the study of composition and properties of the obtained copolymers with the aim of creation of new photosensitive copolymers.

CMCP is the new reactive monomer, formula, and data of synthesis and homopolymerization of which are presented in work [15]. The choice of this monomer for investigation of the copolymerization with GMA has been stipulated by availability of strongly absorbing light energy of groups (carbonyl, cyclopropane, epoxy, and chlorine atom) in the molecule the synthesized copolymer, which in decisive degree influences on such important photolithographic parameters of the resist, as photosensitivity, elasticity, adhesion, etc., which is confirmed by experimental material accumulated to the present time [16, 17]. It has been shown that for the copolymerization of the studied systems, it is important to choose the optimal conditions under which the polymerization would occur only on the vinyl group, and the reactive fragments would remain unchanged in the side chain.

## 9.2 EXPERIMENTAL PART

The synthesis of CMCP was carried out on the methodology described in the work [15]. The copolymerization of the synthesized CMCP with GMA was carried out in ampoules in benzene solution in the presence of 0.5% (from total monomer mass) dinitrilazoisobutyric acid (AIBN) at 70°C. The total concentration of the initial monomers was constant and was 0.2 mol/l, and the ratio of the initial monomers changed in the concentrations shown in Table 9.1. The forming copolymer was purified by twofold precipitation from benzene solution to methanol and dried in vacuum (15–20 mm merc.c.) at 30°C to constant mass. After the specified time (10–20 min), the copolymers of various compositions of the comonomers have been isolated by the addition of the reaction mixture into an excess of methanol. The conversion of the copolymer samples, for which there have been calculated the copolymerization constants, was 10–12%. The elemental analysis: Found % C 75.45; H 6.72; Cl 10.68. $C_{19}H_{23}O_3Cl$. Calculated, %: C 75.33; H 6.87; Cl 10.61.

The copolymers, being a white powder, are highly soluble in aromatic and chlorinated hydrocarbons. The yield of the copolymer reaches 92%. The characteristic viscosity was determined in benzene in Ubbelohde viscometer ($[\eta] = 0.75$–$0.78$ dl/g). The copolymer composition was determined on data of analysis of the functional groups (on epoxide number).

The IR spectra of the copolymers were registered on spectrometer "Agilent Cary 630 FTIR," and PMR spectra-on spectrometer BS-487B Tesla (80 MHz) in solution of deuterated chloroform.

The determination of the copolymerization parameter values presented in Table 9.1 was carried out on methods described in [18].

For investigation of the photochemical structuring of the copolymer there have been prepared 4–12%-s' solutions of the copolymers, which were applied on a glass substrate with size of 60 × 90 mm. The application was carried out by a method of centrifugation at 2500 rev·min$^{-1}$. The thickness of the resist layer after it's drying for 10 minutes at room temperature and for 20 min. at 25°C/10 mm merc.c was 0.15–0.20 mcm.

The mercury lamp DRT-220 (current force: 2.2 A, distance from radiation source: 15 cm, rate of the mobile valve of the exposure meter: 720 mm·h$^{-1}$, exposure time: 5–10 sec) was used as a source of UV-irradiation. The content of insoluble copolymer was calculated on residue mass as a fact of formation of the crosslinked product.

**TABLE 9.1** Parameters of Copolymerization CMCP ($M_1$) with GMA ($M_2$)

| Composition of the Initial Mixture, mol.% | | E.n. | Composition of the Copolymer, mol.% | | $r_1$ | $r_2$ | $Q_1$ | $e_1$ | $r_1 \cdot r_2$ | Microstructure of the Copolymer | | |
|---|---|---|---|---|---|---|---|---|---|---|---|---|
| CMCP $M_1$ | GMA $M_2$ | | CMCP $m_1$ | GMA $m_2$ | | | | | | $L_{M1}$ | $L_{M2}$ | R |
| 10 | 90 | 10.45 | 18.7 | 81.3 | 1.08 ± 0.03 | 0.42 ± 0.02 | 1.82 ± 0.01 | –0.78 ± 0.02 | 0.45 | 1.12 | 4.78 | 33.9 |
| 25 | 75 | 8.03 | 37.5 | 62.5 | | | | | | 1.36 | 2.26 | 55.24 |
| 50 | 50 | 5.22 | 59.35 | 40.65 | | | | | | 2.08 | 1.42 | 57.14 |
| 75 | 25 | 2.74 | 78.68 | 21.32 | | | | | | 4.24 | 1.13 | 37.24 |
| 90 | 10 | 1.15 | 91.05 | 8.95 | | | | | | 10.72 | 1.04 | 17 |

$L_{M1}$ and $L_{M2}$ – average length of blocks of monomers links; R – coefficient of Harwood blocking; E.n. – content of epoxide groups (%).

## 9.3 RESULTS AND DISCUSSION

By a method of delatometry the kinetics of homopolymerization of CMCP has been studied and it has been shown that the reaction order on initiator is 0.5, on monomer-1, similarly to data of work [19]. This regularity means that the copolymerization process of CMCP with GMA should probably proceed according to the same regularities as radical chain polymerization of the vinyl monomers. The data of the elemental analysis of the synthesized copolymer confirm that the polymers can consist of various combinations of the monomeric links.

CMCP is the new multifunctional monomer, in its radical copolymerization with GMA it would be expected the formation of new reactive polyfunctional copolymer.

The fact of the copolymerization behavior in the investigated systems was confirmed by turbidimetric titration data [20]. The availability of one inflection on the turbidimetric titration curves evidences that in the system it has been obtained the copolymer but not a mixture of two homopolymers (Figure 9.1).

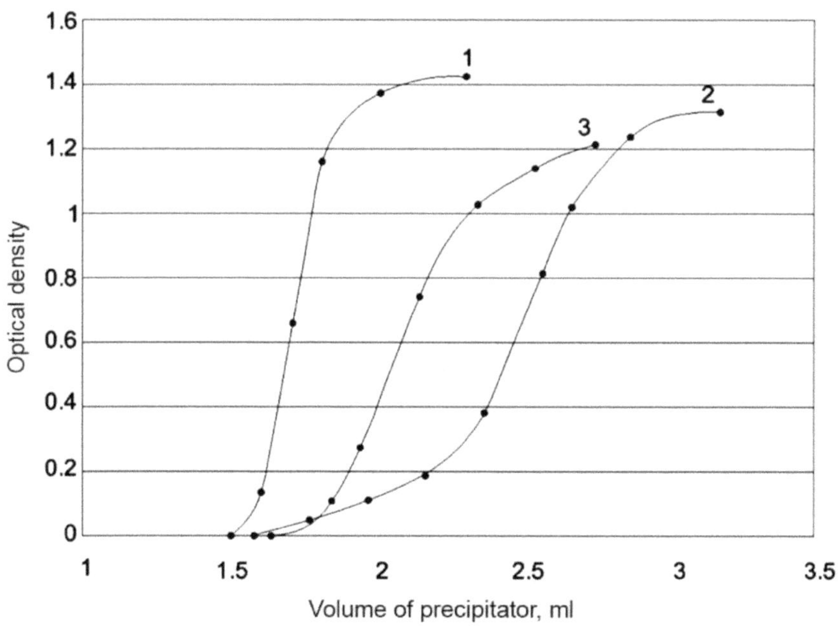


**FIGURE 9.1** Curves of turbidimetric titration of the copolymers of CMCP-co-GMA (1–21.32 mol% of GMA in the copolymer composition, 2–40.65 mol % of GMA in the copolymer composition, 3–62.5 mol % of GMA in the copolymer composition).

In carrying out the copolymerization in any composition of the initial monomer mixture, a content of CMCP and GMA links in the forming copolymers is differed and is from 8.95 to 91.05 mol. Thus, the initial composition is the main factor determining the characteristics of copolymers. It has been revealed that the composition of the forming copolymers depends on the composition of the initial monomer mixture.

The parameters of the microstructure of the copolymers were determined based on copolymerization constants [21]. The obtained data are presented in Table 9.1.

The obtained constant values of relative activity (Table 9.1) indicate a high reactivity of CMCP in comparison with GMA, which has been probably connected with the influence of the substituent in the cyclopropane ring-$CH_2Cl$ on the electron density of the double bond of the vinyl group [22]. It has been revealed that the electron acceptor groups in the cyclopropane cycle are included in the general conjugation system, causing redistribution of the electron density both in the monomer and in the radical center formed from it. As a result, the energy required for the transition state is reduced, leading to an increase in the reactivity of the monomer.

For the determination of activity factors of CMCP, the scheme Q-e proposed by Alfrey and Price has been used [18]. The calculated values Q = 1.82 and e = –0.78 for CMCP show that an introduction of chloromethyl cyclopropane group in styrene in para-position leads to the change of electron-donor properties of the benzene ring.

In the spectrum, the intensive absorption band at 830 $cm^{-1}$, and also the absorption bands in the fields of 3068, 1250, 920 $cm^{-1}$ connected with vibrations of the end epoxide cycle have been detected.

In the spectrum, there are also the absorption bands in the fields of 1030–1040, 1580, and 1600 $cm^{-1}$, characteristic for the three-membered carbon cycle and the benzene ring, respectively. There is an absorption band at 635 $cm^{-1}$, referring to vibrations of the chlorine atom. It has been revealed that the absorption bands at 990 and 1640 $cm^{-1}$ presenting in the IR spectrum of the initial monomers and referring to the deformation and valence vibrations of the double bond of the vinyl group after copolymerization disappear and the absorption bands at 1720 and 1110 $cm^{-1}$, characteristic for vibrations >C = O, and ether bond, respectively, remain unaffected.

In the PMR spectrum of the copolymer, the resonance signals clearly appear, and they can be attributed to the protons of the benzene nucleus ($\delta$ = 6.70–7.30 ppm) and cyclopropane ring ($\delta$ = 0.65–1.68 ppm). The protons of epoxide ring are characterized by signals at 2.30–2.60 ppm (-$CH_2$-) and at 2.96 ppm (-CH-). The signals referring to the protons of the vinyl group ($\delta = 5.10$–$6.66$ ppm) in the polymer sample are completely absent.

Taking into account the above-mentioned one, it has been concluded that the copolymerization of CMCP with GMA proceeds only due to the opening of the double bonds of the vinyl groups with the conservation of other reactive functional fragments of both monomers. Thus, on the basis of the analysis of IR and PMR spectra of the copolymers obtained by copolymerization of CMCP with GMA, the following copolymer structure is assumed:

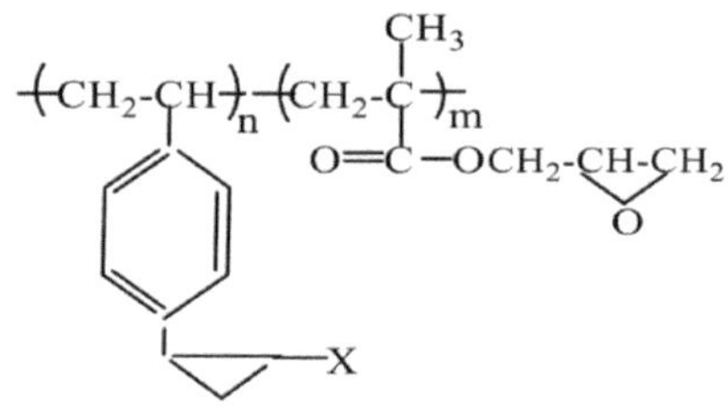

$X = -CH_2Cl$

The investigated copolymers have various chemical natures. Due to the availability of reactive groups of various chemical natures in the links of macromolecule, the investigation of photochemical structuring of the synthesized copolymer is of interest. It has been revealed that under the action of UV-irradiation, a crosslinking occurs, as a result of insoluble material showing itself as photoresist of a negative type is formed.

The obtained copolymers have high photosensitivity (56 $cm^2/J$), good solubility before irradiation, resistance to solvent after crosslinking, and good thermal stability, which is very important for photoresist. This copolymer is capable of the formation of thin films and long-term storage with the conservation of good lithographic properties.

An intensive behavior of the photochemical processes in the copolymers is caused by the presence of such strongly absorbing groups as glycidyl, cyclopropane, carbonyl, and chlorine atoms, etc., in them, which increase the sensitivity of the copolymers to UV-irradiation and are subjected to photochemical conversions leading to crosslinking of the polymer chains. Under the action of UV-irradiation, the copolymer on the basis of CMCP and GMA easily undergoes the structuring, as a result of which the polymer films become insoluble and with low defects.

The UV spectra of the copolymers show the absorption bands around 292 and 300 nm, referring to $\pi \rightarrow \pi$ transitions from suspended photoreactive fragments. The influence of irradiation on photosensitive polymers has been investigated by the measure of the changes in the UV spectrum.

Absorption intensity through various irradiation intervals (Figure 9.2) indicates to changes occurring in the UV spectrum of the copolymer samples with composition of the monomer links in macrochain 59.35: 40.65.

It has been clearly seen from the character of the change of UV spectra of the copolymer that in the first stages of irradiation, the change rate in the band maximum 292–300 nm is linear with irradiation time. However, after 30- and 60-sec, the deceleration of process comes, and almost complete disappearance within 5 min of irradiation occurs.

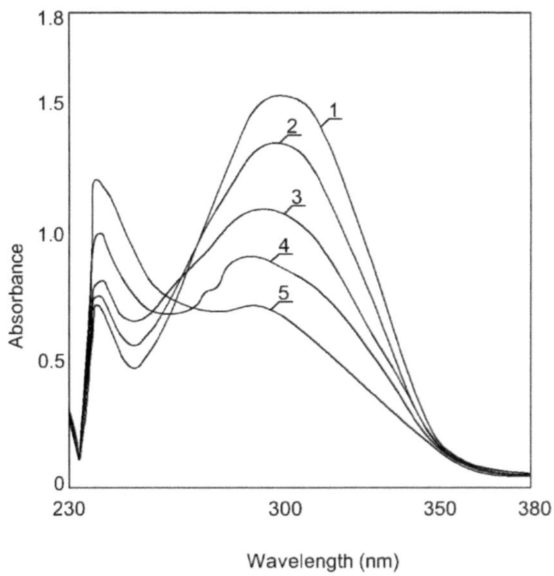


**FIGURE 9.2** Change of UV spectrum of absorption of the prepared film from poly(CMCP-co-GMA) at ratio [59.35:40.65]. 1–5 exposure time, respectively, $\tau = 5, 10, 15, 20, 25$ sec.

It has been revealed that the sensitivity of the samples in the short-wave part of the spectrum strongly depends on layer thickness. In this case, with increase of the layer thickness the three-dimensional structure forming after irradiation has a view of a loose grid with large cells, which swells strongly when manifested and decreases in drying of the polymer layer, causing folds and wrinkles. It has been shown that the good results have been obtained in working with films by thickness of 0.2–0.3 mcm.

For elucidation of ways and mechanisms of the photoreaction behavior we have studied the IR spectra of the copolymer films. The structuring process of the synthesized cyclopropane-containing copolymers has been studied by IR spectroscopic investigations. As follows from data of Figure 9.3 in the IR spectrum of the copolymer after irradiation, a decrease in the peak intensities of some absorption bands is observed. Depending on irradiation duration (1–5 min) an intensity of maxima of the absorption bands characteristic for cyclopropane ring (1030–1035 $cm^{-1}$), chlorine atoms (635 $cm^{-1}$), carbonyl group (1720 $cm^{-1}$), and epoxide ring (830, 1240 $cm^{-1}$) is changed: with an increase of irradiation time these fragments are firstly decreased and then (after ~5 min) are practically disappeared.

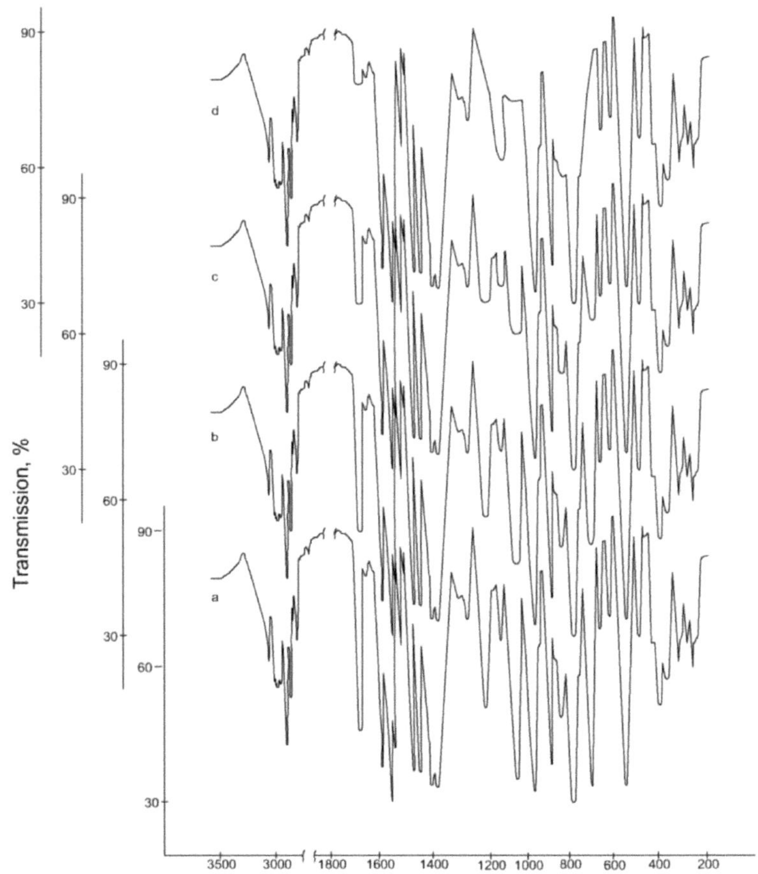


**FIGURE 9.3** IR-spectra of the copolymer film of CMCP-co-GMA: unirradiated (a) and irradiated for 1 (b), 3 (c) and 4 sec. $m_1$: $m_2$ = 78.68: 21.32 mol %.

Thus, it has been synthesized the new copolymer containing cyclopropane ring, carbonyl groups, chlorine atom and also epoxide groups in macromolecule, which allowed to create the polymer materials with high photosensitivity of negative type on the basis of copolymer.

## 9.4 CONCLUSION

1. It has been synthesized and characterized the new polyfunctional CMCP and it has been carried out its radical copolymerization with GMA. The composition and structure of the synthesized copolymer have been established.
2. The copolymerization constants of the copolymerizing monomers ($r_1$ = 1.08; $r_2$ = 0.42) have been found and Q-e parameters values of Alfrey-Price scheme ($Q_1$ = 1.82; $e_1$ = –0.78) have been calculated. It has been studied the process of structuring of the copolymer and it has been established that the synthesized copolymer has relatively high photosensitivity (56 $cm^2 J^{-1}$).

## KEYWORDS

- **copolymerization**
- **epoxy-containing photosensitive copolymers**
- **glycidyl methacrylate**
- **monomers**
- **radical copolymerization**
- **synthesis**

## REFERENCES

1. Bokov, Y. S., (1982). Photo-, electron- and X-ray resists. *Radio and Communications*, 136.
2. Reichmanis, E., Nalamasu, O., Houhihan, F. M., & Novembre, A. E., (1999). Radiation chemistry of polymeric materials: Novel chemistry and applications for microlithography. *Polym. Int.*, *48*, 1053–1059.
3. Ohe, Y., Ito, H., Watanabe, N., & Ichimura, K., (2000). A novel dry photopolymer for volume-phase holograms. *J. Appl. Polym. Sci.*, *77*, 2189–2200.
4. Decker, C., (1997). In: Meijer, H. E. H., (ed.), *Materials Science and Technology* (Vol. 18, p. 615). Weinheim: VCH.
5. Filichkina, V. N., (1985). The use of transparent plastics in the construction of capitalist countries. *Chemical Industry Abroad, 11*, 11–27.
6. Hou, H., Jiang, J., & Ding, M., (1999). Ester-type precursor of polyimide and photosensitivity. *Eur. Polym. J.*, *35*, 1993–2000.
7. Mizoguchi, K., & Hasegaawa, E., (1996). Photoactive polymers applied to advanced microelectronic devices. *Polym. Adv. Technol.*, *7*, 471–477.

8. Vainer, A. Y., Dyumaev, K. M., Likhachev, I. A., Shalatonova, A. D., & Yartsev, Y. A., (2004). Methacrylate derivatives of carboxyl-containing polyimides: Synthesis and photochemical transformations. *Doklady Physical Chemistry. T., 396*(1–3), 115–118.
9. Guliev, K. G., Ponomareva, G. Z., & Guliev, A. M., (2006). Photosensitivity of polymers based on epoxy-substituted vinylphenylcyclopropanes. *Russian Journal of Applied Chemistry, 79*(3), 488–491.
10. Guliev, K. G., Ponomareva, G. Z., & Guliev, A. M., (2007). Synthesis and properties of epoxy-containing poly(cyclopropylstyrenes). *Polymer Science, Series B., 49*, 196–199.
11. Guliev, K. G., Alieva, A. M., & Guliev, A. M., (2013). Synthesis and polymerization of (p-vinylphenyl)cyclopropylmethyl cinnamate. *Russian Journal of Applied Chemistry, 86*(1), 92–95.
12. Subramanian, K., Ramai, R. A. V., & Krishnasamy, V., (1991). Photopolymers: Synthesis and characterization of poly(4-cinnamoylphenyl methacrylate). *Macromol. Chem. Rapid Commun., 12*, 211–214.
13. Guliyev, K. G., Aliyeva, A. M., & Guliyev, A. M., (2015). *Synthesis, Radical Polymerization of 2-Substituted Cyclopropyl Styrenes and Their Lithographic Properties* (pp. 47, 48). All-Russian Scientific-Technical Conference with the participation of young scientists: "Innovative materials and technologies in design." St. Petersburg.
14. Guliyev, K. G., Aliyeva, A. M., Dzhafarov, R. V., Iskenderova, R. M., & Guliyev, A. M., (2015). International Scientific-Technical Conference: "Polymer composites tribology" (Polycomtrib). Gomel, Belarus. *New Photosensitive Cyclopropane and Epoxy-Containing Polymers* (p. 47).
15. Guliyev, K. G., Rzayeva, A. E., Mamedli, S. B., Khamedova, U. A., Nurullayeva, D. R., & Guliyev, A. M., (2017). Synthesis and polymerization of p-(vinyl phenyl)-2-chloromethylcyclo-propane. *World Journal of Research and Review (WJRR), 5*(5), 1–5.
16. Guliev, K. G., Ponomareva, G. Z., & Mamedli, S. B., (2009). Ultraviolet absorption spectra of 2-substituted-1-(*p*-vinylphenyl)cyclopropanes. *Journal of Structural Chemistry, 50*(4ş), 693–695.
17. Guliyev, K. G., Aliyeva, A. M., Ponomaryova, G. Z., Nurullayeva, D. R., & Guliyev, A. M., (2015). Copolymerization of (p-vinylphenyl) cyclopropylmethylcinnamate with methylmethacrylate. *American Journal of Applied Chemistry, 3*(1), 21–24.
18. Odian, J., (1974). In: Korshak, V. M., (ed.), *Fundamentals of Polymer Chemistry* (p. 616).
19. Guliyev, K. G., Aliyeva, A. M., Mamedli, S. B., & Guliyev, A. M., (2015). Synthesis and kinetic investigation of the radical polymerization of (p-vinyl phenyl)cyclopropyl methyl cinnamate. *Azerb. Chem. J., 2*, 40–46.
20. Kantov, M., & Mir, M., (1971). *Fractionation of the Polymers*, p. 335.
21. Zilberman, E. N., (1979). Parameters of microstructure of multicomponent copolymers. *High-Molec. Comp. B, 21*(1), 33–36.
22. Guliyev, K. G., Aliyeva, A. M., Mamedli, S. B., Nurullayeva, D. R., & Guliyev, A. M., (2016). Copolymerization of p-(2-vinylphenyl)cyclopropylmethyl cinnamate with glycidyl methacrylate. *Eur. Chem. Bull., 5*, 108–112.

# CHAPTER 10

# Possibilities of Synthesis of Monodispersed Latex

A. A. HOVHANNISYAN, G. K. GRIGORYAN, A. G. NADARYAN, and N. H. GRIGORYAN

*Scientific-Technological Center of Organic and Pharmaceutic Chemistry NAS Republic of Armenia, 0014, Yerevan, 26 Azatutyan Av., Armenia, E-mail: hovarnos@gmail.com (A. A. Hovhannisyan)*

**ABSTRACT**

Polymer nanodispersed systems are widely used in high technology. Based on these systems, it is possible to obtain modern medicines, and mini-technologies are being created to produce thin films. Using polymer particles, transmission, and other physicochemical properties of biological membranes and films are investigated. Dispersed polymer systems are now accepted in hormone and pulsotherapy for transporting drugs to the internal affected organs. The development of modern supramolecular chemistry is largely associated with the study of dispersed polymer systems. These studies have identified the ability to store nanosystem information. The present work shows the possibility of synthesizing monodisperse latexes in static heterogeneous monomer-water systems.

## 10.1 INTRODUCTION

The main method for the synthesis of polymer dispersions is emulsion polymerization [1–3]. However, the latexes synthesized in emulsions are polydisperse and cannot be used in high technology. Latexes should be monodisperse, with specific particle diameters, must have a certain surface structure, and retain their colloidal properties for a long time.

Doing a search in scientific journals and on the Internet, one can find plenty of publications in which provided recipes are a synthesis of monodisperse latex. After reading these recipes, one can make a conclusion that the monodisperse latex is prepared likely intuitively than using programmed recipes.

For example, in Ref. [4], the formation of monodispersed latex particles in vinyl monomers is result of the presence of the polyamide and epoxy resins (ERs), as well as inorganic salts and oxides in recipes. The authors indicated that particle radius could be adjusted by varying the stirring intensity of the system. Monodisperse lattices can be obtained by polymerizing diene monomers under microgravitation [5]. Latices with a narrow particle size distribution are obtained using suspension polymerizing styrene in the presence of ortho and para divinylbenzene [6].

A list of recipes of monodisperse latex synthesis can be continued further and further, but from the mentioned examples, it is clear that the processing of recipes is an empirical. This approach is justified, since monodisperse lattices are extremely valuable and they have unique applications in various fields of science and technology.

Static mode of polymerization in a heterogeneous monomer-water system can be considered as one of the most promising methods for the synthesis of monodisperse lattices [7–11], since any mixing systems (mechanical, ultrasound, creating convection currents, etc.), expands uncontrollably the particle size distribution. In this work, we investigated the effect of certain factors on the particle size and particle size distribution during static polymerization, and based on the results obtained, developed recipes for the synthesis of latexes with a narrow particle size distribution.

## 10.2 EXPERIMENTAL PART

The photographs and scheme of polymerization plants are shown in Figures 10.1 and 10.2. In these installations, the monomer is layered on the aqueous phase in thermostated tubes. The parameters of the aqueous phase are measured by instruments whose electrodes are immersed in the aqueous phase (Figure 10.2). During the polymerization, samples are taken from the tubes to determine the dry residue of the latexes. Samples are taken using pipettes (Figure 10.2).

Styrene and chloroprene were used as monomer. The polymerization was initiated by potassium persulfate. Which was dissolved in the aqueous phase. In the polymerization of hydrophobic monomers (styrene, chloroprene), the formation of oligomers in water, consisting of two to three monomer groups and a sulfate end group, is possible. Such diphilic oligomers have surface activity and are capable of stabilizing the emulsion.

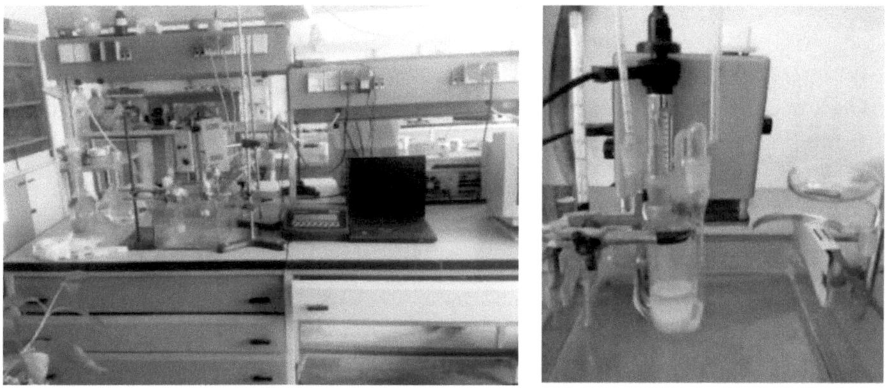

**FIGURE 10.1** Photo of polymerization unit.

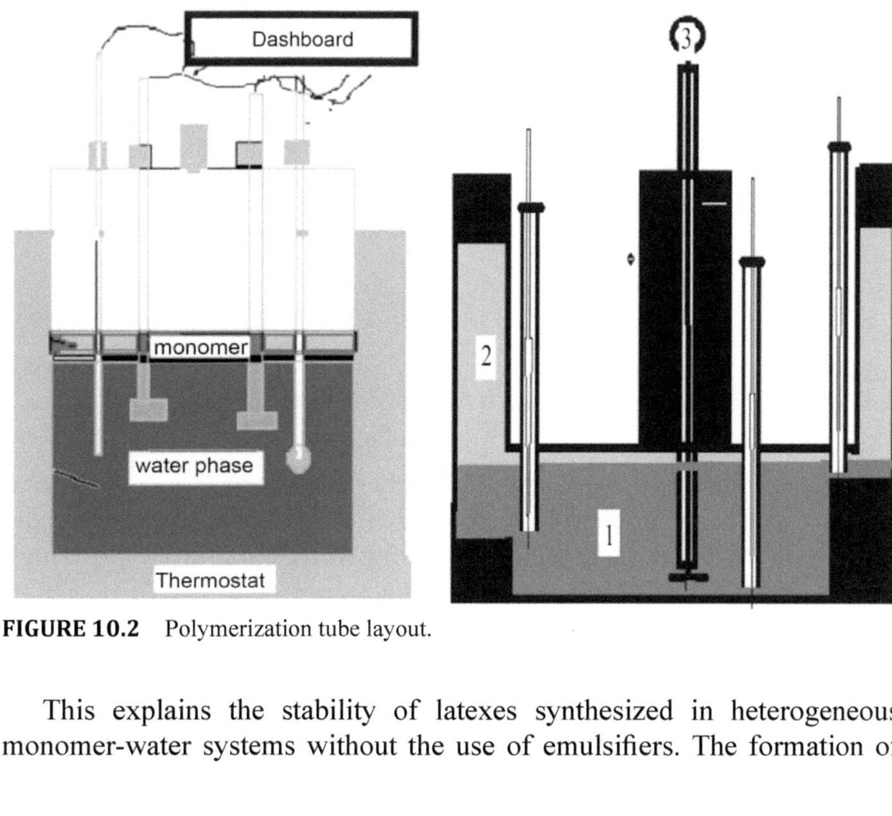


**FIGURE 10.2** Polymerization tube layout.

This explains the stability of latexes synthesized in heterogeneous monomer-water systems without the use of emulsifiers. The formation of diphilic oligomers is recorded by measuring the capacitance of the double electric layer of the platinum electrode (C), which is immersed in the aqueous phase (Figure 10.2). This parameter is very sensitive to the appearance of surface-active molecules in water. The measurement procedure C is detailed, as described in Ref. [12].

Measurement C helps determine the concentration of persulfate in the polymerization recipes. Figure 10.3 shows the change in the relative value of $C/C^0$ at different concentrations of potassium persulfate ($C^0$ is the value of C at the beginning of the process).

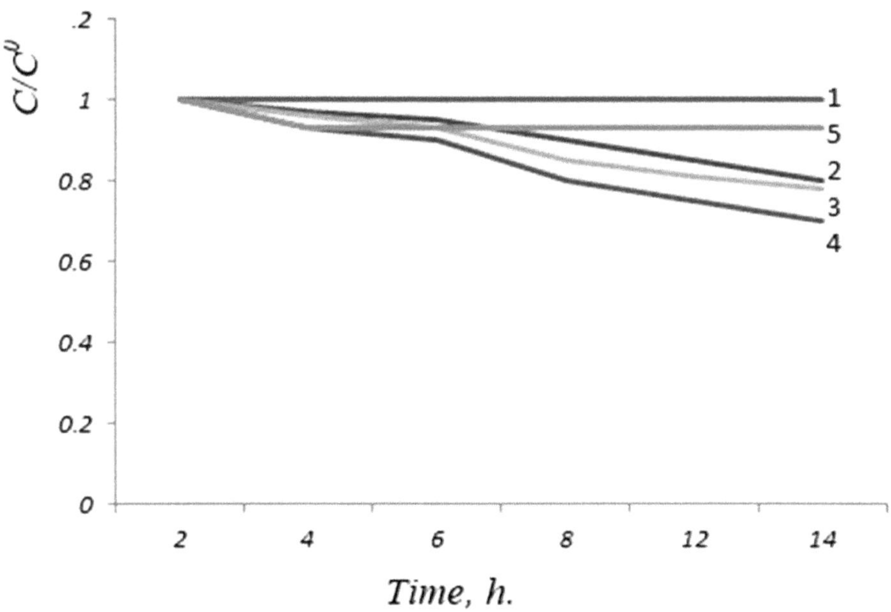


**FIGURE 10.3** The change in $C/C^0$ over time at different concentrations of potassium persulfate in the aqueous phase: 1–0%, 2–0.2%, 3–0.3%, 4–0.4%, 5–0.6%.

It can be seen that this change strongly depends on the concentration of potassium persulfate and has a minimum. This is obviously due to the fact that the formation of surface-active oligomers does not occur at high concentrations of persulfate.

Diphilic molecules have a certain hydrophilic-lipophilic balance (HLB), which is calculated using the Davis formula [13]. The HLB values of a styrene oligomer with a terminal sulfate group and several monomer units (m) were calculated and the results are shown in Table 10.1. Table 10.2 shows the colloidal properties [13] of such oligomers. As can be seen from the table, oligomers acquire stabilizing properties when the number of monomer units is more than 12. Apparently, the proportion of surface-active oligomers in water is greatly reduced at high concentrations of potassium persulfate and, therefore, the latexes that were obtained at concentrations greater than 0.5% were not stable.

**TABLE 10.1** HLB of Growing Styrene Radical

| **m** | 1 | 10 | 11 | 13 | 14 | 15 | 16 | 17 |
|---|---|---|---|---|---|---|---|---|
| **HLB** | 45.1 | 21.7 | 19.1 | 13.9 | 11.3 | 8.7 | 6.1 | 3.5 |

**TABLE 10.2** HLB Scale

| Surfactant Solubility in Water | HLB | Notes |
|---|---|---|
| Do not disperse | 0 | |
| Poor dispersity | 2 | Water/oil type emulsifier |
| | 4 | |
| | 6 | |
| Muddy unstable dispersion | 8 | |
| Muddy stable dispersion | 10 | Wetting reagent |
| Translucent or transparent solution | 12 | Detergent |
| | 14 | |
| Translucent or transparent solution | 14 | Oil/water type emulsifier |
| | 16 | |
| | 18 | |

Determination of the exact time of completion of polymerization processes in a static system is important for the synthesis of latex particles of a given diameter. To establish the ending time of static polymerization of monomer both viscosity of the monomer phase and the dry residue of the aqueous phase have to be measured during the process. After two days withstanding of biphasic system of styrene-0.2 w% $K_2S_20_8$ aqueous solution the viscosity of monomeric phase did not change, and the dry residue of the aqueous phase remained equal to 2% even for longer withstanding. Polymerization was provided at 50°C.

From these results it follows that the determination of the time of the completion of polymerization in a static two-phase system is reduced in reality to establishing the time of completion of growth and the size of the latex particles in the aqueous phase. Dimensions of the latex particles and their variation with polymerization conditions were investigated by electron microscopy.

In Figure 10.4, latex particles obtained at different dwell times of the system are presented. During the first 6 hours of polymerization, the average particle size increases to 300 nm and the particle size distribution is narrow (see Figure 10.4). Particle sizes do not change with the continuation of the polymerization processes in the system, but the particle size distribution undergoes a significant change (Figure 10.5). In Figure 10.5, finely dispersed particles, the electron density of which is significantly different from the electron density of the large ones, are distinguished. The appearance of such small-sized particles may be due to the supersaturation of the aqueous phase by styrene oligomers and their aggregation.

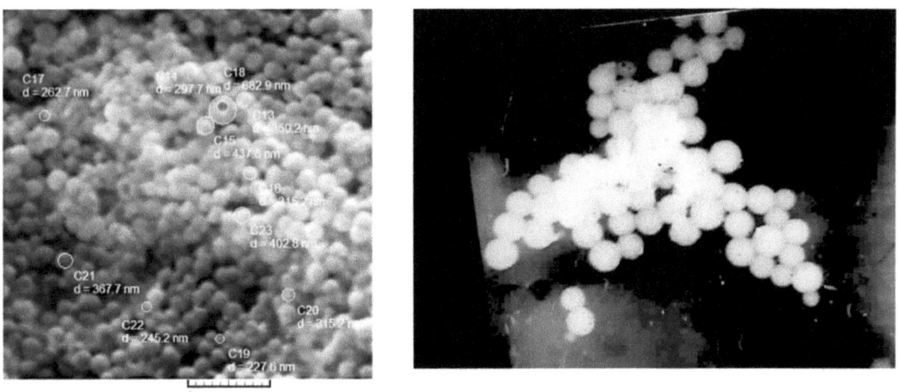


**FIGURE 10.4** Electron microscopic photograph of polystyrene latex obtained after 6 hours withstanding of static system at 50°C.

From the obtained results it follows that if the formation of particles a in the heterogeneous monomer-water system is stopped in time, it will be possible to hope to obtain unimodal latex particles. It is also clear that in static heterogeneous system monodisperse lattices can be most easily obtained by polymerization of such monomers, in which the water solubility and constant chain propagation allow dispersed particles to be nucleated only in one zone of the system. The values of propagation constants and the solubility of chloroprene in water [13, 14] made it possible to suggest that polymerization of this monomer in a static system can be a method for the synthesis of monodisperse lattices.

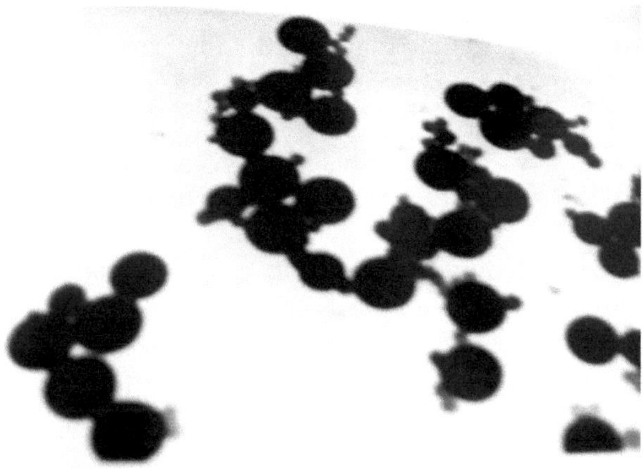

**FIGURE 10.5** Electron microscopic photograph of polystyrene latex obtained at 50°C after withstanding of static system for 150 hours.

Polymerization of chloroprene in a static monomer-water system allowed the synthesis of monodisperse latex. The polymerization time was 6 hours, the latex dry residue was 2%, and the particle diameter was 250 nm. The stability of the obtained latex was investigated by centrifugation: after two hours of centrifugation at a rotation speed of 7500 $min^{-1}$ the precipitate was 0.01 w%. The viscosity of the monomer phase during 6 hours of polymerization increased significantly and the polymerization process was considered to be completed. An electronic photo of latex is shown in Figure 10.6.

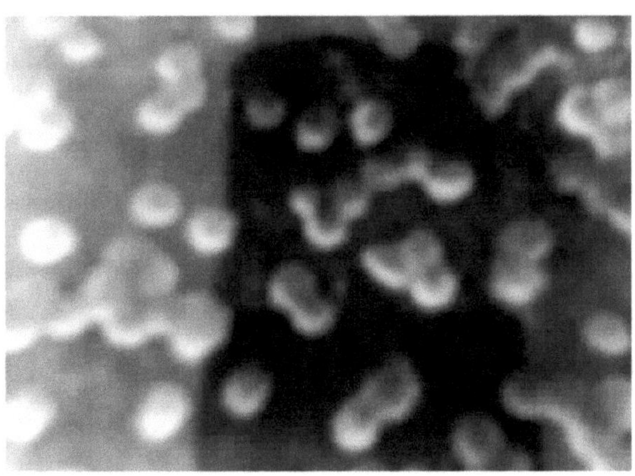

**FIGURE 10.6** Electron photographs of polychloroprene latex obtained after 6 hours of withstanding of the static system.

A feature of polychloroprene latexes synthesized without the use of emulsifiers is that the polymer contains double bonds. This opens up the possibility of changing the chemical structure of the surface of latex particles.

## 10.3 CONCLUSIONS

It was established that polymerization in a static monomer-water system can become a method for the synthesis of monodisperse latexes, without the use of emulsifiers. For the synthesis of stable latexes of a given diameter, an accurate determination of the initiator concentration and the duration of polymerization are necessary.

## KEYWORDS

- **emulsion polymerization**
- **hydrophilic-lipophilic balance**
- **latex**
- **polychloroprene**
- **polymerization**
- **polystyrene**

## REFERENCES

1. Harkins, W. D., (1947). Theory of the mechanism of emulsion polymerization. *J. Amer. Chem. Soc., 69,* 1428–1448.
2. Harkins, W. D., (1950). General theory of mechanism of emulsion polymerization. *J. Polym. Sci., 5,* 217–251.
3. Peter, A. L., & El-Aasser, M. S., (1997). *Emulsion Polymerization and Emulsion Polymers* (p. 218). Hardcover.
4. Matsumoto, T., Wakimoto, S., Miyahara, S., & Heishu, I., (1981). Pat. Japan 55-80402.
5. Lovelace, A. M., Vanderhoff, J. W., Micale, F. J., El-Aasser, M. S., & Kornfeld, D. M., (1981). *Method for Producing Monodisperse Latexes with a Large Particle Size*. Pat. USA 4247434.
6. Dimonie, V., Hagiopol, C., Glorgescu, M., Moraru, G., & Constantinescu, G., (1983). *Method for Producing Monodispersed Polystyrene Latex Used in Immunology*. Pat. SRR 77091.

7. Oganesyan, A. A., Khaddazhb, M., Gritskovac, I. A., Gubinb, S. P., Grigoryana, G. K., Muradyana, G. M., & Nadaryana, A. G., (2013). Polymerization in the static heterogeneous system styrene-water in the presence of methanol. ISSN: 00405795. *Theoretical Foundations of Chemical Engineering, 47*(5), 600–603.
8. Hovhannisyan, A. A., Grigoryan, G. K., Khaddazh, M., & Grigoryan, N. G., (2015). On the mechanism of latex particles formation in polymerization in heterogeneous monomer-water system. *J. Chem. Chem. Eng., USA, 9*(5), 363–368.
9. Hovhannisyan, A. A., & Khaddazh, M., (2016). *On the Theory of Emulsion Polymerization* (pp. 34–38). Lambert Academic Publishing, Germ. ISBN-13: 978-3-659-85137-7.
10. Hovhannisyan, A. A., Khaddazh, M., Grigoryan, N. G., Grigoryan, G. K., & Zhuravleva, O. A., (2017). Topology of formation of latexes in heterogeneous static monomer-water systems. *Chemical Engineering of Polymers* (pp. 301–306). Published by Apple Academic Press, USA.
11. Hovhannisyan, A. A., (2018). *Polymerization in Monomer-Water Heterogeneous Systems* (p. 45). 10th International Conference on Chemistry Education and Research, Oslo, Norway, Advanced Materials, ISSN: 2150-3494.
12. Oganesyan, A. A., (1986). Free radical polymerization and phase formation in heterogeneous monomer/water systems. *Doctoral (Chem.) Dissertation*. Moscow: Inst. of Fine Chemical Technology (in Rus.).
13. Adamson, A., (1979). *Physical Chemistry of Surfaces* (p. 567). (In Russian).
14. Hrabak, F., Bezdek, M., Hynkova, V., & Pelzbauer, Z., (1967). Growth reaction in the radical polymerization of chloroprene. *J. Polym. Sci., C3, 16*, 1345–1353.
15. Gerrens, H., (1964). Radical reactions in polymerization processes. In: *Dechema Monographien* (Vol. 49, No. 859, p. 346S). Frankfurt a/M.

# CHAPTER 11

# Biodegradable Polymer Packaging Materials in Seawater Environment

HELENA JANIK,[1] JUSTYNA KUCIŃSKA-LIPKA,[1] MACIEJ SIENKIEWICZ,[1] and ALICJA KOSAKOWSKA[2]

*[1]Gdansk University of Technology, Chemical Faculty, Polymer Technology Department, Gdansk, Poland, E-mail: helena.janik@pg.edu.pl (H. Janik)*

*[2]The Institute of Oceanology of the Polish Academy of Sciences, Sopot, Poland*

## ABSTRACT

Lack of degradability and the landfill sites closing as well as increasing water and land pollution problems have led to concern about synthetic packaging materials. Awareness of the waste problem and its influence on the environment has awakened a new interest in the field of biodegradable polymers. They are designed to degrade upon disposal by the action of living organisms. Biodegradation is governed by various factors like polymer characteristics, type of microorganism, and the nature of pre-treatment. The polymer features such as its crystallinity, molecular weight, the type of functional groups and substituents present in its structure, and additives added to the polymer all play an important role in their degradation. In nature, the degradation of most synthetic plastics is a very slow process that involves environmental factors and often is followed by the action of microorganisms. Biodegradation of various polymers have been studied under diverse environments like in soil, in sewage sludge, in compost conditions, and in seawater. The aim of this chapter is to present the research on the biodegradation of the available biodegradable polymer packaging materials in seawater and on the trials to overwork the methods to estimate the impact of biodegradable polymers on sea life. Thus, the chapter is unique in its content.

## 11.1 INTRODUCTION

Biodegradable polymers have been developed over a few decades [1–6]. One of the main reasons for that interest is dramatically increasing post-using volume of traditional synthetic polymers used as packaging materials. The second area of interest is the development of biodegradable polymers in medicine [7–11]. The paper will concentrate on the first application.

The definition of biodegradation has undergone gradually some changes. In the beginning, it was only concerned with the decomposition of materials by the action of microorganisms with the generation of carbon dioxide, water, and new biomass. Currently, we know that biodegradation is a complex process and proceeds with several steps [12].

Several standards like ASTM (USA), CEN (Europe), DIN (Germany), and ISO for degradation and biodegradation have been developed to avoid misunderstandings. According to the ASTM standard D5526-94d, biodegradability is the potential of polymer material for decomposition into carbon dioxide, methane, water, inorganic compounds, and biomass under the influence of the enzymatic action of microorganisms. The material should undergo decomposition in 60–90% within 90–180 days. Thus to the previous definition, it was added time. Later the biodegradable conditions were considered as well, and new standardized tests have been overworked (EN ISO 13432:2002, EN 14047:2003, EN 14046:2005, EN ISO 14852:2007, EN ISO 14855-1:2009, EN 14806:2010). The most studied environments to follow biodegradation are soils and composting piles. Starch and starch-based compositions [13–17] PLA [18–21], PHB [22], and PCL [23–25] belong to the polymers studied mostly in those environments.

In this chapter, our interest is directed into sea/ocean/lakes biodegradable environment for polymer packaging materials from two different points of view. It will overview polymer degradability testing in those water environments, and for the first time, it will cover the overview of investigation of the impact of polymer packaging on sea life; the subject is very little described in the literature.

## 11.2 TESTING OF POLYMER DEGRADABILITY IN SEAWATER ENVIRONMENT

### *11.2.1 STARCH AND STARCH-BASED COMPOSITIONS*

Taking into account the very poor biodegradability of polyethylene (PE) in seawater environments, some laboratories and producers obtained starch-based compositions of PE or polypropylene (PP). They were later studied in different sea environments.

Starch ($C_6H_{10}O_5$)$n > 300$) belongs to polysaccharides; it is produced by some plants that accumulate the starch in the roots, rhizomes, tubers, and seeds. Most starch types consist of 25% amylose and 75% amylopectin [26]. The starch obtained from potatoes, corn, wheat, and rice is of industrial importance [27]. The addition of starch to the synthetic non-biodegradable polymers causes susceptibility to attack by microorganisms, which speeds up the degradation [28].

One of the first papers dealing with the biodegradability of starch-based compositions in seawater was that one published in 2002 by Rutkowska et al. [15]. They examined the influence of starch on the biodegradability of starch-PE compositions having 5 and 8% of starch. Biodegradability of those compositions was compared with pure and modified PE with pro-degradant additives in the form of "master batch" in the amount of 20%. The degradation of polymer samples was studied under marine exposure conditions in the Baltic Sea. The temperature of the water was in the range of 3–20°C, pH = 8.1–8.5, and salt content was around 6–7%. The incubation of the samples took 20 months. The experiments were also performed in the laboratory at the ambient temperature, in a liquid medium containing seawater with sodium azide ($NaN_3$) to evaluate the resistance of the samples against hydrolysis. That test showed that the samples are resistant to hydrolysis. In natural conditions, the starch-PE samples did not change much as well in comparison to the PE-master batch. Only small surface changes were noticed. The small changes in the whole volume of the sample could be explained by the fact that in the case of a small amount of starch added to PE, the morphology is not favorable for the degradation of starch. Starch particles are surrounded by PE, and water or microorganisms have limited access to them. Thus analyzing the results for degradability of polymers and polymer blends, the morphology is an important key factor, very often not taken into account in the interpretation of degradation results.

Rutkowska and Heimowska [17] studied as well as thermoplastic starch (TPS) and starch-cellulose materials for around a year in the same Baltic Sea environment as it was described above. It was found that in the natural environment of seawater, the enzymatic hydrolysis of the tested materials occurred, what was demonstrated by clear erosion of the surfaces, and weight loss. The samples of starch-cellulose were more susceptible to attacks of microorganisms living in seawater than the samples of TPS. After 56 days of incubation in natural seawater, only 6.4% of mass loss was observed for TPS in comparison to around 60% for starch-cellulose after 35 days of incubation in the same seawater conditions. The temperature of natural seawater was from 3°C (December) to 20°C (August). The tests carried out in laboratory conditions (18–23°C, pH = 8) with adding of $NaN_3$ to seawater to kill microorganisms showed that the degradation in that environment is very slow what allowed us to conclude that in natural seawater degradation is of enzymatic character [17].

Muthukuma et al. [16] examined the rate of degradation of starch-PP composition in three different environments, namely under direct sunlight, soil, and marine waters in India at a depth of 1 m from the surface (Bay of Bengal, Ennore Port Limited Chennai). It is a pity that no more information was provided about the marine environment. The films were retrieved sequentially after every 30 days till up to 150. For the comparison, pro-oxidant blended and starch blended high-density polyethylene (HDPE) and pro-oxidant blended low-density polyethylene (LDPE) were also studied in the same environments. The percentage of degradation of starch blended with HDPE or PP (starch in the amount of 5%) and pro-oxidant blended LDPE and HDPE in seawater was between sunlight and soil. The exposure to sunlight showed the highest weight loss (>10%), and samples buried in soil showed the lowest (similar to 1%). They concluded that the marine environment is found to be one of the major sites for the dumping of waste especially plastics, and there is a need to develop strategies to degrade them faster than what is observed from their studies on modified polyolefins [16].

Elanmugilan [29] examined the degradation degree of the blends of LDPE with a starch-based additive, namely polystarch N. The samples were tested in seawater in Saudi Arabia. Stress-strain properties and thermal behavior were investigated for the LDPE and LDPE/polystarch N blend, having 40% (w/w) of polystarch N. Rheological studies and scanning electron microscope photomicrographs of the polymer samples after seawater aging showed that the addition of polystarch N enhanced the degradation of LDPE. SEM photomicrographs of the surface indicated that seawater aging caused micro-cracks and crazes in LDPE, while holes on the surface in the case of blends. The latter effect is ascribed to the leaching out of starch from polystarch N. They proved that polystarch N, when present in LDPE, helps in degradation through chain scission when exposed to natural weather [29].

### 11.2.2 POLY(LACTIDE) COMPOSITIONS

PLA is the most technologically advanced biodegradable polymer. This aliphatic polyester is characterized by high transparency and gloss, high rigidity, making it widely used in packaging, agriculture, and medicine [23]. The polymer is easily molded using conventional processing equipment. It can be synthesized by biological and chemical methods [30]. The chemical method used by Cargill Dow is using ring-opening polymerization (ROP) of lactide [31]. In the second method, used by Mitsui Toatsu, PLA is obtained by the polycondensation [32]. In terms of the industrial composting process in the first step of degradation of PLA, hydrolysis takes place by the ingress of water into the amorphous structure. Subsequently disintegrates ester bonds, resulting in the formation of water-soluble oligomers and lactic acid. In the final step of degradation, the microorganisms convert them into simple compounds such as carbon dioxide, water, and biomass [33].

Adamus et al. [24] examined the degradation process in dynamic seawater for six months of poly(L-lactide) (PLLA) and poly[(DL)-lactide] (PLA). Samples of 40 × 40 × 20 mm in size were placed at 2 m depth in Baltic Sea (temp. from 1–20°C, pH = 8, salinity around 6 ppt.). The course of degradation process was monitored by evaluation of molecular weight changes (GPC), macroscopic observation of surface view and mass loss. There were observed no particular changes in monitoring parameters after 6 months of incubation [24].

In 2008, Krasowska et al. [25] showed that PLA degradation rate depended on both environment and polymer type studied. They described the results of environmental degradation of poly(D,L-lactide) (PLA), unmodified or modified with synthetic amorphous poly[(R,S)-3-hydroxybutyrate] (a-PHB), in the Baltic seawater (Gdynia Port, PL) characterized in previous papers [34]. The processes of samples degradation in the applied natural conditions were characterized by macroscopic surfaces' changes as well as by changes of samples weights, molecular weights, and compositions. The tests took place for 24 months. Degradation process of PLA in seawater was no observed. It is consistent with the literature [35, 36] that reports about a few PLA-degrading microorganisms identified so far. Moreover, it was added that PLA is not degrading in the same way as some other aliphatic polyesters, such as polyhydroxyalkanoates (PHA) or poly(epsilon caprolactone) [37, 38].

The presence of a-PHB in the composition with PLA accelerated the degradation process, and the greater modifier content, the faster degradation was noticed. Adding 5% of a-PHB to PLA caused a mass loss in seawater at the level of 65%, while adding 20% of a-PHB led to the increase of mass loss to around 80%. It was found that the degradation of PLA compositions in seawater is the result of chemical hydrolysis of the ester bonds rather than enzymatic degradation. After 24 months of incubation of the a-PHB/PLA compositions in seawater, there was observed complete defragmentation of the samples despite the low temperature of the sea environment (−20°C).

In 2011, Zenkiewicz et al. [39] studied the degradation process in seawater (Baltic Sea, temp. 18–21°C) of PLA crosslinked with the use of both triallyl isocyanurate (TAIC) and electron radiation or using dicumyl peroxide (DCP). Physical or chemical crosslinking of PLA improves its functional qualities, which may be of fundamental practical importance. This mainly concerns PLA intended for thermoforming of packaging or injection molding of disposable articles [39]. Determination of the tensile strength and mass loss of samples incubated in seawater for 9 weeks was performed. No essential changes in either quantity were found. These results were consistent with data reported in the literature [40, 41], which revealed that no essential physical and chemical changes in PLA incubated in seawater at temperatures up to 23°C were detected.

### *11.2.3 POLYHYDROXYALKANOATES (PHA)*

PHA is produced from renewable resources by bacterial fermentation of sugar and lipids. They are thermoplastic or elastomeric polymers, depending on the monomer used in the synthesis. They belong to bio-plastic family, which is developing very quickly. It is predicted that bio-based polymer production capacity will be nearly 12 million tons in 2020, what means that it will triple in comparison to 2011. The expected total polymer production is about 400 million tons in 2020; thus the bio-based share will be around 3% in 2020, meaning that bio-based production capacity will grow faster than overall production [42].

These materials, alone or in a blend with synthetic plastic or starch, give packaging films [43]. The poly(3-hydroxybutyrate) (PHB) is the most common type of bio-plastic polyester, which is obtained in the polymerization of 3-hydroxybutyrate monomer, with properties similar to PP although it is stiffer and more brittle [44]. Scientists from China produce several PHA polymers from food wastes used as carbon sources. Materials had various physical and mechanical properties, like flexibility, tensile strength, melting viscosity. The production of polyhydroxyalkanoates is rather inexpensive, and nowadays, PHB can be found in the market with different grades under Mirel® name [45].

There is not much information on the degradability of PHA in the marine environment. One example is the study (2010) in the tropical environment (India Ocean) of poly(3-hydroxybutyrate), P(3HB), and its copolymer poly(3-hydroxybutyrate-co-3-hydroxyvalerate) P(3HB-co-3HV) by Akmal et al. [46]. The studied materials were obtained by the use of soil microorganism, *Erwinia sp.* USMI-20 from a mixture of palm oil and various second carbon sources [47, 48]. Polyester films measuring 20 × 20 × 0.2 mm with an average weight of 80 mg were immersed into lake or seawater in the depth of 1 m. Unfortunately, there were no reported parameters of those aquatic environments. The experiments were realized during 8 weeks. The investigation showed that P(3HB) and P(3HB-co-3HV) films degraded in the tropical environment. P(3HB) biodegraded at a degradation rate of 1.5% per week in lake water (10% mass loss) and 0.8% per week in seawater (5% mass loss). The degradation rates for P(3HB-co-3HV) were 3.2% per week in lake water (50% mass loss) and 2.7% per week in seawater (15% mass loss), respectively. The following bacteria were identified to degrade P(3HB) or P(3HB-co-3HV): *Alcaligenes* sp. AAC-2205, *Micrococcus* sp. FAAC-2206, *Pseudomonas* sp. FAAC-2208 (lake water) and *Proteus* sp. FAAC-2204 (Indian Ocean water).

The similar PHA samples were studied (2010) in tropical coastal water (at the Marine Research and Testing Station of the Vietnam-Russia Tropical Centre, in the South China Sea-Vietnam) [49]. This time the polymers were synthesized by the use of bacterium *Ralstonia eutropha*. The specimens were submerged to a depth of 120 cm. The average temperature of the water was around 29°C, water pH values were close to neutral (7.0–7.5), and the average salinity in the study period was 34%. Dissolved oxygen concentration varied from 5.4 to 8.3 mg/ml.

The rates of biodegradation of two PHA types (PHB and a PHB/PHV copolymer) in the form of films and compacted pellets were compared. The specimens of different shapes biodegraded at different rates. The most rapid degradation was recorded in films, which had a large surface area. In 160 days after the specimens had been submerged in seawater, the residual mass loss of 3-PHB and 3-PHB/PHV specimens was almost equal (around 50%). Over the first 80 days, the mass of the specimens remained almost unchanged what is almost consistent with the studies of Akmal [46], where after around 60 days, the mass loss was very low. The polydispersity index of the PHAs grew significantly. However, the degree of crystallinity of both PHAs remained almost unchanged, i.e., the amorphous phase of the PHA and the crystalline one are equally disintegrated. Based on the 16S rRNA analysis, the PHA-degrading strains isolated from seawater were identified as *Enterobacter* sp. IBP-VN1, *Bacillus* sp. IBP-VN2, *Gracilibacillus* sp. IBP-VN3, *Enterobacter* sp. IBP-VN4, *Enterobacter* sp. IBP-VN5, and *Enterobacter* sp. IBP-VN6.

Rutkowska et al. [50] examined degradation of blends of atactic poly[(R,S)-3-hydroxybutyrate] (a-PHB) with natural poly(3-hydroxybutyrate-co-3-hydroxyvalerate) (PHBV, 12 mol% of 3HV units), and compared it with plain PHBV degradation process in cold marine exposure conditions in the dynamic water of the Baltic Sea (6 weeks of studies). The degree of blends degradation of a-PHB with PHBV was rather low and depended on the blend composition. The natural PHBV degraded together with its binary blends containing a-PHB; the degree of degradation was dependent on the a-PHB content. The amorphous phase of the binary blend degraded first, resulting in the increased sample crystallinity, which in turn caused higher values of tensile strength and tensile modulus of incubated samples. The weight loss changes of plain PHBV were more significant than changes of the molecular weight [50].

### *11.2.4 POLY(E-CAPROLACTONE) (PCL) COMPOSITIONS*

PCL is synthetic thermoplastic biodegradable polyester produced by chemical conversion of crude oil, followed by ROP. It has good water, oil, solvent, and chlorine resistance. It has a low melting point of 58–60°C and low viscosity. PCL packaging may be mixed with starch to reduce manufacturing costs and is then used for trash bags [28, 44]. According to the available literature, PCL (and its modifications) undergoes biodegradation in the presence of specific microorganisms. Its degradation occurs to different extents, depending on different environments [51–57].

La Carra et al. [53] examined the degradation of pure PCL as a result of the attack by various microorganisms. They observed the raising growth of microorganism (*Pseudomonas*, *Erwinia*, and *Bacillus)* in the presence of the tested polymers. It has also been stated that these microorganisms are capable of biodegrading not only PCL but also other polymers [43–46, 58].

Takayoshi Sekiguchi et al. [59] tested PCL fibers (monofilaments) in deep seawater around Japanese islands (320 m-2°C, 350 m-5°C, 612 m-10°C). After 9 months, PCL fibers lost completely their tensile strength in deep-seawaters. In comparison they examined poly(b-hydroxybutyrate/valerate) (PHB/V), and poly(butylene succinate) (PBS) fibers. The latter did not degrade at all while PHB/V fibers degraded faster than PCL ones and even, in some cases, lost its mechanical properties after 3 months. Moreover, 5 PCL-degrading bacteria were isolated from the deep seawaters belonging to the geneses *Pseudomonas, Alcanivorax,* and *Tenacibaculum.* It was also confirmed that all these bacteria degrade PCL fibers also in vitro [59].

PCL-based composition (90% of PCL) was tested in Baltic seawater (Gdynia Harbor) at 2 m depth under the water surface [60]. It was stated that after 6 weeks (August/September, average seawater temperature: 17°C) of incubation the sample was completely disintegrated. Microscopic observations showed that the degradation process proceeded in two steps. In the first step (1–4 weeks) the degradation of amorphous phase took place. After 2 weeks the mass loss was 8% and decrease in tensile strength was around 15% while after 4 weeks the mass loss was 40% and the material was very week (90% decrease in tensile strength). The second step started when almost all amorphous regions degraded; subsequently the crystalline phase degraded. Then the film became prone to fragmentation and enzymatic erosion proceeded.

The PCL and PCL modified with TPS were tested under laboratory conditions in seawater (Baltic Sea, Sopot) with and without microorganisms [61]. All samples were incubated for 28 weeks. Weight loss, microscopic observation of polymer surfaces and tensile strength tests showed that in each experiment, degradation of tested sample occurred. The degradation process was more effective in seawater with microorganisms. For comparison PCL modified with calcium carbonate was also studied. It appeared that that modifier was not as effective as TPS and in some circumstances inhibited degradation rate of the composition.

For unmodified PCL sample it was noted 8% mass loss in seawater with microorganisms and 6% without microorganisms in comparison to 16 and 12% correspondingly found for PCL-TPS composition.

A similar composition of PCL/TPS like in above described paper was studied in natural seawater (Baltic Sea, Gdynia Harbor, and August/ September). The mass loss observed after 3 weeks was around 30% [62]. Thus dynamic seawater strongly speeds up the degradation process in comparison to laboratory condition for comparable temperatures noted for both sets (around 20°C).

## 11.3 TESTING OF INTERACTION POLYMER-MICROORGANISMS SEA LIFE

As it was shown above large numbers of tests were carried out on polymer biodegradability in seawater environment but only a few take into account the influence of biodegradable polymer packaging on aquatic life. From ecological point of view mostly the papers appeared in which it is pointed out that polymer packaging materials are a constant danger to macro marine flora and fauna [6, 63–69].

Those papers appeared as the consequence of using synthetic polymers in packaging materials for a long time. They inform us that many seabirds and mammals die every year from ingestion of traditional plastic. Worldwide statistics show that 43% of marine mammal species, 86% of sea turtle species, and 44% of sea bird species are susceptible to ingesting marine plastic debris [64]. In Newfoundland, 100 hundred marine mammals are killed annually through suffocation from polymer material [65]. Moreover, marine turtles mistake plastic bags for jellyfish, which can look similar while floating on the water. It is common in every region in the world that turtles ingest plastic [66–68]. It can be found in the literature that plastic is the most reported debris found in marine turtles and other marine animals [69].

With the introduction of biodegradable polymers to the packaging industry, there is a need to overwork the tests estimating the influence of polymer packaging material presence in the marine environment on microorganisms as more and more disintegrated polymer materials into micropieces will appear in sea/lake/ocean water. It will be the part of polymer packaging materials that it was not collected in an organized way of composting. There are some papers trying to overwork new methods to follow that problem. In the chapter, we will describe the pioneered steps of undertaken procedures.

To follow the interaction of sea life-polymer it is a need to choose some bioindicators living in marine environment. It is not an easy task but first trials have recently appeared in the literature and it will be shortly described in subsections. There were proposed for the time being two bio-indicators belonging to diatoms and cyanobacteria.

### 11.3.1 THE TESTS WITH PHAEODACTYLUM TRICORNUTUM DIATOMS

Diatoms are mostly unicellular photosynthetic microalgae of the phytoplankton. Diatoms are producers within the food chain. They are thought to be responsible for close to 40% of marine primary productivity and as much as 25% of global productivity [70]. Diatoms are very diverse and abundant organisms present not only in saltwater but also in freshwater and terrestrial environments. Fossil evidence suggests that they originated during, or before, the early Jurassic Period. Their abundance is contributing to significant levels of carbon dioxide fixation and oxygen production in the oceans [71]. Mostly diatoms are surrounded by a silica cell wall; thus, they play an important role in the biogeochemical cycling of silica [72]. They live in saltwater where the salinity can vary. Diatom communities are a popular tool for monitoring environmental conditions, past, and present, and are commonly used in studies of water quality.

*Phaeodactylum tricornutum*, silica-less diatom, is one of the most widely used diatoms in various investigations. *Phaeodactylum tricornutum* has color from orange-brown to yellow-green. It is very useful model because of its short generation time, ease of genetic transformation and apparently small genome [73]. *Phaeodactylum tricornutum* is excellent for studying the ecology, physiology, biochemistry, and molecular biology of diatoms and moreover it could be an alternative source for the production of EPA (eicosapentaenoic acid) and PUFAs (polyunsaturated fatty acids) [74]. It also found an application in genetics as a system for genes transformation [70]. It is suggested to use this diatom in the studies of sea life balance in case the biodegradable polymers appear in seawater [75, 76].

Seawater for the tests was taken from the Sopot Molo Coastal Station and was transported to the Institute of Oceanology of the Polish Academy of Sciences (IO-PAN) in Sopot. There the pH, temperature, and salinity were measured. To remove all the microorganisms form the water filtration was carried out by a glass filters GF/C with diameter 0.7 μm.

*P. tricornutum* was cultured in liquid. Diatoms were grown in marine enrichment medium f/2. The main component was artificial seawater with salinity of 8.9 PSU. All solutions were made up with distilled water. Enriched seawater f/2 medium was prepared from the primary stock solutions of macronutrients, micronutrients, and vitamins which are added to a large proportion of the final volume of water in order to avoid precipitation. Primary stock solutions were diluted in 100 $cm^3$ $H_2O$ and then 1 $cm^3$ of them were added to working stock solution.

*P. tricornutum* SAG 1090-la (Sammlung von Algenkulturen at the University of Göttingen) culture was maintained in enriched seawater f/2 medium. Diatom cells were grown in sterile laboratory conditions at constant temperature 24 ± 0.5°C and under continuous illumination with 7.63 μM/$m^2$/s irradiance. Time of culturing was 2 weeks. Such conditions were maintained during all the experiments.

Commercially available biodegradable polymer packaging materials having 85% of TPS in their compositions were used for the tests. Part of the polymer strips of 2 cm × 15 cm in size were sterilized in the UV light and in solution of 3% hydrogen peroxide for 30 min. All operations were made under the laminar flow cabinet with a lamp which produces ultraviolet light (UVT) of the wavelength of $\lambda = 254$ nm. Consequently, samples were washed with sterile de-ionized water.

The vessel with known volume was filled with enriched seawater f/2 medium and diatom *P. tricornutum* inoculum. Subsequently, the proper diatom culture was poured to the Erlenmeyer flasks with 500 $cm^3$ volume. Each test flask contained two polymer strip-samples and 300 $cm^3$ solution. Vessels with the control samples were only filled with 250 $cm^3$ of enriched seawater f/2 medium with diatom *P. tricornutum* inoculum. The Erlenmeyer flasks were closed with sterile cotton stopper. Diatom culture was grown under continuous UV illumination and at a constant temperature of 22 ± 0.5°C.

The chlorophyll a content of laboratory cultures of *P. tricornutum* was measured spectrophotometrically. The measurements were carried out with Spectrophotometer UV-VIS HITACH U-2800. The absorbance was measured at four wavelengths (750, 665, 630, 645, and 480 nm) to determine turbidity and chlorophyll a. In case of pheopigment the samples absorbance is measured at 750 and 664 nm before acidification and at 750 and 665 nm after acidification with 1 M HCl. 0.06 cm$^3$ of hydrogen chloride was added to each 5 cm$^3$ of extract. After 2 min from acidification the measurements were made. The glass cell with optical path of 2 cm was used. The equation of Jeffrey and Humphrey was used to calculate chlorophyll a in seawater samples [77, 78]:

$$Chl\, a = (11{,}85\, A_{665} - 1{,}54\, A_{645} - 0{,}08\, A_{630}) \cdot v \cdot l^{-1} \cdot V^{-1} [mg / dm^3] \quad (2)$$

The equation of Lorenzen, which was used to calculate overall chlorophyll in seawater samples containing *P. tricornutum* [5, 18], is the following:

$$Chl\, a = 26{,}7\, (A_{665b} - A_{665a}) \cdot v \cdot l^{-1} \cdot V^{-1}\, [mg / dm^3] \quad (3)$$

Level of Pheopigment in samples calculated with the equation of Lorenzen [21, 79]:

$$Ph\, a = 26{,}7 \left[1{,}7 \cdot (A_{665a} - A_{665b})\right] \cdot v \cdot l^{-1} \cdot V^{-1}\, [mg / dm^3] \quad (4)$$

Carotenoids level in samples calculated with the equation of Parsons and Strickland [80]:

$$Car = 10{,}0 \cdot A_{480} \cdot v \cdot l^{-1} \cdot V^{-1} [mg / dm^3] \quad (5)$$

where; $A_x$: sample absorbance at "x" nm (minus absorbance at 750 nm); V: volume of 90% acetone [cm$^3$]; l: cell (cuvette) length [cm]; and V: volume of filtered water [dm$^3$].

At the same time, macroscopic observations were realized. The photos were taken in known time intervals. The color intensity of samples with polymer strips was compared with control samples. The more intensive color, the higher number of cells in inoculums is observed. The growth can be caused by polymer degradation to the compounds, which may perhaps be a perfect nutrient for diatoms [81–84]. Determination of the cell number gave information about diatoms population and cells conditions. Frequently, the presence of polymer samples changes the cells morphology. Predominantly the fusiform cells occur in control samples and spherical cells in samples with polymer strips [83]. Such observations show about the polymer influence on *P. tricornutum.* What is more, the cells number is correlated with the color intensity of inoculums [84].

The spectrophotometric measurements of pigments provided the information about the processes occurred in the cells. More chlorophyll *a* specify phytoplankton abundance in coastal and estuarine waters, thus it is potential indicator of maximum photosynthetic rate and is a commonly used measure of water quality. Pheophytin *a* is a common degradation product of chlorophyll *a*. It enables to evaluate the amount of death cells. Carotenoids are indicators of cells natural aging and nutrient limitations [85].

Macroscopic observations proved the influence of the TPS polymer on *P. tricornutum* marine diatom growth. During the experiment in system A (with diatom *P. tricornutum)* the change of seawater medium color was observed. The most intensive system color was observed for *P. tricornutum* culture incubated with tested polymer. The more intensive color of the system, the higher number of cells in the inoculum is detected. Throughout incubation period diatom *P. tricornutum* cells adhered into TPS composites polymer surface and its growth was stimulated. It was observed higher degree of diatom growth in the presence of TPS samples in comparison with the blank samples Polymer degradation products could be the reason of that. Further research is going to be taken to follow that observation to more detailed examination.

The increase of diatom biomass was in correlation with the increase of pigment content. More than 100% *chlorophyll-a* gain in samples with polymer material was observed comparing with control ones. It was proved that presence of polymeric sample significantly changed kinetic of diatom growth in seawater after 24 weeks of the incubation.

From the test carried out in system B it was stated that in natural seawater (without *P. tricornutum* diatom) only partial destruction of tested polymer took place, resulting in around 5% polymer weight loss after 36 weeks of the incubation.

### 11.3.2 THE TESTS WITH ANABAENA VARIABILIS

Cyanobacteria are microscopic organisms. They get their name from the bluish pigment phycocyanin, which they use to capture light for photosynthesis. They also contain chlorophyll a, the same photosynthetic pigment that plants use. In fact, the chloroplast in plants is a symbiotic cyanobacterium, taken up by a green algal ancestor of the plants sometime in the Precambrian.

However, not all "blue-green" bacteria are blue; some common forms are red or pink from the pigment phycoerythrin. The Red Sea gets its name from occasional blooms of a reddish species of *Oscillatoria*, and African flamingos get their pink color from eating *Spirulina* [86, 87]. An axcenic culture of cyanobacterium *Anabaena variabilis* (ATCC-29413) obtained from American Type Culture Collection was chosen to study sea life balance in the presence of modified polycaprolactone samples, which were studied before, and at least one of them degraded much [88]. The *Anabaena variabilis* is identified in many waters, including those of Baltic Sea where the water for studies was taken out (Sopot). The chosen polymer materials were polycaprolactone modified with thermoplastic starch (PCL/TPS>85%) or with calcium carbonate (60% PCL/40% $CaCO_3$). The experimental system consisted of natural seawater, the polymer to be tested and a culture of *Anabaena variabilis*. Natural seawater was filtered and sterilized to remove suspended matter and microorganisms, and then enriched with 1.5 g $dm^{-3}$ $NaNO_3$ and 0.04 g $dm^{-3}$ $K_2HPO_4$. *Cyanobacteria inocula* were then added to the nutrient-rich seawater. Cells were grown in original 616 medium [89] and in the nutrient-enriched Baltic seawater. The *A. variabilis* inoculum was incubated for 12 days in the original medium and 16 weeks in the enriched seawater (in the presence and absence of polymer samples). The water-inoculum set-up was stirred magnetically. The medium with defined initial pH, salinity, and chlorophyll-a content was used in the experimental system. Sterilized polymer samples were used. The chlorophyll-a content was determined as the criterion of cyanobacterial growth in the presence of the tested polymers. The spectrophotometric measurements of pigments provided information about the processes occurring in the cells. Chlorophyll-a was measured to estimate the abundance of phytoplankton in the water. Higher content of chlorophyll-a is indicative of phytoplankton abundance in coastal and estuarine waters, so it is a potential indicator of maximum photosynthetic rate and a commonly used measure of water quality.

The polymer surface and color changes in the cyanobacterium culture were also recorded photographically. The experimental results indicated that the addition of polymer samples to the cyanobacterium culture affected its biological balance in different ways depending on the composition studied. Thus the suggested test differentiates the behavior of both materials studied. Cyanobacterial growth was lower in the presence of PCL modified with calcium carbonate than in the presence PCL/TPS blend. There is much to do to verify this procedure with more polymer packaging samples, but the first step has been done.

## 11.4 SUMMARY

From the first part of the overview, it is clear that the testing of polymer degradability is very chaotic and overspread. There is no standardized method to obtain comparable results coming from different laboratories. Very often the authors are not giving even the parameters of seawater. There is a need to establish at least the depth of the water for the test, the size (the weight) of the sample and time of the experiment for particular type of a polymer. Despite of mentioned drawbacks it is possible to put the composition studied in some order. The least degradable in seawater is PLA and the quickest degradable polymer compositions are starch-based. As to PHA it depends on morphology. Amorphous PHB is very vulnerable for degradation in seawater contrary to semicrystalline one.

As to the second part of the overview, it is an interesting approach to estimate the impact of polymer on sea life balance. In case the tests with diatoms and cyanobacterium would be verified, and more popular, the obtained results could help to predict the direction of development of particular polymer packaging materials taking into account the ecological aspect. Moreover, the obtained results could be used in life cycle analysis.

## KEYWORDS


- **biodegradable polymer packaging**
- **biodegradation**
- **high-density polyethylene**
- **microorganisms**
- **seawater environment**
- **testing methods**


## REFERENCES


1. Leja, K., & Lewandowicz, G., (2010). Polymer biodegradation and biodegradable polymers: A review. *Polish Journal of Environmental Studies, 19*, 255–266.
2. Cancan, S., Shuhong, Z., & Mengting, L., (2013). Barrier and mechanical properties of biodegradable poly(epsilon-caprolactone)/cellophane multilayer film. *Journal of Applied Polymer Science, 130*, 1805–1811.

3. Yang, Y., Dayi, P., & Kui, L., (2013). Biodegradable and amphiphilic block copolymer-doxorubicin conjugate as polymeric nanoscale drug delivery vehicle for breast cancer therapy. *Biomaterials, 34*, 8430–43.
4. Leonor, I. B., Kim, H. M., & Balas, F., (2007). Alkaline treatments to render starch-based biodegradable polymers self-mineralize. *Journal of Tissue Engineering and Regenerative Medicine, 1*, 425–435.
5. Bordes, P., Pollet, E., & Averous, L., (2008). Nano-biocomposites: Biodegradable polyester/nanoclay systems. *Progress in Polymer Science, 34*, 125–155.
6. Guzman, A., Gnutek, N., & Janik, H., (2011). Biodegradable polymers for food packaging-factors influencing their degradation and certification types: A comprehensive review. *Chemistry and Chemical Technology, 5*, 115–122.
7. Kucińska-Lipka, J., Gubańska, I., & Janik, H., (2013). Polyurethanes modified with natural polymers for medical application-Part I. Polyurethane/chitosan and polyurethane/collagen. *Polimery, 9*, 37–43.
8. Kucińska-Lipka, J., Gubańska, I., & Janik, H., (2013). Polyurethanes modified with natural polymers for medical application-Part II. Polyurethane/gelatin, polyurethane/starch polyurethane/cellulose. *Polimery, 11*, 678.
9. Caracciolo, P. C., Thomas, V., Vohra, Y. K., Buffa, F., & Abraham, G. A., (2009). Electrospinning of novel biodegradable poly(ester urethane)s and poly(ester urethane urea)s for soft tissue-engineering applications. *Journal of Materials Science-Materials in Medicine, 20*, 2129–2137.
10. Rockwood, D. N., Woodhouse, K. A., Fromstein, J. D., Chase, D. B., & Rabolt, J. F., (2007). Characterization of biodegradable polyurethane microfibers for tissue engineering. *Journal of Biomaterials Science-Polymer Edition, 18*, 743–758.
11. Gibas, I., Janik, H., & Dini, L., (2010). Poly(epsilon-caprolactonediol) and polyethylene glycol-based polyurethanes for medical applications. *Przemysl Chemiczny, 89*, 1622–1626.
12. Lucas, N., Bienaime, C., Belloy, C., Queneudec, M., Silvestre, F., & Nava-Saucedo, J. E., (2008). Polymer biodegradation: Mechanisms and estimation techniques. *Chemosphere, 73*, 429–442.
13. Alvarez, V. A., Ruseckaite, R. A., & Vazquez, A., (2006). Degradation of sisal fiber/Mater Bi-Y biocomposites buried in soil. *Polymer Degradation and Stability, 91*, 3156–3162.
14. Rutkowska, M., Krasowska, K., Steinka, I., & Janik, H., (2004). Biodeterioration of mater-Bi Y class in compost with sewage sludge. *Polish Journal of Environmental Studies, 13*, 85–89.
15. Rutkowska, M., Heimowska, A., Krasowska, K., & Janik, H., (2002). Biodegradability of polyethylene starch blends in seawater. *Polish Journal of Environmental, 11*, 267–271.
16. Muthukumar, T., Aravinthan, A., & Mukesh, D., (2010). Effect of environment on the degradation of starch and pro-oxidant blended polyolefins. *Polymers Degradation and Stability, 95*, 1988–1993.
17. Rutkowska, M., & Heimowska, O., (2008). Degradation of naturally occurring polymeric materials in seawater environment. *Polimery, 11, 12*, 854–863.
18. Kale, G., Aurasa, R., Singha, S. P., & Narayan, R., (2007). Biodegradability of polylactide bottles in real and simulated composting conditions. *Polymer Testing, 26*, 1049–1061.
19. Agarwal, M., Koelling, K. W., & Chalmers, J. J., (1998). Characterization of the degradation of polylactic acid polymer in a solid substrate environment. *Biotechnology Progress, 14*, 517–526.
20. Pranamuda, H., Tokiwa, Y., & Tanaka, H., (1997). Polylactide degradation by an *Amycolatopsis* sp, *Applied and Environmental Microbiology, 63*, 1637–1640.

21. Torres, A., Li, S. M., Roussos, S., & Vert, M., (1996). Poly(lactic acid) degradation in soil or under controlled conditions. *Journal of Applied Polymer Science, 62*, 2295–2302.
22. Kim, M. N., Lee, A. R., Yoon, J. S., & Chin, I. J., (2000). Biodegradation of poly(3-hydroxybutyrate), sky green and mater-bi by fungi isolated from soils. *European Polymer Journal, 35*, 1677–1685.
23. Petersen, K., & Nielsen, P. V., (1999). Potential of bio-based materials for food packaging. *Trends in Food Science and Technology, 10*, 52–68.
24. Adamus, G., Dacko, P., Musiol, M., Sikorska, W., Sobota, M., & Biczak, R., Herman, B., et al., (2006). Degradation of selected synthetic polyesters in natural conditions. *Polimery, 7, 8*, 539–546.
25. Krasowska, K., Brzeska, J., Rutkowska, M., Dacko, P., Sobota, M., & Kowalczuk, M., (2008). The effect of poly(D,L-lactide) modification with poli[(R,S)-3-hydroxybutyrate on the course of its degradation in natural environments. *Polimery, 10*, 730–736.
26. Nayak, P., (1999). Biodegradable polymer: Opportunities and challenges. *J. M. S. Rev. Macromol. Phys., C39*, 481–505.
27. Lucas, N., Bienaime, C., Belloy, C., et al., (2008). Polymer biodegradation: Mechanisms and estimation techniques: A review. *Chemosphere, 73*, 429.
28. Gross, R., & Kalra, B., (2002). Biodegradable polymers for the environment. *Science, 297*, 803–807.
29. Elanmugilan, M., Sreekumar, P. A., Singha, N. K., Al-Harthi, M. A., & De, S. K., (2013). Natural weather, soil burial, and seawater ageing of low-density polyethylene: Effect of starch/linear low-density polyethylene masterbatch. *Journal of Applied Polymer Science, 12*, 9449–457.
30. Nafchi, A. M., & Moradpour, M., (2013). Thermoplastic starches: Properties, challenges and prospects. *Starch, 65*, 61–72.
31. Bhattacharya, R., & Mani, M., (1998). Properties of injection molded starch/synthetic polymer blends III. Effect of amylopectin to amylose ratio in starch. *Eur. Polym. J., 34*, 1461–1475.
32. Wang, J. Y., (2007). Preparation and characterization of thermoplastic starch/PLA blends by one-step reactive extrusion. *Polym. Int., 65*, 1440–1447.
33. Danyluk C., Erickson R., Burrows S., & Auras R., (2010). Industrial composting of poly(lactic acid) bottles. *Journal of Testing and Evaluation, 38*(6), 717–723. https://doi.org/10.1520/JTE102685.
34. Rutkowska, M., Krasowska, K., Heimowska, A., et al., (2002). Degradation of polyurethanes in seawater. *Polymer Degradation and Stability, 76*, 233–239.
35. Shimao, M., (2001). Biodegradation of plastics. *Biotechnol., 12*, 242–247.
36. Tsuji, H., & Suzuyoshi, H., (2002). Environmental degradation of biodegradable polyesters 1. Poly(ε-caprolactone), poly[(R)-3 hydroxybutyrate], and poly(L-lactide) films in controlled static seawater. *Polymer Degradation and Stability, 75*, 347–356.
37. Karjoma, S., Suortti, T., Lempiainen, R., Selin, J. F., & Itavaara, M., (1998). Microbial degradation of poly-(L-lactic acid) oligomers. *Polymer Degradation and Stability, 59*, 333–336.
38. Khabbaz, F., Karlsson, S., & Albertsson, A. C. J., (2000). PY-GC/MS an effective technique to characterizing of degradation mechanism of poly (L-lactide) in the different environment. *Appl. Polym. Sci., 78*, 2369–2378.
39. Zenkiewicz, M., & Richert, J., (2009). Thermoforming of polylactide nanocomposite films for packaging containers. *Polimery, 54*, 299–302.

40. Yew, G., Mohd, Y. A. M., Mohd, I. A. A., & Ishiaku, U. S., (2005). Water adsorption and enzymatic degradation of poly(lactid acid)/rice starch composites. *Polymer Degradation and Stability, 90*, 488–500.
41. Taib, R. M., Ramarad, S., Mohd, I. Z. A., & Todo, M., (2009). Properties of kenaf fibres/polylactid acid biocomposites plasticized with polyethylene glycol. *Polymer Composites, 31*, 1213–1222.
42. http://bioplastic-innovation.com/2013/06/28/market-study-on-bio-based-polymers-in-the-world/ (accessed on 19 October 2020).
43. Tharanathan, R. N., (2003). Biodegradable films and composite coatings: Past, present and future. *Trends in Food Sci. Tech., 14*, 71–78.
44. Siracusa, V., Rocculi, P., Romani, S., & Dalla, R. M., (2008). Biodegradable polymers for food packaging: A review. *Trends in Food Sci. Tech., 19*, 634–643.
45. http://www.plasticstoday.com/mpw/articles/telles-bags-film-customer-partner-its-mirel-bioplastic-03093 (accessed on 19 October 2020).
46. Akmal, D., Azizanc, M. N., & Majida, M. I. A., (2003). Biodegradation of microbial polyesters P(3HB) and P(3HB-co-3HV) under the tropical climate environment. *Polymer Degradation and Stability, 80*, 513–518.
47. Majid, M. I. A., Akmal, D. H., Few, L. L., Agustien, A., Toh, M. S., Samian, M. R., Najimudin, N., & Azizan, M. N., (1999). Production of poly(3-hydroxybutyrate) and its copolymer poly(3-hydroxybutyrate-co-3-hydroxyvalerate) by *Erwinia* sp. USMI-20. *Int. J. Biol. Macromol., 25*, 95–104.
48. Akmal, D. H., Majid, M. I. A., Razip, M. S., Nazalan, M. N., & Azizan, M. N., (1998). Biodegradation of microbial polyesters P(3HB) and P(3HB-co-3HV) under the tropical climate environment. *Proceedings of the International Symposium on Biological Polyhydroxyalkanoates*. Tokyo, Japan.
49. Volova, T. G., Boyandin, A. N., Vasiliev, A. D., Karpov, V. A., Prudnikova, S. V., Mishukova, O. V., Boyarskikh, U. A., Filipenko, M. L., Rudnev, V. P., et al., (2010). Biodegradation of polyhydroxyalkanoates (PHAs) in tropical coastal waters and identification of PHA-degrading bacteria. *Polymer Degradation and Stability, 95*, 2350–2359.
50. Rutkowska, M., Krasowska, K., Heimowska, A., Adamus, G., & Sobota, M., (2008). Environmental degradation of blends of atactic poly[(R,S)-3-hydroxybutyrate] with natural PHBV in Baltic seawater and compost with activated sludge. *Journal of Polymers and the Environment, 16*, 183–191.
51. Jastrzębska, M., & Janik, H., (1998). Biodegradation of polycaprolactone in seawater. *React. Funct. Polym., 38*, 27–30.
52. Kasuya, K., Takagi, K., Yoshida, Y., & Doi, Y., (1998). Biodegradability's of various aliphatic polyesters in natural waters. *Polym. Degrad. Stabil., 59*, 327–332.
53. Carra, F. L., Immirzi, B., Ionata, E., Mazzella, A., Portofino, S., Orsello, G., & De Prisco, P. P., (2003). Biodegradation of poly-ε-caprolactone/poly-β-hydroxybutyrate blend. *Polym. Degrad. Stabil., 79*, 37–43.
54. Oda, Y., Asari, H., Urakami, T., & Tonomura, K., (1995). Microbial degradation of poly(3-hydroxybutyrate) and polycaprolactone by filamentous fungi. *J. Ferment. Bioeng., 3*, 265–269.
55. Cofta, G., Borysiak, S., Doczekalska, B., & Garbarczyk, J., (2006). Resistance of polypropylene-wood composites to fungi. *Polimery, 51*(4), 276–279.
56. Rutkowska, M., Krasowska, K., Heimowska, A., & Steinka, I., (2002). Effect of poly(ε-caprolactone) modification on its biodegradation in natural environments. *Polimery, 47*(4), 262–268.

57. Krasowska, K., Hejmowska, A., & Rutkowska, M., (2006). Enzymatic and hydrolytic degradation of poly(ε-caprolactone) in natural environment. *Polimery, 51*(1), 21–26.
58. Cacciari, I., Quatrini, P., Zirletti, G., Mincione, V., Vinciguerra, V., Lupattelli, P., & Giovannozzi, S. G., (1993). Isotactic polypropylene biodegradation by a microbial community-physicochemical characterization of metabolites produced. *Environ. Microbiol., 59*, 3695–3700.
59. Sekiguchi, T., Saika, A., Nomura, K., Watanabe, T., et al., (2011). Biodegradation of aliphatic polyesters soaked in deep seawaters and isolation of poly(epsilon-caprolactone)-degrading bacteria. *Polymer Degradation and Stability, 96*, 1397–1403.
60. Rutkowska, M., Krasowska, K., Heimowska, A., Steinka, I., Janik, H., Haponiuk, J., & Karlsson, S., (2002). Biodegradation of modified poly(caprolactone) in different environments. *Polish Journal of Environmental Studies, 11*(4), 413–420.
61. Guzman-Sielicka, A., Janik, H., & Sielicki, P., (2012). Degradation of polycaprolactone modified with TPS and $CaCo_3$ in biotoc/abiotic seawater. *J. Polym. Environ., 20*(2), 353–360.
62. Rutkowska, M., Heimowska, A., Krasowska, K., & Janik, H., (2000). Biodegradation of thermoplastic starch-polyepsilon caprolactone) in different environments. *Proc. 3rd International Symposium on Natural Polymers and Composites–ISNaPol/2000* (pp. 553–556). Brazil, Sao Carlos.
63. Siracusaa, V., Rocculib, P., Romanib, S., & Dalla, R. M., (2008). Biodegradable polymers for food packaging: A review. *Trends in Food Science and Technology, 19*, 634–643.
64. Derraik, J., (2002). The pollution of the marine environment by plastic debris: A review. *Mar. Pollut. Bull., 44*, 842.
65. Brown, S., (2003). *Seven Billion Bags a Year: As Landfill or Litter, the Plastic Bag is Increasingly Unpopular and Dangerous to our Wildlife*. Australian Conservation Foundation.
66. Mrosovsky, N., et al., (2009). Leatherback turtles: The menace of plastic. *Mar. Pollut. Bull., 58*, 287.
67. Mascarenhas, R., et al., (2004). Plastic debris ingestion by sea turtle in Paraíba, Brazil. *Mar. Pollut. Bull., 49*, 354.
68. Tourinho, P., et al., (2010). Is marine debris ingestion still a problem for the coastal marine biota of southern Brazil? *Mar. Pollut. Bull., 60*, 396.
69. Tomas, J., Guitart, R., Mateo, R., & Raga, J., (2002). Marine debris ingestion in loggerhead sea turtles (*Caretta caretta*), from the Western Mediterranean. *Mar. Pollut. Bull., 44*, 211.
70. Scala, S., & Bowler, C., (2001). Molecular insights into the novel aspects of diatom biology. *Cellular and Molecular Life Sciences, 58*, 1666–1673.
71. Apt, K. E., Kroth-Pancic, P. G., & Grossman, A. R., (1996). Stable nuclear transformation of the diatom *Phaeodactylum tricornutum. Mol. Gen. Genet., 252*, 572–579.
72. Domergue, F., Lerchl, J., Zahringer, U., & Heinz, E., (2002). Cloning and functional characterization of *Phaeodactylum tricornutum* front-end desaturases involved in eicosapentaenoic acid biosynthesis. *Eur. J. Biochem., 269*, 4105–4113.
73. Scala, S., Carels, N., Falciatore, A., Chiusano, M. L., & Bowler, C., (2002). Genome properties of the diatom *Phaeodactylum tricornutum. Plant Physiology, 129*, 993–1002.
74. Apt, K. E., & Behrens, P. W., (1999). Commercial developments in microalgae biotechnology. *J. Phycol., 35*, 215–226.
75. Guzman, A., Janik, H., Mastalerz, M., & Kosakowska, A., (2011). Pilot study of the influence of thermoplastic starch-based polymer packaging material on the growth of diatom population in seawater environment. *Polish J. Chemical Technology, 13*(2), 57–61.

76. Gnutek, N., Janik, H., Guzman, A., & Kosakowska, A., (2010). Biodegrdable polymer packaging materials-influence on environment. *Proc. 13th International Symposium of Students and Young Mechanical Engineers: Advances in Chemical and Mechanical Engineering* (pp. 114–120.). Gdansk.
77. Lorenzen, C. J., & Jeffrey, S. W., (1980). *Determinations of Chlorophyll in Seawater.* UNESCO, Paris.
78. Jeffrey, S. W., & Humphrey, G. F., (1975). New spectrophotometric equations for determining chlorophylls a, b, c, c1, c2 in higher plants, algae, and natural phytoplankton. *Biochem. Physiol. Pflanzen., 167*, 191–194.
79. Lorenzen, C. J., (1967). Determination of chlorophyll and pheopigments: Spectrophotometric equations. *Limnol. Oceanogr.,* 12.
80. Strickland, J. D. H., & Parsons, T. R., (1968). A practical handbook of seawater analysis. Pigment analysis. *Bull. Fish. Res. Bd.,* (p. 167). Canada.
81. Janik, H., Mycio, P., Guzman, A., & Kosakowska, A., (2009). *Influence of New Polymer Packaging Degradation Products on the Growth of Cyanobacteria Toxic Population of Baltic Sea.* Lvivska Politechnika, Lviv.
82. Mycio, P., (2008). *Study on Biodegradation of Selected Polyesters and Their Modification in Seawater and Analysis of the Effect of Biodegradation Products on Cyanobacterial Populations.* Master Thesis, Gdansk University of Technology/ Sopot Institute of Oceanology of the Polish Academy of Sciences, Poland.
83. Mastalerz, M., (2009). *The Effect of Degradation Products of the "Carrefour Bag" Polymer Packaging on Diatom Populations in the Baltic Sea.* Master Thesis, Technical University of Gdańsk/Institute of Oceanology of the Polish Academy of Science. Gdansk.
84. Deroine, M., Le, D., Corre, Y. M., Le, G. P. Y., Davies, P., Cesar, G., & Bruzaud, S., (2014). Accelerated aging of polylactide in aqueous environments: Comparative study between distilled water and seawater. *Polymer Degradation and Stability, 108*, 319–329.
85. Matile, P., Hörtensteiner, S., & Thomas, H., (1999). Chlorophyll degradation. *Annu. Rev. Plant Physiol. Plant Mol. Biol., 50*, 67–95.
86. Ungerer, J., Pratte, B., & Thiel, T., (2008). Regulation of fructose transport and its effect on fructose toxicity in *Anabaena spp. Journal of Bacteriology, 190*(24), 8115–8125.
87. Pliński, M., & Komarek, J., (2007). *Sinice-cyanobakterie (Cyanoprokaryota).* Flora Zatoki Gdańskiej i wód przyległych (Bałtyk Południowy) (Blue-green bacteria-cyanobacteria (Cyanoprokaryota) of the Gulf of Gdańsk and adjacent waters (southern Baltic) flora) (University of Gdansk Publishing House, Gdansk).
88. Guzman, A., Janik, H., & Kosakowska, A., (2010). Preliminary testing of the influence of modified polycaprolactones on *Anabaena variabilis* growth in seawater. *J. Polym. Environ., 18*(4), 679–684.
89. Stanier, R. Y., Kunisawa, R., Mandel, M., & Bazire, C. G., (1971). Purification and properties of unicellular blue-green algae (Order: Chroococcales). *Bact. Rew., 35*, 171–205.

## CHAPTER 12

# Water Sorption in Polyester/Dust/Glass Polyester Recyclate Composites with Nanofillers

M. JASTRZĘBSKA and M. RUTKOWSKA

*Department of Industrial Commodity Science and Chemistry,*
*Faculty of Entrepreneurship and Quality Science,*
*Gdynia Maritime University, 83 Morska Str., 81–225 Gdynia, Poland,*
*E-mail: m.jastrzebska@wpit.umg.edu.pl (M. Jastrzębska)*

## ABSTRACT

Glass fiber reinforced plastics dominate the composites market accounting for approx. 95% of its total volume and create the problem of how to dispose of after-production waste and utilized product waste in a sustainable way in the future. The best way of solving the problem with composite after-production waste is mechanical recycling via collection, shredding, and milling down into fractions. The recyclate can then be reused as filler in new formulations and has other industrial uses. This case study was carried out on the mechanical recycling of glass reinforced polyester waste shredded into smaller fragments. The recyclate can be added into polyester composites with dolomite dust. Nanofillers have been applied into composites with the glass-reinforced polyester recyclate. The purpose of the presented work has been observed the change of water uptake in the composites.

Water uptake can lead to a change in the physical, chemical, mechanical, and dimensional properties of the matrix (hydrolysis, swelling as a function of its affinity with water). It has been found that there is a correlation between water absorption and discontinuities structure of composites [1]. The higher absorption indicates a greater number of defects in composites. It is therefore of great importance to characterize water absorption in the complicated polyester/dolomite dust composites with glass polyester recyclate and nanofillers. Sorption curves were obtained for composites to determine the maximum water uptake. The obtained results have shown that water uptake was clearly reduced in a result of the addition of nanofillers. The glass-reinforced polyester recyclate can be applied only as filler in building materials which do not require high strength. The composites with the recyclate can be used as countertops, parking curbs, construction barricades, insulation material, and garden ornaments.

## 12.1 INTRODUCTION

These days' composites are applied more extensively, especially for transporting structures, the construction industry, electronics, sport, and leisure. Fiber-reinforced plastics are very strong and durable. They have been used as boat materials. In the world, there are more than 23 million pleasure boats. In Poland, there are three producers of unsaturated polyester resins and gel-coats, one producer of glass fiber and reinforcements, and several hundred composites manufacturers with an overall production estimated at 80,000 tonnes/year. Production is mainly focused on marine applications, ranging from sporting boats and yachts to parts for big ships. It creates the problem of how to dispose of after-production waste and utilized product waste in a sustainable way in the future. Several studies have been reported on mechanical recycling of glass reinforced polyester waste in recent years [2–4]. Recycling process reduces environmental impact by reusing materials in a more sustainable way. Polymer composites made from thermoplastics can be recycled directly by remelting and remolding into new materials. However, recycling polymer composites based on thermoset matrix is more difficult because the material is fully cured and contains incorporated glass or carbon reinforcement. Currently, mechanical recycling the most widely conducted research, aimed at partially replacing traditional aggregate and fillers with ground glass fiber reinforced polyester recyclate.

Polyester composites are used in application in marine environments. Water, therefore, plays a significant role in the long-term performance of glass reinforced composite material, particularly in outdoor applications. A disadvantage of using polyester-based composites in water is that the polymer matrix and fiber/matrix interphase can be degraded by a hydrolysis reaction of unsaturated groups within the resin. Water penetration into the composition occurs through mechanisms: diffusion-directly into the matrix, capillary flow-to interface the fiber/matrix and transporting water molecules through microcracks [5]. The osmosis phenomenon was put forward to explain the crack propagation. It was also pointed out that unreacted glycols present inside the matrix may be responsible for cracking nucleation in that they are water-soluble inclusions which attract water and therefore increase the pressure inside the micropockets [6]. Several authors have studied the absorption process and showed the water absorption in distilled water following Fick's law. In Fickian diffusion, water could diffuse into the materials through atomic motion. Spherical water molecules of approximately 0.096 nm radius can move randomly (Brownian motion) and site in the voids (sub-microscopic network and free volume) of the material network by a concentration gradient. Debonding may occur at the fiber/matrix interface. Furthermore, water sorbed in voids or cavities may cause cracking as a result of the freeze-thaw phenomenon. Studies on the impact of water absorption on the properties of glass composites have been conducted for many years [5–8].

Fan et al. [8] observed that one of the probable solutions to close the pores/voids and improvement of interface/interphase strength is by adding nanofillers in glass-reinforced polymer composites. Among the different nanofillers, inorganic nanofillers are the most promising because of their availability, low fabrication cost, and readily optimization of mechanical and thermal properties at the design stage.

In our earliest study, has been shown that glass reinforced polyester waste from boat parts can be recycled [9–11]. The original material composed of polyester resin and dolomite dust with glass reinforced polyester recyclate has been made. The process of recycling glass fiber reinforced unsaturated polyester composites by disintegration and (then their) 10% by mass application to the production of new building material was developed and evaluated from the economical point of view. It was also ascertained that the use of nanofiller for composites with glass polyester recyclate may improve the mechanical properties of the finished products. The studies have shown that the new materials with glass polyester recyclate can be used as window-sills, prefabricated flagstones, and wall panels. However, the mechanical properties of composites with recyclate can be very sensitive to environmental parameters, such as temperature, humidity, light or other corrosive environments, especially those controlled by the properties of matrix and fiber/matrix interface. The stability of polyester depends on the cross-linked structure. However, ester linkages in cured polyester are prone to hydrolysis. The observation of the changes of water uptake in the composites was the purpose of the presented work. The changes of water uptake in the polyester composites were observed during the experiment. The polyester composites contained glass reinforced polyester recyclate, dolomite dust and nanofillers (modified montmorillonite). Then, the influence of nanofillers on water uptake of composites containing glass reinforced polyester recyclate has been tested.

## 12.2 EXPERIMENTAL

### *12.2.1 MATERIALS*

The recyclate of glass fiber reinforced cold-cured polyester laminates and dolomite dust (marketed by Kambud Sp. z o.o.) were used in this work as reinforcement of polyester matrix. The recyclate contains 45 wt.% of glass fiber in polyester. The composites were prepared using recyclate 20 wt.% concentration which were added to the unsaturated polyester at room temperature. The matrix of composites unsaturated ortophthalic polyester resin Polimal 109-32 K manufactured in "Organika-Sarzyna" Chemical Works S.A. (Poland).

Two kinds of modified montmorillonite were used as organic nanofillers:

- Nanofiller NanoBent® ZR2 delivered by ZGM. Zębiec S.A. (Poland) and synthesized by scientists from Rzeszow Technical University, according to the technology being implemented under a grant deliberate No.03933/C ZR7-6/2007, bentonites modified with quaternary ammonium salts BARDAC®2270 (70% solution of didecylodimethylammonium chloride), cation exchange capacity CEC minimum 80 meq/100g, d 001 = 1.84 nm [12].
- Nanofiller NanoBent® ZW1 manufactured in ZGM Zębiec S.A. with cooperation with Wroclaw Technical University, organophilized montmorillonite, by changing the hydrophilic character bentonite special (BS) into organophilic using a quaternary ammonium salts 3-dimethyloaminopropylamides fatty acid, cation exchange capacity CEC minimum 85 meq/100g, d 001 = 1.8 nm [12].

The compositions (Table 12.1) were mixed with metyl ethyl ketone peroxide as a initiator in the amount of 0.01 wt.% and cobalt naphthenate as an accelerator in the amount of 1 wt.% 22°C in our laboratory. The composites were prepared using a mixer and the mixture was then poured into suitable mold prepared from stainless steel. The interior surface of the mold was coated with paraffin wax and polyvinyl acetate to prevent sticking of the sample with the mold.

In this work polyester composites with 20 wt.% of polyester resin and 12 wt.% of glass polyester recyclate and different amount of organic modified montmorillonite NanoBent® ZW1 or modified montmorillonite NanoBent® ZR2 (1, 2, 3 wt.%) were prepared and water sorption have been studied.

**TABLE 12.1** Composition of Polyester Composites with Nanofillers

| Content / Formulation | Nanofiller [%] | Dolomite Dust [%] | Glass Polyester Recyclate [%] | Polyester Resin [%] |
|---|---|---|---|---|
| ZW1-1 | 1 | 67 | 12 | 20 |
| ZW1-2 | 2 | 66 | 12 | 20 |
| ZW1-3 | 3 | 65 | 12 | 20 |
| ZR2-1 | 1 | 67 | 12 | 20 |
| ZR2-2 | 2 | 66 | 12 | 20 |
| ZR2-3 | 3 | 65 | 12 | 20 |

**Note:** Samples ZW1-1 is a composite with nanofiller NanoBent® ZW1 in amount of 1%, samples ZR2-2 is a composite with nanofiller NanoBent® ZR2 in amount of 2%.

### *12.2.2 MEASUREMENT*

The moisture absorption test in this study was complied with the ASTM D 5229:2004: Standard Test Method for Moisture Absorption Properties and Equilibrium Conditioning of Polymer Matrix Composite Materials.

## 12.3 RESULTS AND DISCUSSION

In polyester/dust/glass polyester recyclate composites the empty spaces have been observed, and they have an influence on weakening and increasing of water absorption. The low water resistance is one of the disadvantages of composites with the recyclate. The closed pores present in these composites play the initial absorption of water into the composites and resulting in reduction in mechanical properties. When the glass/polyester composites are immersed in the water, water uptake would happen. This is the results of capillarity of the materials and the water absorption of the hydrophilic groups in the unsaturated polyester and the glass fiber in the recyclate. The weight uptake would increase with prolonged immersion time as far as the composite is unsaturated. The reaction between the water molecules and the matrix would deteriorate the interphase resulting in a weaker material. Water can cause matrix swelling, interphase debonding, physical damage of the interphase and hydrolysis of the material; these are the main reasons of the deteriorated tensile strength.

The adding nanofillers to these composites may be attributed to a well-dispersed reinforcing phase creating a large interfacial surface area and unique properties may be obtained. Nanoclays such montmorillonite, a hydrated sodium calcium aluminum magnesium silicate hydroxide, (Na,Ca)(Al,Mg)$_6$(Si$_4$O$_{10}$)$_3$·nH$_2$O, after modification, in addition to their primary function as high aspect ratio reinforcements, also have the important functions in absorption properties.

Preparation of composites with the nanofiller NanoBent® ZW1 proved to be an effective method of improving the mechanical properties. When the nanofiller was added in 1, 2, 3 wt.%, to the composites with 20 wt.% of resin and 12 wt.% glass reinforced polyester recyclate the compressive strength and the flexural strength increased comparing to the properties of the sample without the nanofiller (Figure 12.1).

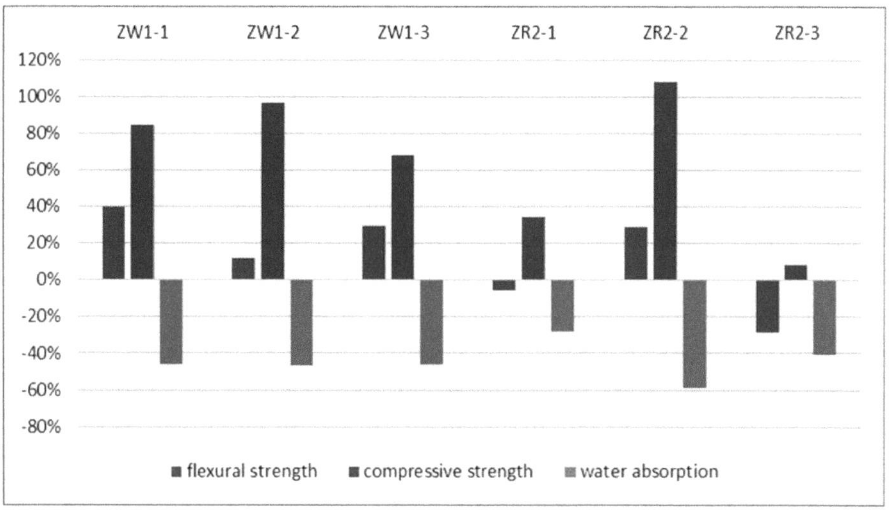


**FIGURE 12.1** Changes of some properties of composites containing 12 wt.% of glass polyester recyclate and nanofillers compared with properties of composites without nanofillers.

This was a result of the high surface area attained by adding the nanofiller to the polyester with the recyclate. The polyester interacts with the filler surface forming an interphase of absorbed polymer and the overall polymer-filler adhesion increases due to the high surface area and thereby improves the strength. After adding the nanofiller NanoBent® ZW1 to composites the flexure strength and the compressive strength increased (Figure 12.1). In contrast after adding 1 wt.% or 3 wt.% the nanofiller NanoBent® ZR2 only compressive strength increased.

Figure 12.2 presents water absorption of composites with 12 wt.% recyclate with different the nanofiller NanoBent® ZW1 content. All samples studied followed Fick's law since water absorption quickly increased. The trends are for all composites with the same slope for the first region and the same time of plateau start. However, the water absorption of the composite without nanofiller increased bigger than the others. It should be mentioned that all composites reached the equilibrium values of absorbed water during the experiment. The use of the nanofiller reduced the water uptake. This can be explained by the improved interfacial adhesion, which reduced the water accumulation in the interfacial gaps.

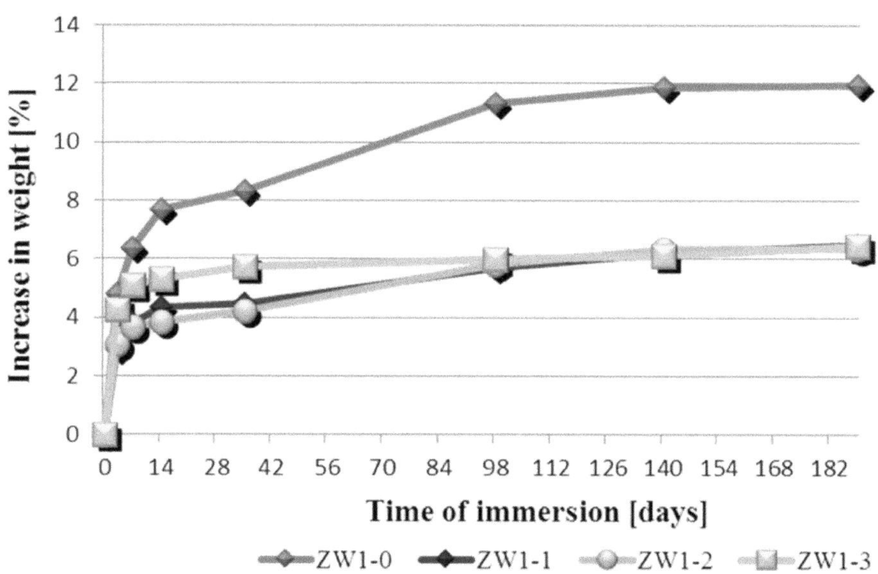


**FIGURE 12.2** Water absorption of composites with 12 wt.% recyclate with different the nanofiller NanoBent® ZW1 content.

The amount of absorbed water was different: about 12 and 6% for the unmodified and with the nanofiller NanoBent® ZW1 composites, respectively. The difference in the content of absorbed water for the modified, different amount of the nanofiller, composites was negligible. This is due to a good fiber/matrix interfacial bond.

Figure 12.3 reports the water absorption of composites with 12 wt.% recyclate with different the nanofiller NanoBent® ZR2 content as a function of the time. Identically as in the previous case the greatest increase in the water uptake occurred for unmodified composites with 12 wt.% recyclate. It is possible to define two regions: absorption and stabilization, according to Fick's law. Initially, the absorption increases with a constant slope until the value reaches a plateau at about 12% for the unmodified and 5% for 2 wt.% nanofiller modified composite. The presence of the nanofiller NanoBent® ZR2 greatly reduced the water uptake by composites.

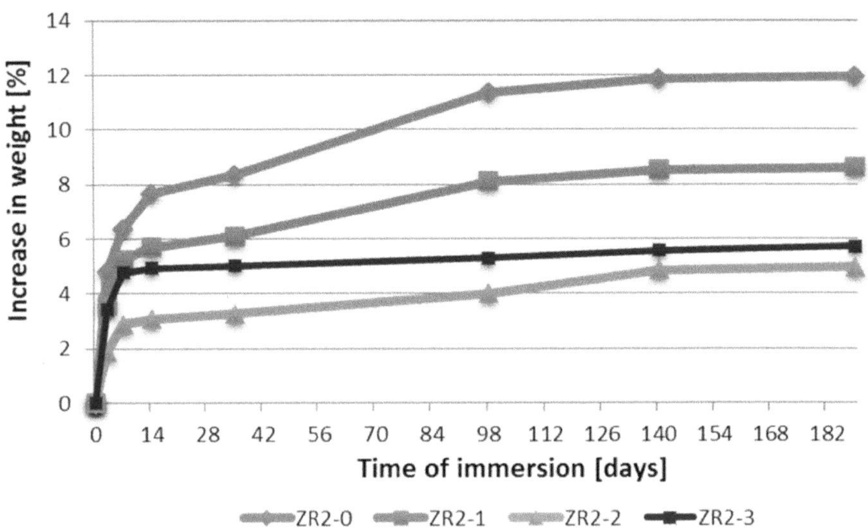


**FIGURE 12.3** Water absorption of composites with 12 wt.% recyclate with different the nanofiller NanoBent® ZR2 content.

The low densely packed polyester network is due to the presence of long polystyrene chains rather than styrene monomer units. The polystyrene chains during the cross-linking reaction with the polyester chains leave a high amount of free volume between them. This aspect, together with the nonlinearity of the polyester repeating unit, reduces the bulk compactness. The higher hydrophilic character can be explained by considering that the water can enter its polymeric network, so the degradation reactions easily occur. In particular, the water molecules can solvate easily the low monomers and/or oligomers which are present in the network. Besides, the chemical reactivity of the ester groups present in the polyester resin allows its degradation. Great defects appear on its surface confirming that water has the possibility to enter the material bulk and decrease the mechanical and chemical features of the material.

Adding the nanofillers to composites with the recyclate has an influence on the water uptake behavior of the composites.

In summary, a new polyester composite with the nanofillers was developed with high compressive strength, low water absorption; using 12% glass reinforced polyester recyclate, thus contributing to sustainability.

## 12.4 CONCLUSIONS

Polyester composites with 12 wt.% glass reinforced polyester recyclate and the dolomite dust were obtained. The glass polyester recyclate, used as component, is often imperfectly bonded, and micro-voids along interfaces are common, resulting in free spaces which provide a pathway for high water absorption. The water uptake increased with the time of exposure for all specimens in water, ultimately reaching saturation. The Fickian nature of water absorption for all specimens was noticed. Composites with 12 wt.% of glass reinforced polyester recyclate without the nanofillers have high water absorption (after 190 days, values reached 12%).

When the curves increased rapidly for all composites, the water uptake was lower for the composites with the nanofillers. The results suggest that there are probably lower void contents in composites with the recyclate and the nanofillers.

The obtained results have shown that water uptake was clearly reduced by the addition of the nanofillers. It is appropriate to use the nanofiller NanoBent® ZR2 in the amount of 2 wt.% or the nanofiller NanoBent® ZW1 in the amount of 1–2 wt.% to composites with 12 wt.% glass-polyester recyclate in order to improve the mechanical and sorption properties of the final products.

## KEYWORDS

- **bentonite special**
- **glass polyester recyclate**
- **material recycling**
- **nanofillers**
- **recyclate**
- **water sorption**

## REFERENCES

1. Woo, M., & Piggott, M., (1988). *Journal of Composites Technology and Research, 10*, 16–19.
2. Goodship, V., (2010). *Management, Recycling and Reuse of Waste Composites* (pp. 281–300). Woodhead Publishing, Oxford.
3. Overcash, M., Twomey, J., Asmatulu, E., Vozzola, E., & Griffing, E., (2018). *Journal of Composite Materials,* 52(8), 1033–1043.
4. Meira, C. A. C., Ribeiro, M. C. S., Santos, J., Meixedo, J. P., Silva, F. J. G., Fiúza, A., Dinis, M. L., Alvimet, M. R., (2013). *Construction and Building Materials, 45*, 87–94.
5. Gautier, L., Mortaigne, B., & Bellenger, V., (1999). *Composites Science and Technology, 59*, 2329–2337.
6. Sonawala, S. P., Spontak, R. J., (1996). *J. Mater. Sci., 31*, 4745–56.
7. Gu, H., & Hongxia, S., (2007). *Materials and Design, 28*, 1647–1650.
8. Fan, X. J., Lee, S. W. R., & Han, Q., (2009). *Microelectron Reliab, 49*, 861–71.
9. Jastrzębska, M., & Blokus-Roszkowska, A., (2015). *Przemysł Chemiczny, 94*(6), 1003.
10. Jastrzębska, M., & Rutkowska, M., (2019). In: Mukbaniani, O. V., Tatrishvili, T. N., & Abadie, M. J. M., (eds.), *Science and Technology of Polymers and Advanced Materials: Applied Research Methods* (pp. 35–45). Apple Academic Press.
11. Jastrzębska, M., Janik, H., & Paukszta, D., (2014). *Polimery, 59*(9), 656–661.
12. Spychaj, T., Heneczkowski, M., Pigłowski, J., Oleksy, M., Kowalczyk, K., Kiersnowski, A., & Galina, H., (2006). *Inżynieria Materiałowa [Materials Engineering], 27*(6), 1296–1302.

# CHAPTER 13

# Sorption Properties of Hydrogels of Rare-Crosslinked Functional Polyacids and Polybases in Relation to Rare Earth Metal Ions

T. K. JUMADILOV, R. G. KONDAUROV, and A. M. IMANGAZY

*JSC "Institute of Chemical Sciences after A.B. Bekturov," Almaty, the Republic of Kazakhstan, E-mail: jumadilov@mail.ru (T. K. Jumadilov)*

## ABSTRACT

This research is devoted to the sorption features of lanthanum (La), cerium (Ce), neodymium (Nd), samarium (Sm) ions from the solutions by polymer hydrogels of polyacrylic acid (hPAA), polymethacrylic acid (hPMAA), poly-4-vinylpyridine (hP4VP), and poly-2-methyl-5-vinylpyridine (hP2M5VP). We found that the sorption of the above-mentioned rare-earth metals (REMs) ions by individual hydrogels was increasing in time.

1. **PAA Hydrogel:** The maximum values of extraction degree of La, Ce, Nd, Sm ions by PAA hydrogel were 68.16%, 62.83%, 60.90%, and 65.80%, respectively. The highest values of the polymer chain binding degree values 55.90% (La), 51.80% (Ce), 49.90% (Nd), and 52.80% (Sm) of hPAA were achieved after 48 hours of interaction. The effective dynamic exchange capacity maximum values of hPAA were observed after 48 hours of interaction: 5.12 (La), 4.25 (Ce), 4.17 (Nd), and 4.62 (Sm) (mmol/g).
2. **PMAA Hydrogel:** Maximum sorption of REMs ions by hPMAA occurs after 48 hours of polymer interaction with saline solution. The extraction degree values of lanthanum, cerium, neodymium,

samarium were 65.80%, 59.90%, 58.10%, and 65.30%, respectively; polymer chain binding degree values were 54.80% (La), 49.90% (Ce), 46.80% (Nd), and 53.10% (Sm); effective dynamic exchange capacity values were 4.85 (La), 4.05 (Ce), 3.92 (Nd), and 4.37(Sm) (mmol/g).

3. **P4VP Hydrogel:** The maximum values of the hP4VP sorption parameters: extraction degree values were 66.12% (La), 57.07% (Ce), 55.10% (Nd), and 63.10% (Sm); polymer chain binding degree values were 55.20% (La), 46.90% (Ce), 46.10% (Nd), and 51.90% (Sm); effective dynamic exchange capacity values were 4.95 (La), 3.78 (Ce), 3.67 (Nd), and 4.30 (Sm) (mmol/g).
4. **P2M5VP Hydrogel:** Maximum sorption of the REMs ions by hP2M5VP occurs after 48 hours of interaction. The sorption degree values were 64.10% (La), 50.50% (Ce), 49.10% (Nd), and 58.10% (Sm); polymer chain binding degree values were 52.90% (La), 41.50% (Ce), 39.00% (Nd), 48.20% (Sm); and effective dynamic exchange capacity values were 4.67 (La), 3.45 (Ce), 3.10 (Nd), and 4.20 (Sm) (mmol/g).

## 13.1 INTRODUCTION

Rare earth metals (REMs) in alloys are useful in many devices that people use in everyday settings, such as cell phones with rechargeable batteries, computer memory, catalytic converters, magnets, fluorescent lighting, etc. [1–4]. In recent years, there has been an explosion in demand for many items that require these elements [5, 6].

REMs also play an essential role in defense equipment. For example, the military uses night-vision goggles, precision-guided weapons, communication equipment, batteries, and other defense electronics, which give the military an enormous advantage. REMs are key ingredients for making the very hard alloys used in armored vehicles and projectiles that shatter upon impact [7, 8].

Demand for these metals is expanding as a sustainable power source turns out to be progressively significant over the globe. REMs such as neodymium, which is significant in "green" technologies and innovative industries, is in the spotlight, especially as hybrid vehicles and electric cars gain their popularity nowadays. This fact makes the production of REMs and their compounds are highly profitable in the near future.

## 13.2 EXPERIMENTAL PART

### *13.2.1 EQUIPMENT*

The REMs ions concentration in solutions we determined by KFK-3KM (Russia) and Jenway-6305 (UK) spectrophotometers.

### *13.2.2 MATERIALS*

This research was carried out in 0.005 M solutions of lanthanum (III), cerium (III), neodymium (III), and samarium (III) nitrate hexahydrate. Polyacrylic (hPAA) and polymethacrylic acid (hPMAA) hydrogels were synthesized in the presence of N,N-methylene-bis-acrylamide as crosslinking agent and redox system of $K_2S_2O_8$-$Na_2S_2O_3$ in aqueous medium. The synthesized hydrogels were crushed into tiny dispersions and washed with distilled water until constant conductivity value of aqueous solutions was achieved. Poly-4-vinylpyridine (hP4VP) and poly-2-methyl-5-vinylpyridine (hP2M5VP) hydrogels (linear polymers cross-linked by divinylbenzene, Sigma-Aldrich) was used as polybasis.

### *13.2.3 METHODOLOGY OF REMS IONS DETERMINATION*

Methodology of REMs ions determination in solution is based on colored complex compound formation of organic analytical reagent arsenazo III with REMs ions [9].

Extraction (sorption) degree was calculated by equation:

$$\eta = \frac{C_{initial} - C_{residual}}{C_{initial}} \times 100\%$$

where; $C_{initial}$-initial concentration of REMs in solution, g/L; $C_{residual}$-residual concentration of REMs in solution, g/L.

Polymer chain binding degree was determined by calculations in accordance with equation:

$$\theta = \frac{\nu_{sorb}}{\nu} \times 100\%$$

where; $\nu_{sorb}$ is the quantity of polymer links with sorbed REM, mol; and ν is the total quantity of polymer links, mol.

Effective dynamic exchange capacity was calculated by the following formula:

$$Q = \frac{\nu_{sorbed}}{m_{sorbent}}$$

where; $\nu_{sorbed}$ is the amount of sorbed metal, mol; and $m_{sorbent}$ is the mass of the sorbent, g.

## 13.3 RESULTS AND DISCUSSIONS

As known, there is an electrochemical equilibrium in any solution. Any intervention in the system provides changes in this balance. Polymer hydrogels are weak electrolytes. However, despite this fact, macromolecules interact with present ions of metals in solutions. Polyacids are able to dissociation of carboxylic groups and sorption of REMs by binding them to oppositely charged carboxylate anions. Poylbases (polyvinylpyridines) undergo ionization of heteroatoms (nitrogen atoms), which provides changes in their conformation [10–14].

### *13.3.1 STUDY OF SORPTION PROPERTIES OF PAA HYDROGEL*

Figure 13.1 shows the dependence of extraction degree values of lanthanum, cerium, neodymium, and samarium ions by PAA hydrogel in time. As seen from Figure 13.1, an increase of this rare-earth metals ions extraction degree occurs in time. The main amount of REMs 41.43% (La), 39.00% (Ce), 37.80% (Nd), and 40.55% (Sm) was sorbed during first 6 hours of the polymer interaction in the saline solutions. The results showed that the further increase of the sorption degree was not so intense [10–14]. In our opinion, the reason for this in decreasing of carboxylic groups dissociation in time. Further interaction of hPAA with REMs nitrates provides reaching of extraction degree maximum values of 68.16% (La), 62.83% (Ce), 60.90% (Nd), and 65.80% (Sm) over 48 hours of interaction.

The dependence of polymer chain binding degree (in relation to La, Ce, Nd, Sm ions) of polyacrylic acid hydrogel is shown in Figure 13.2. The most intensive binding of the mentioned REMs by PAA hydrogel occurs during 6 hours, at this time binding degree values are 34.49% (La), 32.34% (Ce), 30.37% (Nd), and 33.40% (Sm). Further interaction of the polymer macromolecule with saline solution provides further increase of this parameter. After this time, raise of binding degree occurred not so fast, and the final values of the polymer chain binding degree values 55.90% (La), 51.80% (Ce), 49.90% (Nd), and 52.80% (Sm) were reached over 48 hours.

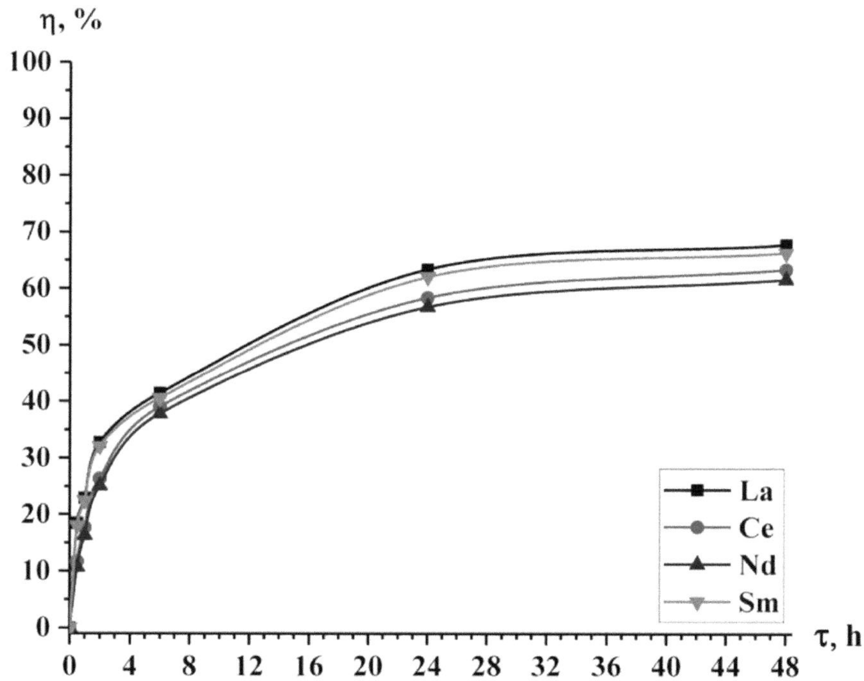


**FIGURE 13.1** The dependence of extraction degree values of lanthanum, cerium, neodymium, and samarium ions by PAA hydrogel in time.

Figure 13.3 shows the dependence of effective dynamic exchange capacity values for La, Ce, Nd, and Sm ions by PAA hydrogel in time. The maximum values 5.12 (La), 4.25 (Ce), 4.17 (Nd), and 4.62 (Sm) (mmol/g) were reached over 48 hours of the polymer interaction in the corresponding saline solutions.

### *13.3.2 STUDY OF PMAA HYDROGEL SORPTION PROPERTIES*

Figure 13.4 presents the dependence of REMs ions sorption degree values by PMAA hydrogel in time.

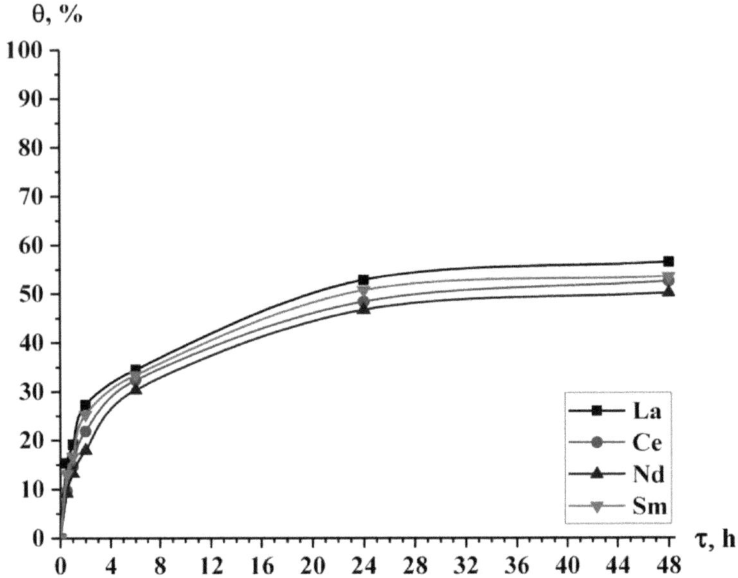


**FIGURE 13.2** The dependence of polymer chain binding degree for La, Ce, Nd, Sm ions by PAA hydrogel in time.\

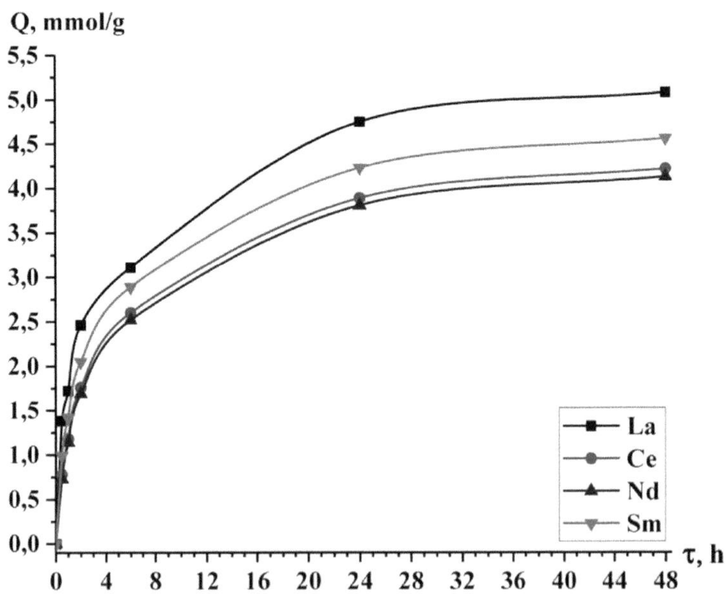


**FIGURE 13.3** The dependence of effective dynamic exchange capacity (in relation to La, Ce, Nd, Sm ions) of PAA hydrogel in time.

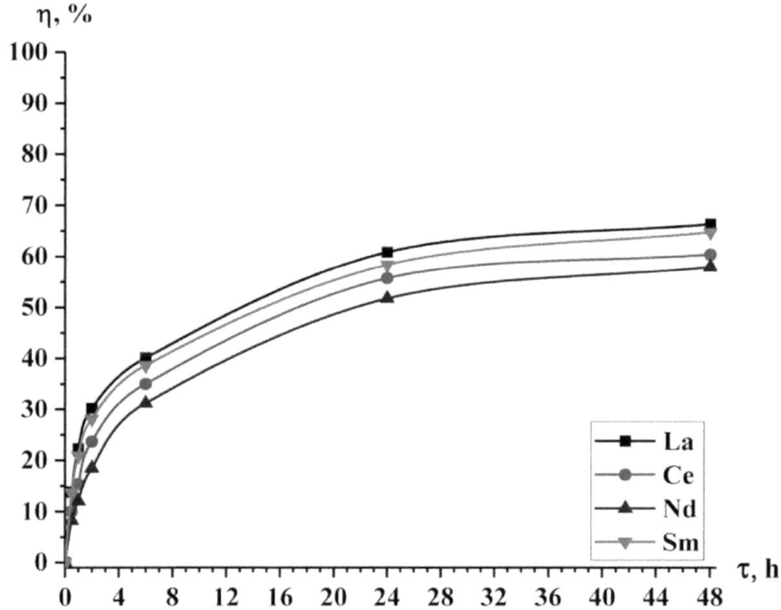


**FIGURE 13.4** The dependence of extraction degree of La, Ce, Nd, and Sm ions by PMAA hydrogel in time.

As seen from Figure 13.4, the overwhelming majority (~40%) was extracted by PMAA hydrogel during the first 6 hours. Further increase (after 24 hours) of sorption degree occurred less intensively which indicated that the "polymer-saline solution" system was reaching equilibrium state. The maximum sorption of REMs occurs over 48 hours of polymer interaction in saline solutions with ions extraction degree values: 65.80% (La), 59.90% (Ce), 58.10% (Nd), and 65.30% (Sm).

Figure 13.5 shows the dependence of polymer chain binding degree (in relation to La, Ce, Nd, Sm ions) of PMAA hydrogel in time. The most intensive sorption occurred during 6 hours of interaction. At this time, the binding degree values were 33.50%, 29.03%, 26.80%, and 28.40% in relation to La, Ce, Nd, Sm ions. Further interaction provided increase of this parameter to 50.67% (La), 46.18% (Ce), 43.80% (Nd), and 48.10% (Sm). The maximum binding was observed over 48 hours, the polymer chain binding degree values were 54.80% (La), 49.90% (Ce), 46.80% (Nd), and 53.10% (Sm).

The dependence of effective dynamic exchange capacity (in relation to La, Ce, Nd, Sm ions) of PMAA hydrogel in time is presented in Figure 13.6. This parameter was increasing over time, wherein the most intense growth

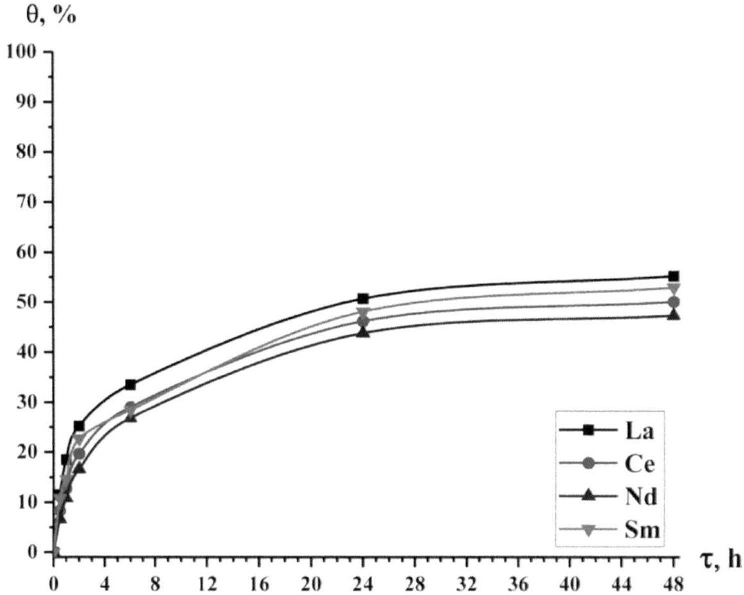


**FIGURE 13.5** The dependence of polymer chain binding degree (in relation to La, Ce, Nd, Sm ions) by PMAA hydrogel in time.

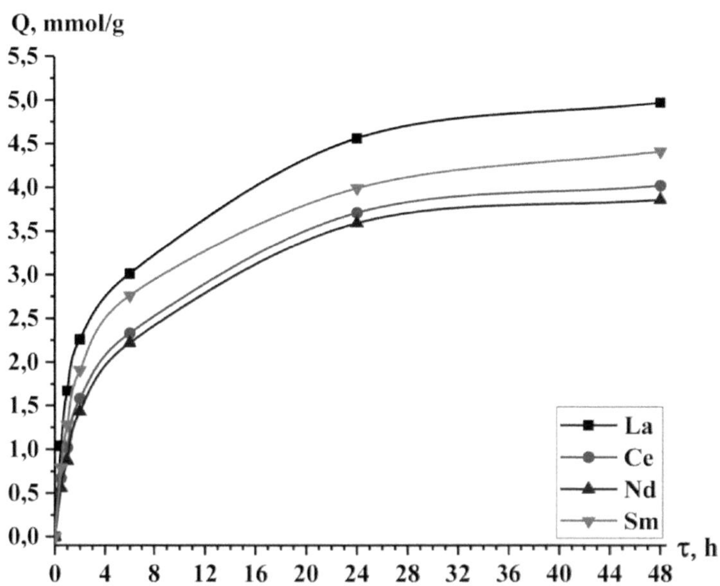


**FIGURE 13.6** The dependence of effective dynamic exchange capacity (in relation to La, Ce, Nd, Sm ions) of PMAA hydrogel over time.

was observed during the first 6 hours. The maximum of exchange capacity values 4.85 (La), 4.05 (Ce), 3.92 (Nd), and 4.37 (Sm) (mmol/g) were observed over 48 hours of interaction.

### *13.3.3 STUDY OF P4VP HYDROGEL SORPTION PROPERTIES*

Figure 13.7 shows the dependence of sorption degree values of La, Ce, Nd, and Sm ions by P4VP hydrogel in time. Strong increase of extraction degree was observed during 6 hours. The sorption degree values for studied ions were 38.43% (La), 30.00% (Ce), 28.30% (Nd), and 36.59% (Sm). The highest values of this parameter were observed over 48 hours of interaction; the extraction degree values for ions were 66.12% (La), 57.07% (Ce), 55.10% (Nd), and 63.10% (Sm).

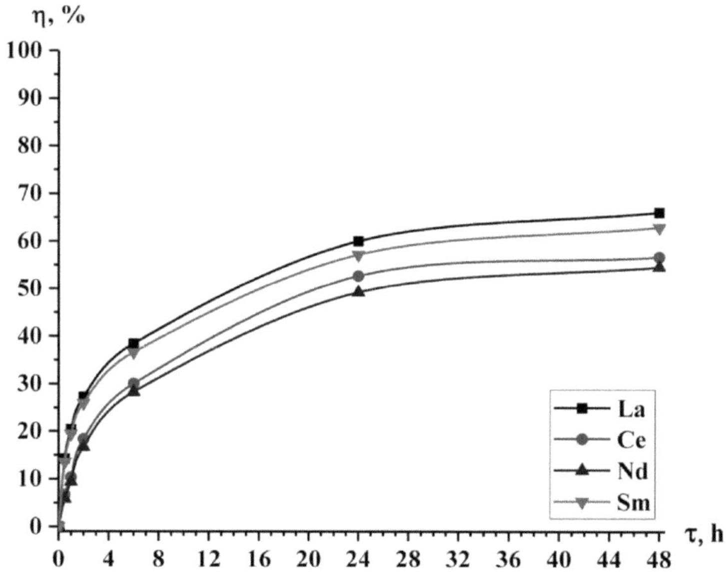


**FIGURE 13.7** The dependence of extraction degrees of lanthanum, cerium, neodymium, and samarium by P4VP hydrogel in time.

The dependence of polymer chain binding degree (in relation to La, Ce, Nd, and Sm ions) by P4VP hydrogel in time is presented in Figure 13.8. As seen from Figure 13.8, the binding degree values increased in time. The obtained results show that the most intense binding of the above-mentioned

REMs by P4VP hydrogel occurred during 6 hours, the polymer chain binding degree values were 32.00% (La), 24.88% (Ce), 22.60% (Nd), and 27.50% (Sm). Further sorption of REMs provided increase of binding degree values; over 24 hours 49.99% (La), 43.69% (Ce), 41.80% (Nd), and 46.70% (Sm) were bind. The maximum values of binding degree were observed over 48 hours of interaction between poly-4-vinylpyridine hydrogel with saline solutions, the binding degree values for ions were 55.20% (La), 46.90% (Ce), 46.10% (Nd), and 51.90% (Sm).

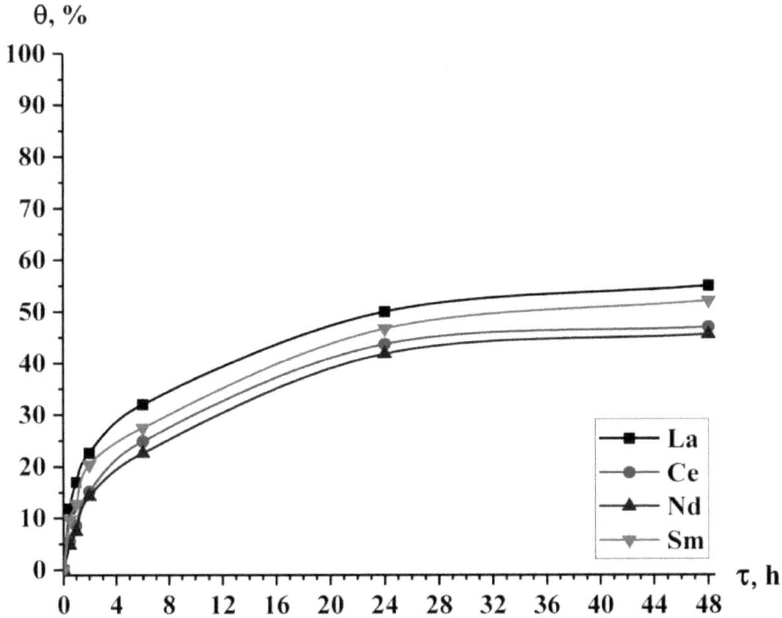


**FIGURE 13.8** The dependence of polymer chain binding degree (in relation to La, Ce, Nd, Sm ions) of P4VP hydrogel in time.

Figure 13.9 shows the dependence of effective dynamic exchange capacity (in relation to La, Ce, Nd, Sm ions) of P4VP hydrogel in time. The most REMs ions were sorbed over 6 hours. The exchange capacity values were 2.89 (La), 2.00 (Ce), 1.92 (Nd), and 2.31 (Sm) (mmol/g). The further weak increase (after 24 hours) was due to the nature of the polybasis (P4VP is a weak basis and it is weakly ionized), and the equilibrium was rather reached fast in the saline solution. The maximum exchange capacity values 4.95 (La), 3.78 (Ce), 3.67 (Nd), and 4.30 (Sm) (mmol/g) were reached over 48 hours of interaction.

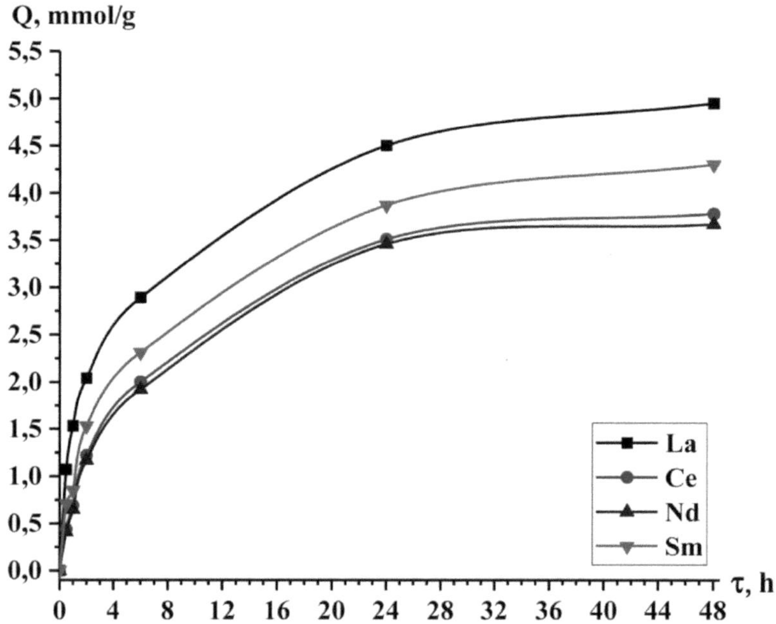


**FIGURE 13.9** The dependence of effective dynamic exchange capacity (in relation to La, Ce, Nd, Sm ions) of P4VP hydrogel in time.

### *13.3.4 STUDY OF SORPTION PROPERTIES OF P2M5VP HYDROGEL*

Figure 13.10 shows the dependence of extraction degrees of La, Ce, Nd, and Sm ions in time. As seen from Figure 13.10, the sorption degree of the above-mentioned REMs ions increased over time. After 6 hours of interaction, the sorption degree values for ions were 36.95% (La), 25.33% (Ce), 22.90% (Nd), and 31.40% (Sm). After this time, sorption occurred slower. The maximum values of this parameter were reached over 48 hours of interaction. The extraction degree values for ions were 64.10% (La), 50.50% (Ce), 49.10% (Nd), and 58.10% (Sm).

The polymer chain binding degree (in relation to La, Ce, Nd, Sm ions) by P2M5VP hydrogel is presented in Figure 13.11. Similar to data on binding degrees of PAA, PMAA, P4VP hydrogels, from Figure 13.11 it can be seen, that the most amount of the REMs were binded during 6 hours, at this time-binding degree values were 30.83% (La), 21.01% (Ce), 18.70% (Nd), and 27.80% (Sm). After this time, there was further increase of these parameters values, at 24 hours binding degree was 49.17% (La), 33.73% (Ce), 30.80%

(Nd), and 41.90% (Sm). A further increase was very slow, which indicated that the equilibrium state was reached. The final polymer chain binding degree values 52.90% (La), 41.50% (Ce), 39.00% (Nd), and 48.20% (Sm) were observed at 48 hours.

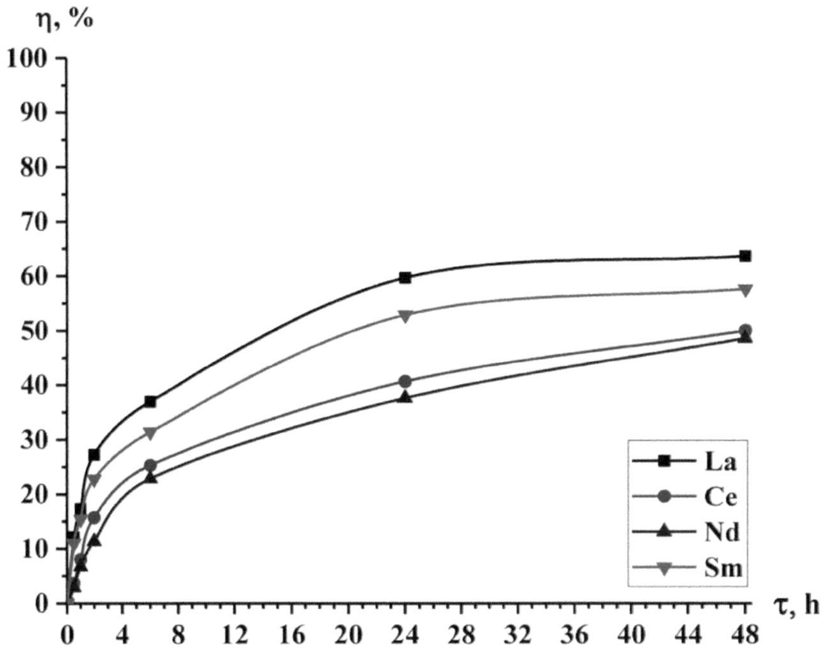


**FIGURE 13.10** The dependence of extraction degree values of La, Ce, Nd, and Sm ions by P2M5VP hydrogel in time.

The dependence of effective dynamic exchange capacity (in relation to La, Ce, Nd, Sm ions) of P2M5VP hydrogel in time is shown in Figure 13.12.

The highest sorption of the REMs ions was observed over 6 hours of the interaction of P2M5VP hydrogel with La, Ce, Nd, and Sm nitrate solutions. At this time of interaction, the parameter had values of 2.77 (La), 1.69 (Ce), 1.43 (Nd), and 2.05 (Sm) (mmol/g). Further increase of effective dynamic exchange capacity was not so intensive. At 24 hours of interaction, the exchange capacity values were 4.47 (La), 2.71 (Ce), 2.56 (Nd), and 3.13 (Sm) (mmol/g). The maximum values of exchange capacity 4.67 (La), 3.45 (Ce), 3.10 (Nd), and 4.20 (Sm) (mmol/g) were reached over 48 hours.

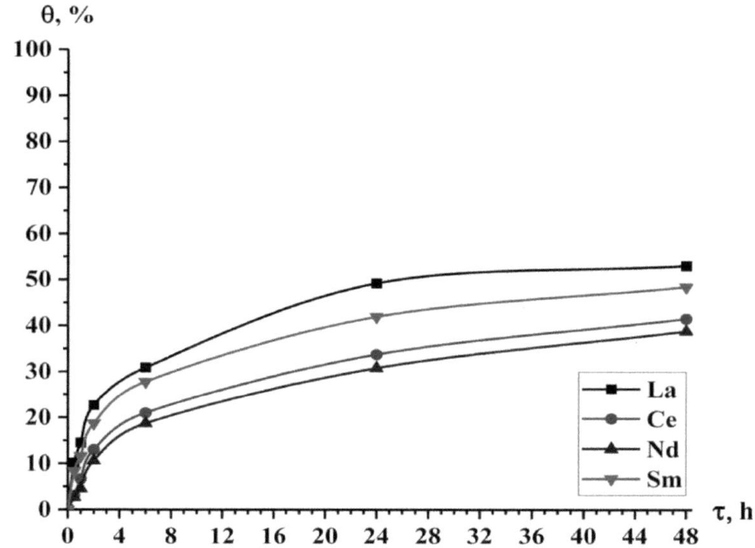


**FIGURE 13.11** The dependence of polymer chain binding degree (in relation to La, Ce, Nd, Sm ions) by P2M5VP hydrogel in time.

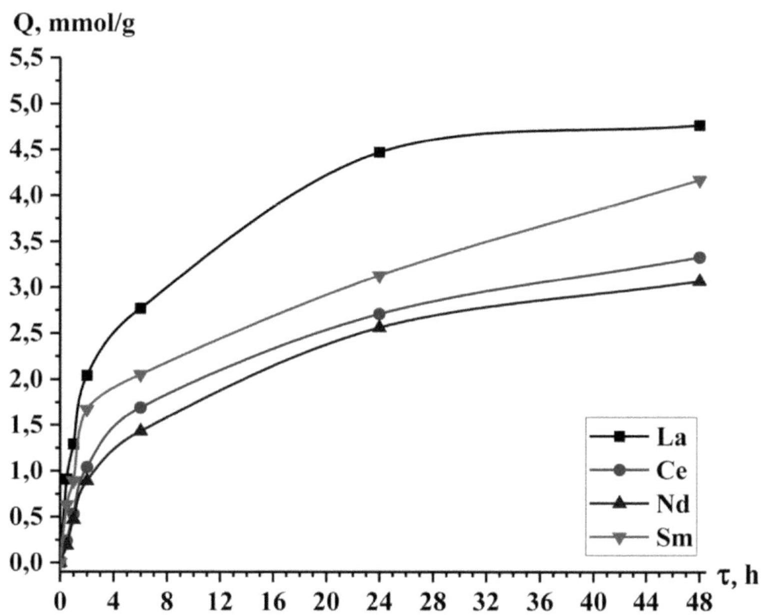


**FIGURE 13.12** The dependence of effective dynamic exchange capacity (in relation to La, Ce, Nd, Sm ions) of P2M5VP hydrogel in time.

## 13.4 CONCLUSIONS

1. The process of La, Ce, Nd, and Sm ions sorption by PAA, PMAA, P4VP, and P2M5VP hydrogels occurred over time. It should be noted that the main sorption parameters (sorption degree, polymer chain binding degree, effective dynamic exchange capacity) increased in time. Maximum sorption of the metal ions occurred at 48 hours of interaction between the polymers and the saline solutions.
2. The maximum values of La, Ce, Nd, and Sm extraction degree by PAA hydrogel were 68.16%, 62.83%, 60.90%, and 65.80%, respectively. The highest values of La, Ce, Nd, and Sm ions extraction degree by PMAA hydrogel were 65.80%, 59.90%, 58.10%, and 65.30%, respectively. The maximum extraction degree values of La, Ce, Nd, and Sm ions by hP4VP were 66.12%, 57.07%, 55.10%, and 63.10%, respectively. The maximum sorption of the REMs ions by hP2M5VP occurred at 48 hours of interaction. The sorption degree values for ions were 64.10% (La), 50.50% (Ce), 49.10% (Nd), and 58.10% (Sm).
3. The highest values of the polymer chain binding degree values 55.90% (La), 51.80% (Ce), 49.90% (Nd), and 52.80% (Sm) of hPAA were reached at 48 hours. The highest values of polymer chain binding degree of hPMAA were 54.80% (La), 49.90% (Ce), 46.80% (Nd), and 53.10% (Sm). The maximum values of the binding degree of hP4VP were 55.20% (La), 46.90% (Ce), 46.10% (Nd), and 51.90% (Sm). The maximum values of polymer chain binding degree values of hP2M5VP were 52.90% (La), 41.50% (Ce), 39.00% (Nd), and 48.20% (Sm).
4. Effective dynamic exchange capacity maximum values 5.12 (La), 4.25 (Ce), 4.17 (Nd), and 4.62 (Sm) (mmol/g) of hPAA were observed at 48 hours of interaction. The highest values of effective dynamic exchange capacity of hPMAA were 4.85 (La), 4.05 (Ce), 3.92 (Nd), and 4.37 (Sm) (mmol/g). The highest values of effective dynamic exchange capacity values of hP4VP were 4.95 (La), 3.78 (Ce), 3.67 (Nd), and 4.30 (Sm) (mmol/g). The maximum values of effective dynamic exchange capacity values of hP2M5VP were 4.67 (La), 3.45 (Ce), 3.10 (Nd), and 4.20 (Sm) (mmol/g).
5. The difference in the sorption degree values is due to the difference in atomic radius, charge density, and polarizability of the La, Ce, Nd, and Sm ions.

## ACKNOWLEDGMENT

This work was financially supported (Grant funding of two Projects: AP05131302 and AP05131451) by the Science Committee of the Ministry of education and science of the Republic of Kazakhstan.

## KEYWORDS

- **$Ce^{3+}$ ions**
- **$La^{3+}$ ions**
- **$Nd^{3+}$ ions**
- **polyacids**
- **polybases**
- **$Sm^{3+}$ ions**
- **sorption**

## REFERENCES

1. Greenwood, N. N., & Earnshaw, A., (1997). *Chemistry of the Elements* (2nd edn., p. 1341). Oxford: Butterworth-Heinemann.
2. Evans, C. H., (1996). *Episodes from the History of the Rare Earth Elements* (Vol. 15, p. 268). Kluwer Academic *Publishers.*
3. Evans, C. H. (1990). *Biochemistry of the Lanthanides* (Vol. 8, p. 2). Springer US.
4. Atwood, D. A., (2012). *The Rare Earth Elements: Fundamentals and Applications* (p. 606). Chichester, West Sussex: John Wiley & Sons, Ltd.
5. Gschneidner, Jr. K. A., & Eyring, L. R., (1988). *Handbook of the Physics and Chemistry of Rare Earths* (Vol. 11, p. 46). Elsevier Science Publishers B.V.
6. Möller, P., Černý, P., & Saupé F., (1989). *Lanthanides, Tantalum and Niobium* (Vol. 7, pp. 3–26). Springer Berlin Heidelberg.
7. Edelmann, F. T., (2014). Lanthanides and actinides: Annual survey of their organometallic chemistry covering the year 2012. *Coordination Chemistry Reviews, 261*, 73–155.
8. Edelmann, F. T., (2015). Lanthanides and actinides: Annual survey of their organometallic chemistry covering the year 2013. *Coordination Chemistry Reviews, 284*, 124–205.
9. Petrukhin, O. M., (1987). Methodology of physico-chemical methods of analysis. *M.: Chemistry*, 77–80.
10. Jumadilov, T. K., Kondaurov, R. G., & Imangazy, A. M., (2019). Phenomenon of remote interaction and sorption ability of rare cross-linked hydrogels of polymethacrylic acid and poly-4-vinylpyridine in relation to erbium ions. *Chemistry and Chemical Technology, 13*(4), 451–458.

11. Jumadilov, T. K., Kondaurov, R. G., & Imangazy, A. M., (2019). Comparison of sorption properties of polyacids and polybases and intergel systems on their basis in relation to neodymium ions. *Chemical Journal of Kazakhstan, 1*, 201–213.
12. Jumadilov, T. K., Kondaurov, R. G., & Imangazy, A. M., (2019). Comparative characteristics of sorption properties of poly-4-vinylpyridine and poly-2-methyl-5-vinylpyridine in relation to rare earth elements ions. *Chemical Journal of Kazakhstan, 2*, 40–48.
13. Imangazy, A. M., Jumadilov, T. K., Kondaurov, R. G., & Zhora, A. D., (2019). "Remote interaction" effect of polymer hydrogels on samarium ions sorption. *Proceedings of XXI Mendeleev Congress on General and Applied Chemistry* (Vol. 2b, Section: 2, p. 154). Saint-Petersburg, Russian Federation.
14. Jumadilov, T. K., & Kondaurov, R. G., (2019). Features of selective sorption of lanthanum ions from solution containing ions of lanthanum and cerium by intergel system hydrogel of polymethacrylic acid-hydrogel of poly-2-methyl-5-vinylpyridine. In: Haghi, A. K., Pogliani, L., & Ribeiro, A. F., (eds.), *Research Methodologies and Practical Applications of Chemistry* (pp. 167–192). AAP press.

## CHAPTER 14

# Change in Rheological Properties of Pitch-Thermoplastic Under the Effect of Polar Polymers

I. KRUTKO, V. KAULIN, I. DANYLO, K. YAVIR, and K. SATSYUK

*Donetsk National Technical University, Department of Chemical Technologies, 85300 Pokrovsk, Ukraine, E-mail: poshukdoc@gmail.com*

## ABSTRACT

The study results of the influence of the polar polymer: polyvinyl chloride (PVC), maleized ethylene vinyl acetate (MEVA), and polymethyl methacrylate (PMMA) on the rheological properties of coal tar pitch (CTP) are presented in the paper. It has been shown that CTP heat treatment with polar polymers promotes thermochemical transformations that significantly change of the modified pitches structure. Due to the stitching processes and the enhancement of intermolecular bonds, the viscosity of the obtained material increases, the viscosity anomaly appears, and the strength of the structure increases.

It has been proven that the heat treatment of CTP at 170°C with PVC and PMMA allows changing and controlling the pitch-thermoplastic viscosity. Low-temperature modification of CTP by polar polymers at temperatures up to 170°C provides obtaining the plastic material-the pitch-thermoplastic with target rheological properties, which will allow it to be used as a precursor in the production of carbon foam, carbon composites, and others.

The advantages of the CTP modification with polar polymers to change its rheological properties are the use of sufficiently low temperatures and the absence of any waste, which allows solving the environmental and energy-saving issues a pitch precursor obtaining.

## 14.1 INTRODUCTION

Coal tar pitch (CTP) is a heavy thermoplastic residue formed during the distillation of coal tar. It is a heterogeneous system of highly condensed carbo- and heterocyclic compounds that are capable of exhibiting polymeric properties. Due to its unique properties, CTP is a promising and inexpensive raw material for the production of various carbon materials: carbon composites, solid foams, carbon fibers, adsorbents, etc. The main problem is that its rheological properties do not meet the requirements for the production of carbon materials [1–7].

CTP refers to non-Newtonian liquids, the anomaly degree of which increases with decreasing temperature and with small shear stress. It has been established that when the temperature of the CTP varies from the softening temperature to 125°C, a transition is observed from elastic-brittle bodies to hard-like plastic ones and then to structured and Newtonian liquids. At temperatures above 125°C, the CTP exhibits a Newtonian flow, due to the destruction of its coagulation structure; that is, it has a very low viscosity. This does not allow its use in the processes of obtaining carbon materials, which are carried out at higher temperatures.

It is known that highly condensed aromatic compounds presented in the pitch are characterized by a low strength of chemical bonds at the periphery. In addition, components with various active functional groups are contained in the pitch. Due to this, the CTP can, under certain conditions, actively interact with various chemical additives, changing its properties [3, 7–9].

By its nature, the pitch refers to weakly polar materials. Therefore, polar polymers were chosen for research: polyvinyl chloride (PVC), maleized ethylene vinyl acetate (MEVA), and polymethyl methacrylate (PMMA). This chapter presents the study results of the effect of the polar polymers as active chemical additives on the rheological properties of CTP.

## 14.2 EXPERIMENTAL

Laboratory investigations were carried out with a medium-temperature CTP. The medium-temperature CPT had the METTLER softening point of 108°C and a coke yield of 56 wt.%. The CPT had $\alpha$-fraction of 35%, $\alpha_1$-fraction of 9%, $\alpha_2$-fraction of 26%, $\beta$-fraction of 33–34%, $\gamma$-fraction of 31–32%. The softening temperature was 86–88°C using the ring and core method. A viscosity of CPT at 140°C and at 160°C was 9.3 and 1.3 Pa·s, respectively.

### *14.2.1 CHARACTERISTICS OF POLYMER ADDITIVES*

Suspension polyvinyl chloride PVC-S (CAS: 9002), powder with a particle size of 100–200 microns, density 1350 kg/m$^3$. Fikentcher constant 63; the melting temperature and the transition to a viscous flowing state are 150–170°C (with decomposition). It is an amorphous polar polymer with a high intermolecular interaction. According to the solubility parameter (20.2 MPa$^{1/2}$), PVC refers to a moderately polar polymer. This indicates the sufficient simplicity of its combination with other polymers. Therefore, PVC is considered an easily compounded polymer.

PMMA $[C_5O_2H_8]_n$ (CAS: 9011-14-7). PMMA is a powder with a density of 1190 kg/m$^3$ and a melting temperature of 150°C. MEVA is a polymer with low polarity, thanks to the grafted maleic groups has high reactivity and the ability to form strong bonds at the phase boundary. The effect of active additives on the viscosity of the pitch was researched in a laboratory reactor at 170°C, stirring the pitch for up to 2 hours with a certain amount of additive.

The method of CTP processing was as follows. The initial CTP, which is in the form of granules, was crushed in a mill to a powder state. The pitch powder was then mixed with a certain amount of polymer or mixtures of polymers. The purpose of the mixing is a uniform distribution of all the components in the mixture. The number of additives was (wt. parts per 100 wt. parts of CTP): PVC from 3 to 20, PMMA from 1 to 5, MEVA 5. The resulting mixture was heated in a reactor equipped with a mechanical stirrer. After melting the pitch-polymer mixture, the stirrer was switched on to follow the conditions of the mixture uniformity and to intensify the chemical reactions. The molten mass was mixing in the reactor for up to 2 hours at a temperature of 170°C.

The viscosity of the obtained pitches modified with PVC (3 wt.p.) and MEVA (5 wt.p.) was determined using a cone-plate rotational viscometer RHEOTEST-2.1. The viscosity of the pitch samples modified with a high PVC content (5 wt.p. and above) was characterized by a melt flow rate (MFR), which was determined using the equipment of the brand IIRT-AM (according to the test methods ASTM D1238, ASTM D3364, GOST 11645). The essence of the method was to determine the mass of the material in grams, extruded from the device for 10 minutes at a temperature of 150°C and at a load of 49, 98 N (5, 10 kgf).

## 14.3 RESULTS AND DISCUSSION

Studies have shown that at 135°C and above, CTP shows a Newtonian flow (Figure 14.1, sample 1), that is, a completely destroyed structure. The viscosity of heat-treated pitch without chemical additives (Figure 14.1, sample 2) is about three times greater than the viscosity of CTP, but in these conditions, its flow also corresponds to the completely destroyed structure.

Viscosity is known to depend on the chemical composition and molecular kinetic characteristics of the system. The interaction of molecules leads to the formation of supramolecular structures, the number and size of which affect the viscosity. Thus, the viscosity indirectly characterizes the intermolecular interaction in the pitch, as in the case of any other condensed system. The most complete idea of the change in pitch structure gives a study of the viscosity anomaly phenomenon, that is, the decrease of viscosity with increasing shear rate.

Addition to the CTP of the PVC (3 wt.p.) leads to a change in the chemical composition of the modified CTP and to an increase in intermolecular interaction, which is expressed in the increase of viscosity and viscosity anomaly (Figure 14.1, sample 3) comparing to pitch and heat-treated pitch. Addition to the mixture of pitch and PVC (3 wt.p.) the MEVA modifier (5 wt.p.) enhances the effect of structuring the pitch (Figure 14.1, sample 4).

The nature of the change in the effective viscosity of the modified pitch depending on the shear rate (Figure 14.1) shows the intensification of the structuring processes under the action of the active additives. Moreover, the strength of the modified pitches structure, which is determined by the intermolecular bonds strength, depends on the type of modifier and the temperature.

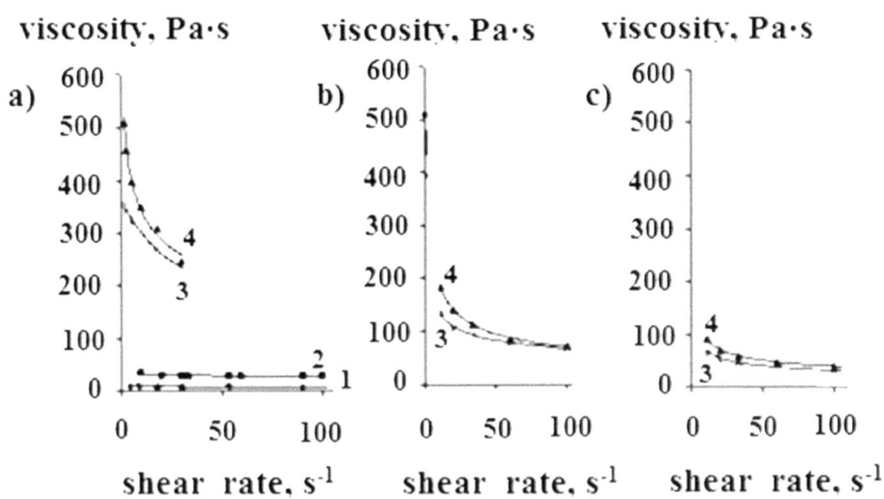


**FIGURE 14.1** The dependence of viscosity on the shear rate: (a) T = 135°C; (b) T = 145°C; (c) T = 155°C.

The feature of the rheological behavior of the modified pitch is due to the fact that the active additives PVC and MEVA significantly change the chemical composition and the pitch structure. In fact, it is no longer a pitch, but a pitch-polymer having a higher viscosity and viscosity anomaly even at 155°C (Figure 14.1). The viscosity is determined by the ratio of the energy of interaction and thermal motion. Under these conditions, the intermolecular interaction energy exceeds the energy of thermal motion, because even the application of a mechanical field ($\gamma = 100\ s^{-1}$) at 155°C does not cause complete destruction of the modified pitches structure.

CTP is attributed to concentrated dispersed systems. The high concentration of the dispersed phase leads to the interaction of its particles. In our opinion, the structure of the obtained pitch-polymers corresponds to the coagulation structure, which is characterized by chemical bonds formed by the chemical interaction of the coal pitch components with active functional groups of PVC and MEVA with the formation of macromolecules, and the intermolecular interaction forces that result of the formation of stronger hydrogen bonds due to the presence of such functional groups as -OH, -COOH, $-NH_2$.

As the temperature increases, the kinetic energy of macromolecules and supramolecular structures increases, and when the attraction energy is reached, the Van der Waals bonds are broken. This results in the ability of macromolecules and supramolecular structures to move. Pitch-polymers with coagulation structure flow with increasing temperature, exhibiting high elasticity.

The analysis of the flow curves-the dependence of shear stress on the shear rate $P = f(\gamma)$ (Figure 14.2) – for the pitches and pitches modified with PVC and MEVA made it possible to estimate the strength of the modified pitch structure.

Figure 14.2 shows that, at 135°C, the flow curves for the initial pitch (sample 1) and the heat-treated pitch without additives (sample 2) correspond to straight lines, which are extrapolated to zero. This means that under these conditions, these samples are characterized by a Newtonian flow, which can only be realized in the practical absence of structure.

The study of the patterns of rheological behavior of CTP and modified pitches in the temperature range 125–155°C showed that the intensification of thermal motion in the system of the modified pitch with increasing temperature when applying a mechanical field does not ultimately lead to the complete destruction of system (Figure 14.1). This fact confirms the significant influence of the introduced polar polymers on the processes of structure formation and the emergence of stronger intermolecular bonds in the pitch-modifier system.

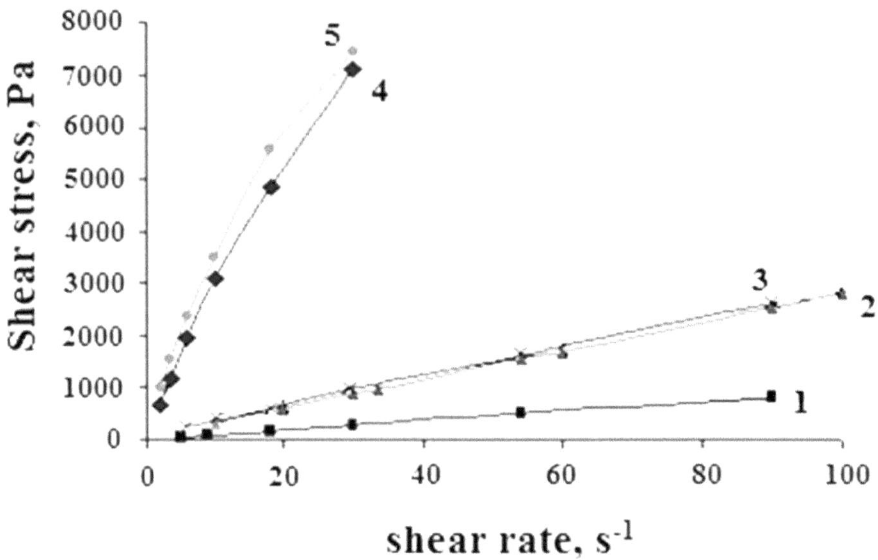


**FIGURE 14.2** The dependence of shear stress on the shear rate at t = 135°C of CTP and pitches, modified with PVC and MEVA: (1) CTP; (2) heat-treated CTP; (3) CTP: MEVA = 100: 5 wt.p.; (4) CTP: PVC = 100: 3 wt.p.; (5) CTP: PVC: MEVA = 100: 3: 5 wt.p.

The value of the ultimate shear stress $P_y$ (the yield point in the Bingham equation) characterizes the effort required to destroy the structure, that is, the structure strength. Calculation of the yield points at *t* = 135°C of the pitches (Table 14.1) showed that pitches modified with PVC and MEVA have greater strength of structure compared to the initial and heat-treated pitches. The highest strength of the structure has a pitch, modified by a mixture of PVC and MEVA ($P_y$ = 2900 Pa). When the temperature rises to 155°C (Figure 14.2), the yield point of the modified pitches is reduced by 1.5–2 times. In spite of this, the modified pitches structure retains strength and is not completely destroyed.

The temperature dependence of CTP viscosity is subordinated to the Arrhenius equation [3, 6]:

$$\eta = A \cdot e^{E/RT}$$

The activation energy of a viscous flow is the energy required to move molecular complexes from one position to another. It is connected with overcoming the energy barrier in the process of viscous flow. The conditional activation energy of the viscous flow (E) of CTP, which shows how strongly the viscosity depends on temperature, decreases with the introduction of active additives (Table 14.1). Rheological studies of the behavior of pitches modified with PVC and MEVA have shown that with the introduction of active additives to CTP, the conditional activation energy of the viscous flow decreases in the temperature range 125–155°C.

The flow of pitch is related to the movement and orientation of macromolecules and supramolecular structures, mainly in one direction (in the force direction). Large complex structural units of such material as pitch are unable to make an instant transition from one state to another. The flow in such a system is due to the successive movements of individual segments, and further, the whole structural unit. It can be assumed that during the flow, the straightening of macromolecules and supramolecular structures occurs, that is, highly elastic deformation will happen. If the flexibility of macromolecular chains increases, the potential barrier needed to move them decreases.

**TABLE 14.1** Rheological Characteristics of Pitches Modified with PVC and MEVA

| No. Sample | Sample Name | Sample Composition, wt. p. | | | Yield Point Py, Pa | Activation Energy E, kJ/mol |
|---|---|---|---|---|---|---|
| | | CTP | PVC | MEVA | | |
| 1. | The initial pitch | 100 | 0 | 0 | 0 | 177 |
| 2. | Heat-treated pitch | 100 | 0 | 0 | 0 | 176 |
| 3. | Modified pitch | 100 | 0 | 5 | 250 | 109 |
| 4. | Modified pitch | 100 | 3 | 0 | 1500 | 123 |
| 5. | Modified pitch | 100 | 3 | 5 | 2900 | 100 |

The obtained values of E for CTP, heat-treated pitch, and modified (PVC and MEVA) pitches lead us to believe that the conditional activation energy of a viscous flow is determined not only by the value of the intermolecular interaction forces but also by the construction of macromolecular and supramolecular structures. The less correct and orderly the construction of the structural units is, the more difficult they are to move relative to each other, as in the case of unmodified pitches characterized by the highest E = 176–177 kJ/mol.

The introduction of a PVC polymer modifier into the CTP whose polar groups interact with the polar active functional groups of the pitch leads to the ordering of the newly formed supramolecular structures, which is evidenced by a decrease in E = 123 kJ/mol compared to a 177 kJ/mol of the pitch.

A further decrease of E to 100 kJ/mol is observed if the pitch is modified by a mixture of PVC and MEVA polymer additives. Most likely, this is due to the further ordering of the formed structural units and the increased flexibility of the macromolecular chains. Thus, this additive has a plasticizing effect without reducing the strength of the structure. Therefore, the temperature interval of the highly elastic state of the modified pitch is expanded. Extension of the range of highly elastic state improves the technological properties of the pitch-polymers, which is very important for their further processing.

The viscosity of the pitch samples modified with high PVC content (5 wt.p. and above) increased significantly, so the MFR was used to determine it. In the interaction of pitch with PVC (10 to 20 wt.p.) the formation of the strongest structures is observed, as indicated by the sharp decrease in the MFR from 23.8 to 1.1 g/10 min. (Figure 14.3), which is the increase in viscosity.

Rheological studies have shown that the addition of PVC (10–20 wt.p.) during the heat treatment (170°C) of CTP promoted thermochemical transformations, which significantly changed the structure of the modified pitch. Due to the processes of crosslinking and enhancement of intermolecular bonds, the obtained material viscosity has increased significantly. It is found that the strength of the modified pitches structure, which is determined by the strength of intermolecular bonds, increases with the amount of PVC.

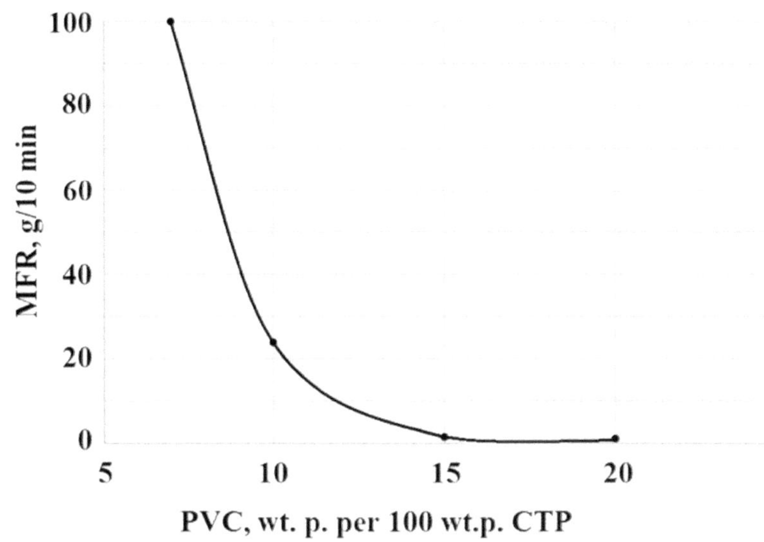


**FIGURE 14.3** The dependence of MFR of modified pitch on the amount of PVC. Load weight 10 kg, T = 150°C.

The use of PVC as a CTP modifier has made it possible to regulate the viscosity of the material obtained. However, the samples of modified pitch obtained were characterized by considerable fragility. PMMA (1 and 5 wt.p.), known as impact strength modifier, was used to give modified pitch samples of higher strength.

Kinetic studies have shown (Figure 14.4) that with increasing the heat treatment time of the pitch at 170°C, regardless of the PVC and PMMA content, the MFR of the modified pitch decreases increases, which indicates an increase of viscosity.

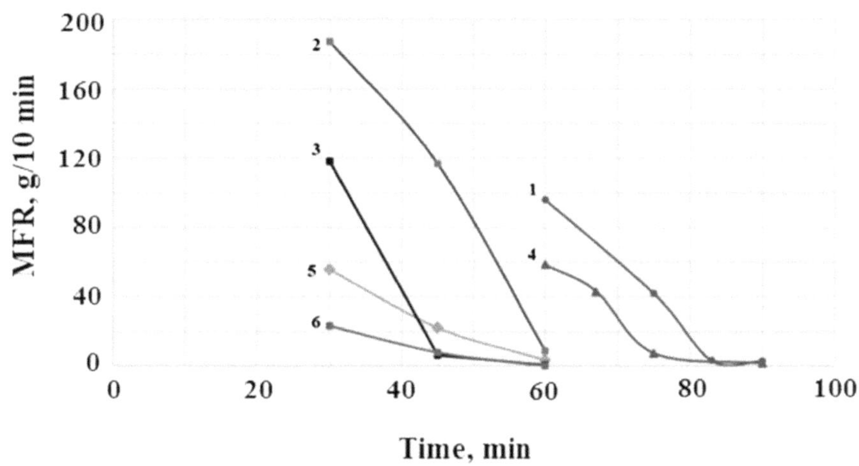


**FIGURE 14.4** Kinetic curves of modification at T = 170°C of the CTP:PVC:PMMA mixture (wt.p.): for CTP:PMMA = 100:1 the amount of PVC: (1) 10; (2) 15; (3) 20. For CTP:PMMA = 100:5 the amount of PVC: (4) 10; (5) 15; (6) 20.

Rheological studies (Figure 14.5) showed that PMMA addition significantly intensifies the thermochemical transformations that occur during the modification of CTP by PVC, significantly affecting its structure, resulting in a significant increase in viscosity.

By increasing the amount of PMMA from 1 to 5 wt.p., with the constant content of PVC the rate of change of MFR decreases (Table 14.2), which can be explained by the decrease in the rate of interaction of pitch with PVC and PMMA due to the higher viscosity of the reaction medium. Table 14.2 shows that the amount of PVC and PMMA affect the viscosity and duration of heat treatment of the mixture at 170°C. It takes less time to achieve the same MFR value of modified pitch with increasing PVC content.

Thus, by varying the CTP:PVC:PMMA ratio and the mixture heat treatment time at 170°C, it is possible to regulate the viscosity of the modified pitch to achieve the desired value.

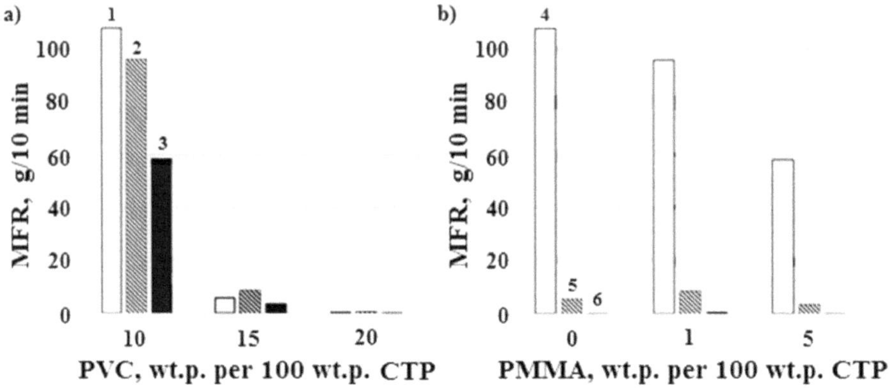


**FIGURE 14.5** The dependence of MFR (load weight 5 kg, t = 150°C) of modified pitch on the amount of PVC and PMMA (wt.p. per 100 wt.p. CTP): (a) the amount of PMMA: (1) 0; (2) 1; (3) 5; (b) the amount of PVC: (4) 10; (5) 15; (6) 20.

**TABLE 14.2** The Dependence of Viscosity and Change Rate of MFR on the Composition of the Mixture CTP: PVC: PMMA and Heat-Treat Time of the Mixture at 170°C

| No. Mixture | The Mixture Composition for Modification, wt.p. | | | Heat Treatment Time of Mixture at 170°C, min. | Achieved Viscosity MFR, g/10 min | The Average Rate of Change MFR, g/10 min per Minute |
|---|---|---|---|---|---|---|
| | CTP | PVC | PMMA | | | |
| 1. | 100 | 5 | 5 | 105–112 | 11.82 | 3.46 |
| 2. | 100 | 10 | 1 | 60–83 | 3.48 | 4.02 |
| 3. | 100 | 10 | 5 | 60–75 | 7.41 | 3.40 |
| 4. | 100 | 15 | 1 | 30–60 | 8.62 | 5.95 |
| 5. | 100 | 15 | 5 | 30–60 | 3.52 | 1.75 |
| 6. | 100 | 20 | 1 | 30–45 | 6.43 | 7.44 |
| 7. | 100 | 20 | 5 | 30–45 | 7.94 | 1.03 |

## 14.4 CONCLUSIONS

Experimental studies showed that at 135–155°C, CTP, and heat-treated pitch without chemical additives behaves like a Newtonian liquid, do not have a viscosity anomaly and have the lowest viscosity of all samples. The slight increase in the viscosity of the heat-treated pitch can be explained by the increase in the number of condensed products that occurred during the heat treatment due to polycondensation reactions. The activation energy of the viscous flow of the initial and heat-treated pitch is highest, which is explained by the least ordered structure of the pitch material.

The addition of PVC polymer during the heat treatment of CTP promotes thermochemical transformations that significantly change the modified pitch structure. Due to the stitching processes, the enhancement of intermolecular bonds, the viscosity of the material obtained increases, the viscosity anomaly appears, and the strength of the structure increases. The decrease in activation energy indicates a higher degree of structure orderliness of the pitch-polymer. The addition of PVC and MEVA to the pitch intensifies the processes of structure formation, as evidenced by the increase in viscosity, a higher viscosity anomaly, and an almost twice-greater yield point, which indicates the greater strength of the structure of such a pitch-polymer. The greater orderliness of the structural elements of the pitch-polymer is evidenced by the further decrease in the activation energy of the viscous flow.

It has been proven that the heat treatment of CTP at 170°C with PVC and PMMA allows changing and controlling the viscosity of the pitch-thermoplastic. By changing the modification time, the amount of PVC and PMMA, it is possible to increase the viscosity, that is, to change the yield point of the obtained pitch-thermoplastic from 0.5 to 100 g/10 min (at 150°C). Modification of CTP by polar polymers at a treatment temperature up to 170°C provides obtaining of plastic material-pitch-thermoplastic with specified rheological properties, which will allow it to be used as a precursor in the production of carbon foam, carbon composites, and others. The advantages of CTP modification with polar polymers to change its rheological properties are the use of sufficiently low temperatures and the absence of any waste, which allows solving the environmental and energy-saving issues of obtaining a pitch precursor.

## KEYWORDS

- **coal tar pitch**
- **maleized ethylene vinyl acetate**
- **pitch-thermoplastic**
- **polymethyl methacrylate**
- **polyvinylchloride**
- **rheology**
- **viscosity**

## REFERENCES


1. Granda, M., Blanco, C., Alvarez, P., Patrick, J. W., & Menendez, R., (2014). *Chemical Reviews, 114*(3), 1608.
2. Liu, X., Wang, Y., & Zhan, L., (2018). *Chinese Journal of Chemical Engineering, 26*(2), 415.
3. Krutko, I., Kaulin, V., & Satsiuk, K., (2015). *High Performance Polymers for Engineering-Based Composites*, 265.
4. Chen, C., Kenne, E. B., Stiller, A. H., Stansberry, P. G., & Zondlo, J. W., (2006). *Carbon, 44*, 1535.
5. Wang, Y., He, Z., Zhan, L., & Liu, X., (2016). *Materials Letters, 169*, 95.
6. Liu, H., Li, T., Wang, X., Zhang, W., & Zhao, T., (2014). *Journal of Analytical and Applied Pyrolysis, 110*, 442.
7. Krutko, I., Danylo, I., & Kaulin, V., (2019). *Petroleum and Coal, 61*(1), 150.
8. Blanco, C., Santamaria, R., Bermejo, J., & Menendez, R., (2000). *Carbon, 38*(4), 517.
9. Krutko, I., Yavir, K., & Kaulin, V., (2018). Issues of Chemistry and Chemical Technology, *4*, 97.

## CHAPTER 15

# Investigation of Complex Formation Process of Zinc with Fulvic Acids, Isolated from Natural Waters at pH = 9

T. MAKHARADZE

*Ivane Javakhishvili Tbilisi State University, Tbilisi, Georgia,*
*E-mail: makharadze_tako@yahoo.com*

### ABSTRACT

Natural macromolecular organic substances-fulvic acids (FAs) take an active part in complex formation processes and stipulate migration forms of heavy metals in natural waters.

In spite of researches, experimental data on stability constants of complex compounds of FAs with heavy metals (among them zinc) are heterogeneous, and they differ in several lines from each other. One of the reasons for such a condition is ignoring an average molecular weight of the associates of FAs, which finally causes the wrong results. The complex formation process between zinc(II) and FAs was studied by the solubility method at pH = 9.0. ZnO suspension was used as a solid phase. FAs were isolated from Paravani lake by the adsorption-chromatographic method. This chapter shows that during the complex formation process, every 1/5 part of an associate of FAs, inculcates into zinc's inner coordination sphere, as an integral ligand, so it may assume, that the average molecular weight of the associate of FAs which takes part in complex formation process equals to 1674. This part of the associate of FAs was conventionally called an *active associate*. The average molecular weight of the *active associate* was used for determination the composition of zinc fulvate complex, the concentration of free ligand and stability constant, which equals to $5.56 \times 10^4$.

## 15.1 INTRODUCTION

Fulvic acids (FAs) are one of the first macromolecular organic substances, which were discovered in natural waters. They take an active part in complex formation and sorption processes and stipulate migration forms of heavy metals and radionuclides in natural waters and soils [1–14]. In spite of researches, experimental data on stability constants (β) of complex compounds of FA with zinc (II) are heterogeneous, and they differ in several lines from each other [3, 5, 6, 11, 12]. This condition is mainly stipulated by ignoring the average molecular weight of the associates of FA, which value in its turn depends on pH and finally causes the wrong result. Therefore, it's difficult to investigate complex formation processes taking place in natural waters, identify migration forms of zinc, and evaluate and assess the chemical-ecological condition of natural waters.

The objective of the work was to obtain the pure samples of FA, to investigate complex formation processes between the pure samples of FA, isolated from natural water and Zn(II), and to calculate β of zinc fulvate complex. The complex formation process was studied by the solubility method. Zinc oxide suspension was used as a solid phase. Complex formation process was studied at pH = 9.0.

## 15.2 MATERIALS AND METHODS

For obtaining pure samples of FA, after filtration through membrane filters (0.45 μm pore size), the water of Paravani lake was concentrated by the frozen method. The concentrated water samples were acidified with 6 M HCl to pH 2 and were put for 2 hours on water bath at 60 °C for coagulation of humin acids. Then the solution was centrifuged for 10 min at 8000 rpm (Centrifuge T-23). For the isolation of FA from centrifugate was used the adsorption-chromotographic method. Charcoal was used as a sorbent. Desorption of amino acids and carbohydrates were performed by means of 0.1 M HCl. For desorption of polyphenols was used 90% acetone water solution. The eluation of the fraction of FA was performed with 0.1 N NaOH solution [10, 15]. The obtained alkaline solution of FA, for the purification, was passed through a cation-exchanger (KU-2-8). For determination, the concentration of FA in obtained solution was used gravimetric method, the part of the solution was dried under vacuum until the constant weight was obtained. Then, model solutions of FA were prepared. The solution of fulvic complexes was obtained by the solubility method. 0.1 ml suspension of zinc oxide and increasing quantity of standard solution of FA were placed in 15 ml capacity fluroplastic cylinders. pH = 9.0, ionic strength $\mu = 0.01$ ($KNO_3$), v = 10 ml. The concentration of hydrogen ions was regulated by the addition of 0.01 M $HNO_3$ acid or 0.01 M NaOH(pH meter pH 2006). Then, it was stirred in a mechanical mixer for 100 hours, until the balance was achieved and then the suspension was filtered through the membrane filters (0.45 μm pore size). In filtrates, zinc was measured by atomic absorption spectrophotometer (Perkin Almer 200).

## 15.3 RESULTS AND DISCUSSION

It was taken into consideration the fact that FA form associates in water solutions. It was established that in the interval pH 4–11, there is line dependence between Mw and the value of pH, which is expressed in the following way: Mw = 1350, pH-4540 [10]. Molar concentrations of FAs at pH = 9.0(Mw = 7610) could be calculated.

The data show that in line with the increase of concentration of FA in the solution, the concentration of zinc increases for several times as well due to formation of fulvic complex.

If it is not taken into consideration charges of ions, the reaction of formation of zinc fulvate complexes could be written in the following way:

$$Zn(II)_{free} + mFA_{total} = ZnFA_m \quad (1)$$

$$\beta = [ZnFA_m]/([Zn(II)_{free}]\ [FA_{total}]^m) \quad (2)$$

In balanced solutions, correlation $[Zn(II)_{total}]:[FA_{total}]$on average equals to 1:0.22. This means that during the complex formation process, the associate of FA, which Mw at pH = 9 equals to 7610, divides, and every 1/5 part of this associate inculcates into zinc's (II) inner coordination sphere, as an integral ligand. Therefore, it may assume that Mw of the associate of FA, which takes part in the complex formation process, equals to 1674.

This part of the associate of FA was conventionally called an *active associate.* The meaning of Mw of the *active associate* of fulvic acid (Mw = 1674) was used for determination the composition of zinc fulvate complexes, the concentration of free ligand ($[FA_{free}]$) and β [16]. It should be noted that in the case of using Mw of the associate (7610), it will be impossible to calculate $[FA_{free}]$. The calculation of $[FA_{free}]$ is impossible. Without it, it's impossible to calculate β of zinc fulvate complex.

In solution, the concentration of fulvate complex equals to the difference between total and free concentrations of zinc (II) received after formation the complex:

$$[ZnFA_m] = [Zn(II)_{total}] - [Zn(II)_{free}] \quad (3)$$

$$\beta = ([Zn(II)_{total}]-[Zn(II)_{free}])/([Zn(II)_{free}][FA_{total}]^m) \quad (4)$$

From (4) equation:

$$\beta[Zn(II)_{free}] = ([Zn(II)_{total}] - [Zn(II)_{free}])/[FA_{total}]^m \quad (5)$$

At the fixed pH, the left part of the equation is a permanent value and was marked it as K'

$$K' = ([Zn(II)_{total}] - [Zn(II)_{free}])/[FA_{total}]^m \quad (6)$$

The logarithm of this equation is:

$$\log K' = \log([Zn(II)_{total}]-[Zn(II)_{free}])-m\log[FA_{total}] \quad (7)$$

The numeral value (m) of thestoichiometric coefficient or the number of ligands in the inner coordination sphere of complex equals to tangens of tilt angle of straight line built in coordinates.

$$\log([Zn(II)_{total}]-[Zn(II)_{free}])-m\log[FA_{total}] \quad (8)$$

To calculate the exact value of tangens tilt angle of straight line, for this purpose was used the least square method. After the calculation, was obtained the numeral value of m(Mw(FA) = 7610) and m(Mw(FA) = 1674), which equal to 0.82 and 1.11 (Tables 15.1–15.4).

**TABLE 15.1** Experimental Data Necessary for Calculation the Composition of Zinc Fulvate Complex by the Solubility Method pH = 9.0; Mw(FA) = 7610; $[ZnFA_m] = [Zn(II)_{total}] - [Zn(II)_{free}]$; $[Zn_{free}] = 2.52 \times 10^{-5}$ Mol/L

| Mol/L | | | $[Zn(II)_{total}]$: $[FA_{total}]$ | Log $[FA_{total}]$ | Log $[ZnFA_m]$ |
|---|---|---|---|---|---|
| $[FA_{total}]$ | $[Zn_{total}]$ | $[ZnFA_m]$ | | | |
| $1.14 \times 10^{-5}$ | $6.08 \times 10^{-5}$ | $3.56 \times 10^{-5}$ | 1:0.19 | –4,9431 | –4,4485 |
| $1.48 \times 10^{-5}$ | $6.80 \times 10^{-5}$ | $4.28 \times 10^{-5}$ | 1:0.22 | –4,8297 | –4,3685 |
| $1.77 \times 10^{-5}$ | $8.28 \times 10^{-5}$ | $5.76 \times 10^{-5}$ | 1:0.21 | –4,7520 | –4,2396 |
| $2.11 \times 10^{-5}$ | $9.60 \times 10^{-5}$ | $7.08 \times 10^{-5}$ | 1:0.22 | –4,6757 | –4,1499 |
| $2.39 \times 10^{-5}$ | $10.59 \times 10^{-5}$ | $8.07 \times 10^{-5}$ | 1:0.22 | –4,6260 | –4,0931 |
| $2.96 \times 10^{-5}$ | $12.67 \times 10^{-5}$ | $10.15 \times 10^{-5}$ | 1:0.23 | –4,5287 | –3,9935 |
| $3.59 \times 10^{-5}$ | $14.86 \times 10^{-5}$ | $12.34 \times 10^{-5}$ | 1:0.24 | –4,4449 | –3,9087 |
| $4.16 \times 10^{-5}$ | $17.00 \times 10^{-5}$ | $14.48 \times 10^{-5}$ | 1:0.24 | –4,3809 | –3,8392 |

$[Zn(II)_{total}]$: $[FA_{total}]$ = 1:0.22

**TABLE 15.2** The Calculation of Zinc Fulvate Complex by the Least Square Method; Mw(FA) = 7610

| Xi | Yi, | XiYi | $Xi^2$ |
|---|---|---|---|
| –4,9431 | –4,4485 | 21,9894 | 24,4342 |
| –4,8297 | –4,3685 | 21,0998 | 23,8737 |
| –4,7520 | –4,2396 | 20,1466 | 22,5815 |
| –4,6757 | –4,1499 | 19,4037 | 21,8662 |
| –4,6260 | –4,0931 | 18,9347 | 21,3999 |
| –4,5287 | –3,9935 | 18,0854 | 20,5091 |
| –4,4449 | –3,9087 | 17,7507 | 19,7571 |
| –4,3809 | –3,8392 | 16,8191 | 19,1923 |

pH = 9.0; Xi = log $[FA_{total}]$, Yi = log$[ZnFA_m]$

$\sum$ Xi = –37,181; $(\sum Xi)^2$ = 1382,4267; $\sum$Yi = –33,041; $\sum Xi^2$ = 173,614; $\sum$XiY = 154,2294

Y = ax+b, m = a = tga = (nSxiyi-SxiSyi)/(nSxi²-(Sxi)²) = 0.82

**TABLE 15.3** Experimental Data Necessary for Calculation the Composition of Zinc Fulvate Complex by the Solubility Method pH = 9.0; Active Associate Mw(FA) = 1674

| Mol/L | | | Log $[FA_{total}]$ | Log $[ZnFA_m]$ |
|---|---|---|---|---|
| $[FA_{total}]$ | $[Zn_{total}]$ | $[ZnFA_m]$ | | |
| $5.19 \times 10^{-5}$ | $6.08 \times 10^{-5}$ | $3.56 \times 10^{-5}$ | –4,2848 | –4,4485 |
| $6.75 \times 10^{-5}$ | $6.80 \times 10^{-5}$ | $4.28 \times 10^{-5}$ | –4,1707 | –4,3685 |
| $8.04 \times 10^{-5}$ | $8.28 \times 10^{-5}$ | $5.76 \times 10^{-5}$ | –4,0947 | –4,2396 |
| $9.60 \times 10^{-5}$ | $9.60 \times 10^{-5}$ | $7.08 \times 10^{-5}$ | –4,0177 | –4,1499 |
| $10.90 \times 10^{-5}$ | $10.59 \times 10^{-5}$ | $8.07 \times 10^{-5}$ | –3,9626 | –4,0931 |
| $13.49 \times 10^{-5}$ | $12.67 \times 10^{-5}$ | $10.15 \times 10^{-5}$ | –3,8700 | –3,9935 |
| $16.35 \times 10^{-5}$ | $14.86 \times 10^{-5}$ | $12.34 \times 10^{-5}$ | –3,7865 | –3,9087 |
| $18.94 \times 10^{-5}$ | $17.00 \times 10^{-5}$ | $14.48 \times 10^{-5}$ | –3,7226 | –3,8392 |

$[ZnFA_m] = [Zn(II)_{total}] - [Zn(II)_{free}]$; $[Zn_{free}] = 2.52 \times 10^{-5}$ Mol/L

**TABLE 15.4** The Calculation of Zinc Fulvate Complex by the Least Square Method; Active Associate

| Xi | Yi, | XiYi | $Xi^2$ |
|---|---|---|---|
| –4,2848 | –4,4485 | 19,0609 | 18,3595 |
| –4,1707 | –4,3685 | 18,2197 | 17,3947 |
| –4,0947 | –4,2396 | 17,3599 | 16,7665 |
| –4,0177 | –4,1499 | 16,6730 | 16,1419 |
| –3,9626 | –4,0931 | 16,2193 | 15,7022 |
| –3,8700 | –3,9935 | 15,4548 | 14,9769 |
| –3,7865 | –3,9087 | 14,8003 | 14,3376 |
| –3,7226 | –3,8392 | 14,2918 | 13,8577 |

Mw (FA) = 1674 pH = 9.0. Xi = log $[FA_{total}]$, Yi = log$[ZnFA_m]$

$\sum$ Xi = –31,9096; $(\sum Xi)^2$ = 1018,2225; $\sum$Yi = –33,041; $\sum Xi^2$ = 127,537; $\sum$XiY = 132,0797

Y = ax+b, m = a = tga = (nSxiyi-SxiSyi)/(nSxi²-(Sxi)²) = 1.11

So in ZnO(solid) —Zn(II)(solution) —FA—$H_2O$ system at pH = 8.0, dominates zinc fulvate complex, with the structure 1:1. For the calculation of β of zinc fulvate at pH = 8.0 was used Leden function F(L) [17]. The necessary data, for calculation are given in Tables 15.5 and 15.6.

**TABLE 15.5** Experimental Data Necessary for the Calculation of the Conditional Stability Constant by the Leden Method pH = 9.0; Active Associate Mw(FA) = 1674

| Mol/L | | | | F (FA) |
|---|---|---|---|---|
| **[$FA_{total}$]** | **[$Zn_{total}$]** | **[ZnFA]** | **[$FA_{free}$]** | |
| $5.19 \times 10^{-5}$ | $6.08 \times 10^{-5}$ | $3.56 \times 10^{-5}$ | $1.63 \times 10^{-5}$ | $8.68 \times 10^{4}$ |
| $6.75 \times 10^{-5}$ | $6.80 \times 10^{-5}$ | $4.28 \times 10^{-5}$ | $2.47 \times 10^{-5}$ | $6.88 \times 10^{4}$ |
| $8.04 \times 10^{-5}$ | $8.28 \times 10^{-5}$ | $5.76 \times 10^{-5}$ | $2.28 \times 10^{-5}$ | $10.03 \times 10^{4}$ |
| $9.60 \times 10^{-5}$ | $9.60 \times 10^{-5}$ | $7.08 \times 10^{-5}$ | $2.52 \times 10^{-5}$ | $11.15 \times 10^{4}$ |
| $10.90 \times 10^{-5}$ | $10.59 \times 10^{-5}$ | $8.07 \times 10^{-5}$ | $2.83 \times 10^{-5}$ | $11.32 \times 10^{4}$ |
| $13.49 \times 10^{-5}$ | $12.67 \times 10^{-5}$ | $10.15 \times 10^{-5}$ | $3.34 \times 10^{-5}$ | $12.05 \times 10^{4}$ |
| $16.35 \times 10^{-5}$ | $14.86 \times 10^{-5}$ | $12.34 \times 10^{-5}$ | $4.01 \times 10^{-5}$ | $12.22 \times 10^{4}$ |
| $18.94 \times 10^{-5}$ | $17.00 \times 10^{-5}$ | $14.48 \times 10^{-5}$ | $4.46 \times 10^{-5}$ | $12.88 \times 10^{4}$ |

[$Zn_{free}$] = $2.52 \times 10^{-5}$ Mol/L; F(FA) = [ZnFA]/([$Zn_{free}$][$FA_{free}$]); [$FA_{free}$] = [$FA_{total}$] – [ZnFA]; [ZnFA] = [$Zn_{total}$] – [$Zn_{free}$]

$\beta = 5.56 \times 10^{4}$; lg β = 4.74;

**TABLE 15.6** The Calculation of Conditional Stability Constant of Zinc Fulvate Complex by the Least Square Method: Active Associate Mw(FA) = 1674; pH = 9.0; Xi = [$FA_{free}$]; Yi = F (FA)

| Xi | Yi | XiYi | Xi² |
|---|---|---|---|
| $1.63 \times 10^{-5}$ | $8.68 \times 10^{4}$ | 1,4148 | $2{,}6569 \times 10^{-10}$ |
| $2.47 \times 10^{-5}$ | $6.88 \times 10^{4}$ | 1,6994 | $6{,}1009 \times 10^{-10}$ |
| $2.28 \times 10^{-5}$ | $10.03 \times 10^{4}$ | 2,2868 | $5{,}1984 \times 10^{-10}$ |
| $2.52 \times 10^{-5}$ | $11.15 \times 10^{4}$ | 2,898 | $6{,}3504 \times 10^{-10}$ |
| $2.83 \times 10^{-5}$ | $11.32 \times 10^{4}$ | 3,2036 | $8{,}0089 \times 10^{-10}$ |
| $3.34 \times 10^{-5}$ | $12.05 \times 10^{4}$ | 4,0247 | $11{,}1556 \times 10^{-10}$ |
| $4.01 \times 10^{-5}$ | $12.22 \times 10^{4}$ | 4,9002 | $16{,}0801 \times 10^{-10}$ |
| $4.46 \times 10^{-5}$ | $12.88 \times 10^{4}$ | 5,7445 | $19{,}8916 \times 10^{-10}$ |

$$\sum Xi = 23.54 \times 10^{-5};\ (\sum Xi)^2 = 554{,}1316 \times 10^{-10};\ \sum Yi = 85.21 \times 10^{4};$$
$$\sum Xi^2 = 75{,}4428 \times 10^{-10};\ \sum XiYi = 26{,}172$$

$$Y = ax+b,\ b = \beta = (Syi{-}^{a}Sxi)/n = 5.56 \times 10^{4}([Zn(II)_{total}] - [Zn(II)_{free}])/([Zn(II)_{free}][FA_{free}]) = \beta_1+\beta_2[FA_{free}]$$

where; a = (nSxiyi-SxiSyi)/(nSxi²-(Sxi)²).

$$\text{Function } F(L) = F(FA) = ([Zn(II)_{total}] - [Zn(II)_{free}])/([Zn(II)_{free}][FA_{free}]) = \beta_1+\beta_2[FA_{free}] \quad (9)$$

where;

$$[FA_{free}] = [FA_{total}]-[ZnFA] = [FA_{tota}l-[ZnFA] = [FA_{total}] - ([Zn(II)_{total}] - [Zn(II)_{free}]) \quad (10)$$

when; $[FA_{free}]$ aspires to zero, β could be found by the graphical method. The section which is cut on the ordinate by the straight line built in coordinates F(FA)— $[FA_{free}]$ equals to the stability constant. The value of β was calculated by the square method.

$$a = (nSxiyi\text{-}SxiSyi)/(nSxi^2\text{-}(Sxi)^2);\ xi = [FA_{free}] \text{ and } yi = F(FA).$$

$$b = \beta = (xi^2Syi\text{-}SxiSxiSyi)/(nSxi^2\text{-}(Sxi)^2) \quad (11)$$

where;

$$x_i = [FA_{free}] \text{ and } y_i = F(FA) \quad (12)$$

$$\beta = 5.56 \times 10^4$$

## 15.4 CONCLUSION

The complex formation process between Zn(II) and FA isolated from water Paravani lake was studied by the solubility method at pH = 9.0. It was shown that, during the complex formation process, an associate of FA, which Mw at pH = 9 equals to 7610 divides, and every 1/5 part of this associate inculcates into zinc's inner coordination sphere, as an integral ligand. Therefore, it may assume that Mw of the associate of FA, which takes part in the complex formation process, equals 1674. This part of the associate of FA was conventionally called the *active associate.*

Mw of the *active associate* (Mw = 1674) was used for determination $[FA_{free}]$, the composition of zinc(II) fulvate complex and β. It was established, that in the ZnO(solid) – Zn(II)(solution) –FA–$H_2O$ system at pH = 9.0, dominates zinc fulvic complex with the structure 1:1, which $\beta = 5.56 \times 10^4$.

## ACKNOWLEDGMENTS

The work was done by supporting the World Federation of Scientists and the World Laboratory.

## KEYWORDS


- **fulvate complexes**
- **fulvic acids**
- **ionic strength**
- **Leden function**
- **ligand**
- **solubility**


## REFERENCES


1. Rey-Castro, C., Mongin, S., Huidobro, C., David, C., Salvador, J., Garces, J., Galceran, J., Mas, F., & Puy, J., (2009). Effective affinity distribution for the binding of metal ions to a generic fulvic acid in natural waters. *Environ. Sci. Technol., 43*, 7184–7191.
2. Town, R. M., Van, L. H. P., & Buffle, J., (2012). Chemodynamics of soft nanoparticulate complexes: Cu(II) and Ni(II) complexes with fulvic acids and aquatic humic acids. *Environ. Sci. Technol., 46*, 10487–10498.
3. Shizuko, H., (1981). Stability constants for the complexes of transition-metal ions with fulvic and humic acids in sediments measured by gel-filtration. *Talanta, 28*, 809–815.
4. Sasaki, T., Yoshida, H., Kobayashi, T., Takagi, I., & Moriyama, H., (2012). Determination of apparent formation constants of Eu(III) with humic Substances by ion selective liquid membrane electrode. *American Journal of Analytical Chemistry, 3*, 462–469.
5. Ephraim, J., (1992). Heterogeneity as a concept in the interpretation of metal-ion binding by humic substances—the binding of zinc by an aquatic fulvic acid. *Anal. Chim. Acta, 267*, 39–45.
6. Schnitzer, M., & Skinner, S. I. M., (1966). Organo-metallic interactions in soils: 5. Stability constants of Cu++-, Fe++-, and Zn++-fulvic acid complexes. *Soil Sci., 102*, 361–365.
7. Bertoli, A. C., Garcia, J. S., Trevisan, M. G., Ramalho, T. C., & Freitas, M. P., (2016). Interactions fulvate-metal ($Zn^{2+}$, $Cu^{2+}$, and $Fe^{2+}$): Theoretical investigation of thermodynamic, structural, and spectroscopic properties. *Biometals, 29*, 275–285.
8. Kirishima, A., Ohnishi, T., Sato, N., & Tochiyama, O., (2010). Simplified modeling of the complexation of humic substance for equilibrium calculations. *J. Nucl. Sci. Technol., 4*, 71044–1054.
9. Tochiyama, O., Nibbori, Y., Tanaka, K., Yoshino, H., Kubota, T., Kirishima, A., & Setiawan, B., (2004). Modeling of the complex formation of metal ions with humic acids. *Radio Chim. Acta, 92*, 559–565.
10. Varshal, G. M., (1994). *Migration Forms of Fulvic Acids and Metals in Natural Waters: Dissertation.* Vernadsky Institute of Geochemistry and Analytical Chemistry of Russian Academy of Sciences.

11. De Castro, R. T., Da, C. E. E. F., De Alencastro, R. B., & Espınola, A., (2007). Differential complexation between $Zn^{2+}$ and $Cd^{2+}$ with fulvic acid: A computational chemistry study. *Water Air Soil Poll., 183*, 467–472.
12. Wang, J., Lü, C., He, J., & Zhao, B., (2016). Binding characteristics of $Pb^{2+}$ to natural fulvic Acids extracted from the sediments in lake Wuliangsuhai, inner Mongolia plateau, P.R. China. *Environmental Earth Science, 75*, 768–779.
13. Linnik, P. N., Zhezherya, V. A., Linnik, R. P., & Ivanechko, Y. S., (2013). Influence of the component composition of organic matter on relationship between dissolved forms of metals in the surface waters. *Hydrobiological Journal, 49*, 91–108.
14. Makharadze, G., & Makharadze, T., (2014). Method of calculation of stability constants of fulvic complexes on the example of copper. *Journal of Chemistry and Chemical Engineering, 8*, 108–111.
15. Reviaand, R., & Makharadze, G., (1999). Cloud-point preconcentration of fulvic and humic acids. *Talanta, 48*, 409–413.
16. Makharadze, G., Supatashvili, G., & Makharadze, T., (2018). New version of calculation of stability constant of metal-fulvate complexes on the example of zinc fulvate. *International Journal of Environmental Science and Technology, 15*, 2165–2168.
17. Beck, M. T., & Nagypal, I., (1990). *Chemistry of Complex Equilibria.* Chichester, Horwood, Halsted Press, New York.

## CHAPTER 16

# Effects of Dysprosium Addition on the Superconducting Properties of Hg-1223 HTS

I. R. METSKHVARISHVILI,[1,3] T. E. LOBZHANIDZE,[2] G. N. DGEBUADZE,[1] B. G. BENDELIANI,[1] M. R. METSKHVARISHVILI,[3] and V. M. GABUNIA[1,4]

[1]*Ilia Vekua Sukhumi Institute of Physics and Technology, Department of Cryogenic Technique and Technologies, Tbilisi–0186, Georgia*

[2]*Ivane Javakhishvili Tbilisi State University, Department of Chemistry, Faculty of Exact and Natural Sciences, 0179 Tbilisi, Georgia*

[3]*Georgian Technical University, Faculty of Informatics and Control Systems, 0175 Tbilisi, Georgia, E-mail: i.metskhvarishvili@gtu.ge*

[4]*Petre Melikishvili Institute of Physical and Organic Chemistry of the Iv. Javakhishvili Tbilisi State University, Jikia Str., 5, Tbilisi–0186, Georgia*

### ABSTRACT

The influence of dysprosium (III) oxides on the physical and chemical properties of Hg-1223 material has been examined. Dysprosium-free Hg-1223 and dysprosium doped $HgBa_2Ca_2Cu_3Dy_xO_{8+\delta}$ (x = 0.0–1.0 wt.%) superconductors were synthesized by sealed quartz tube technique. Our results showed that the presence of dysprosium oxide made the system more reactive and enhanced the kinetics of the reaction, as well as the promotion of the high-Tc phase and enhances the transport critical current densities $J_c$.

### 16.1 INTRODUCTION

Preparations of Hg-based cuprate superconductors are labor-consuming, expensive, and multi-stage [1, 2]. Particularly, these samples are prepared in closed quartz ampoules [3, 4] or by high-pressure methods [5, 6]. Such as the other high-temperature superconductors, the Hg-based superconductors also have the problems of the weak links. Additionally, the worsening of the weak links is related with the excitation of $O_2$ gas and Hg steam. Particularly, the HgO dismisses on $O_2$ gas and Hg steam in the process of syntheses and sintering. The second problem is that the synthesis of the Hg-1223 phase in a pure state is possible at high pressures, but at low pressures is difficult, and it is feasible only in case when mercury oxide will be substituted by high valency atoms partly. In this case, the technique of the sealed quartz tube is useful for receiving of high purity high-temperature phase. Additives or substituent of the Hg-based superconductors by the appropriate metals or oxides can improve the nature of the grain boundaries, as well as promote the formation of the superconducting phase or introduce the effective pinning centers. Several scientific groups have studied the effect of the partial substitution of Hg by Sb ($Hg_{1-x}Sb_x$) [7–10] in Hg-1223. They have founded that a small substitution of Hg by Sb improves the intergrain critical current density in Hg-1223. Besides, partial substitution of Hg by Pb ($Hg_{1-x}Pb_x$) expedites the formation of nearly single-phase of Hg-1223 compound and enhances the irreversible magnetic fields [11–13]. Li et al. [14] have studied and optimized the influence of the synthesis and processing parameters on the phase purity and grain growth of Pb-doped $HgBa_2Ca_2Cu_3O_{8+\delta}$ cuprate superconductor. Goto [15] showed that in the case of F-doped $HgBa_2Ca_2Cu_3Re_{0.2}O_y$ prepared by the diffusion method, the maximal significance of critical current density $J$ can reach $J_c \approx 104$ A/cm$^2$ at 77 K in zero fields. Partial substitution of Hg by Re greatly enhances the stability of the Hg-1223 [16–18]. The substitution of Hg by Pb and zBa by Sr showed essentially improved flux pinning properties in high magnetic fields [19]. In works [20, 21] was found that partial substitution by Bi allows the synthesis of Hg-1223 at lower temperatures under ambient atmosphere. The effects of Sn doping in $Hg_{1-x}Sn_xBa_2Ca_2Cu_3O_{8+\delta}$ was investigated in ref. [22], they showed that Sn doping is favorable for synthesizing the Hg-1223 superconductor. As we see, nowadays, additives or substitutions by the appropriate metals or oxides on the Hg-1223 phase, are very important subjects in order to improvements the physical and chemical properties of these materials.

For this aim, the subject of this chapter is to investigate the effects of doping of Dysprosium (III) oxides on the physical and chemical properties of Hg-1223 material. Samples with composition $HgBa_2Ca_2Cu_3Dy_xO_{8+\delta}$ (x = 0.0–1.00 wt.%) were prepared by the sealed quartz tube technique. The prepared samples were characterized by x-ray diffraction (XRD); physical properties were investigated by ac susceptibility and high harmonic. As far as we know, the influence of

$Dy_2O_3$ oxides on superconductivity properties of Hg-1223 high-temperature superconductors has still not been studied.

## 16.2 EXPERIMENTAL PART

For synthesis dysprosium free and dysprosium doped $HgBa_2Ca_2Cu_3Dy_xO_{8+\delta}$ superconductors, we used the two-step method. In the first stage was synthesized dysprosium-free and dysprosium doped $Ba_2Ca_2Cu_3Dy_xO_y$ (x = 0.0–1.0 wt.%) precursors before proceeding to the second stage where HgO was added to the precursor before final sintering. For starting, materials were utilized powders materials $BaCO_3$ (99.0% Oxford Chem Serve), $CaCO_3$ (99.98% Oxford Chem Serve), CuO (99.999% Sigma-Aldrich), $Dy_2O_3$ (99.99% Sigma-Aldrich), and $Hg_2O_3$ (99.99% Sigma-Aldrich).

As we noted, in the first phase was synthesized separately dysprosium-free and dysprosium doped precursors in the stoichiometric ratio Ba:Ca:Cu: = 2:2:3:X where X is concentrations of $Dy_2O_3$ in the range of x = 0.000, x = 0.0025 x = 0.005, and x = 0.01 identified as 0.0, 0.25, 0.5, and 1.0 wt.%. Powders were mixed in this stoichiometric ratio (4 stoichiometric ratios) and then independently of each other were mixed and ground carefully in an agate mortar. The resulting powders mixture was calcined in an alumina crucible in the air with four intermediate grindings at 900°C for 60 h. The resulting powders were ground and pressed into the pellets by a hydraulic press with about 500 MPa. The obtained pellet is annealed in tube type furnace at 900°C in flowing oxygen partial pressure of 0.3 bar for 8 h.

In the second step, the precursor was mixed with HgO according to the composition Hg:Ba:Ca:Cu:Dy = 1:2:2:3:X and after the final grinding, the powder was pressed into a disc-shaped tablet with a diameter of 6 mm and a thickness of 4 mm, by using a hydraulic press under a pressure of 400 MPa. The pellets were put separately in a quartz tube and from quartz tube was evacuated up to $10^{-3}$ Torr and sealed. Thereafter, a quartz tube was inserted into a programmed muffle furnace. The temperature of the furnace was raised at a rate of 300°C/h up to 700°C and thereafter at a rate of 120°C/h up to 860°C and held at this temperature for 15 h. The furnace was cooled at the rate of 60°C/h to room temperature. Finally, the samples were oxygenated in tube type furnace at 300°C in flowing oxygen for 20 h.

The prepared patterns were characterized by XRD (Dron-3M) with CuK$\alpha$ radiation. The phase method was used to study the real parts of the linear susceptibility [23]. The errors in the determination of $\chi'$ at higher frequencies than 1 kHz does not exceed 1%. For the measurements of intergranular critical current densities, we used the method of high harmonics [24].

## 16.3 RESULTS AND DISCUSSION

Figure 16.1 shows the XRD patterns of the un-doped and doped samples with different addition of dysprosium (III) oxides. The results show that the Hg-1223 phase has a tetragonal structure, and it is good to agree with PDF faille with the lattice parameters $a$ = 0.2858 nm and $c$ = 1.524. From Figure 16.1(a), we can see that the diffraction pattern of the un-doped sample (X = 0) consists of impurity $HgCaO_2$ and un-reacted $BaCuO_2$ precursor phases. According to our estimate from this XRD data, the un-doped sample containing about 62% of the superconducting phase of Hg-1223. One will notice that the low $Dy_2O_3$ concentration sample (0.25 wt.%) was dominated by the Hg-1223 phase. It can be seen that a sample with a low concentration of $Dy_2O_3$ (0.25 wt.%) impurity phases also exists, but their volume fractions decrease, and the superconducting phase of Hg-1223 increases to 78–80% in the sample. However, in higher doping concentration (0.5 wt.% and 1.0 wt.%), the impurity phase increases and the superconducting phase decrease, we also must note that in 1.0 wt.% concentration of $Dy_2O_3$ the Hg-1212 phase appeared.

Figure 16.2 presented the temperature dependences of the real ($-4\pi\chi'$) part of *ac* susceptibility for un-doped and doped samples. Measurement was carryout in zero magnetic fields ($H$ = 0) at $h$ = 1 Oe and $f$ = 20 kHz. The diamagnetic onset temperature of the superconducting transition for 0.0 wt.% sample is about $T_c \approx 133$K and $T_c$ small different for Dysprosium concentration of 0.25 wt.% and 0.50 wt.% samples. As we see, un-doped (0.00 wt.%) and high doped (0.50 wt.%) samples clearly show a two-step decrease with $T$, which reflects the flux shielding from and between the grains and full screening of applied ac magnetic fields for samples observed at T~ 110 K and Tc≈97 K, respectively. Unlike this, low doped (0.25 wt.%) do not show two steps, and full diamagnetism occurs at T~ 114 K. The absence of two-step behavior could be explained in terms of smaller grain size in the 0.25 wt.% doping sample, so that grains are fully penetrated at a lower field value.

Figure 16.3 presents dependence of the transport critical current densities on temperature $j_c$ value measured by high harmonics method $\chi_3(T)$ [24], at $h$ = 1 Oe and $H$ = 0. As we see the largest value of critical current densities observed in 0.25 wt.% doping sample (220 $A/cm^2$), whereas for $x$ = 0.00 and $x$ = 0.50 samples are at 87 $A/cm^2$ and 24 $A/cm^2$, respectively.

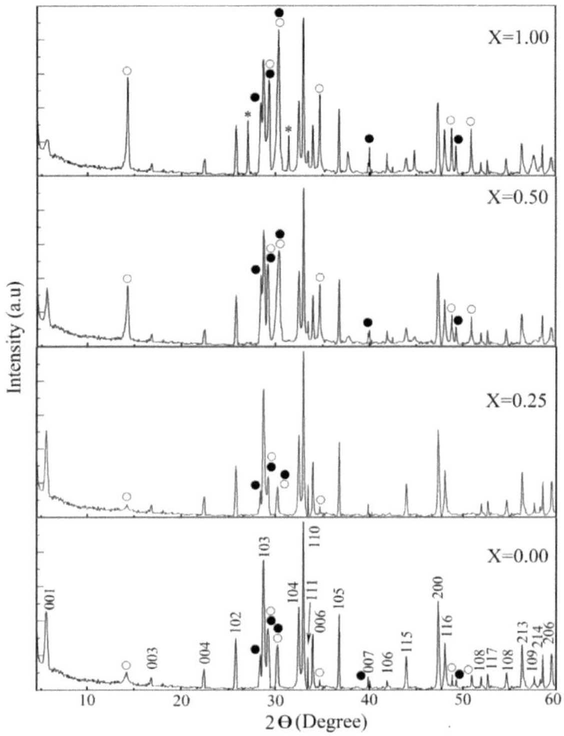


**FIGURE 16.1** XRD patterns of the un-doped and doped samples $HgBa_2Ca_2Cu_3Dy_xO_{8+\delta}$, with $x = 0.00$, $x = 0.25$, $x = 0.50$ and $x = 1.00$. o-for $HgCaO_2$ phase, •-$BaCuO_2$ phase and *-Hg-1212 phase.

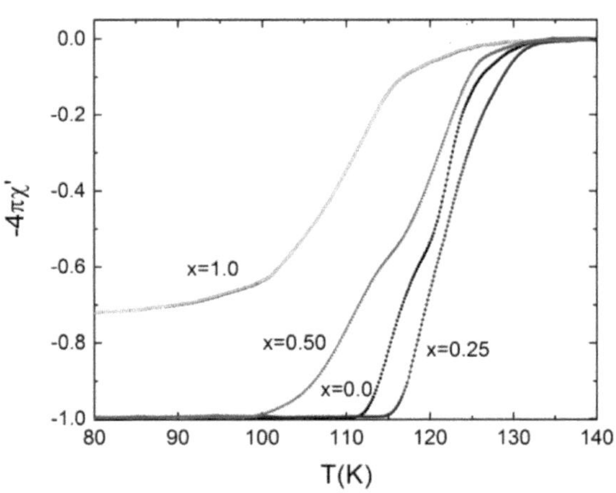


**FIGURE 16.2** Temperature dependences of the real χ' parts of ac susceptibility for un-doped and doped samples.

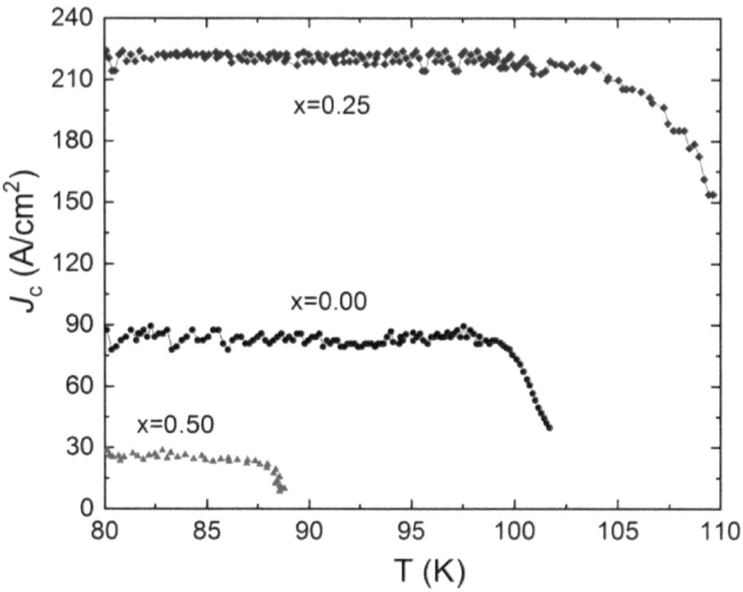


**FIGURE 16.3** Dependence of the critical current densities $J_c(T)$ on temperature for un-doped and doped samples.

## 16.4 CONCLUSIONS

The doping effect of $Dy_2O_3$ as an additive and not as a substituent of any elements from $HgBa_2Ca_2Cu_3O_{8+\delta}$ this set has been studied. For the sintering investigated samples, we used the two-step method. In the first stage were synthesized dysprosium-free and dysprosium doped precursors and the second stage where HgO was added and prepared $HgBa_2Ca_2Cu_3Dy_xO_{8+\delta}$ superconducting materials. X-ray results show that the un-doped sample contains about 62% of the superconducting phase of Hg-1223, whereas at a low concentration of dysprosium (0.25 wt.%), the superconducting phase sample constitutes 78–80%. Temperature dependences of $\chi'(T)$ show a two-step process for un-doped and high doped (0.5 wt.%) samples. It is important to note that the low-doped sample does not show two steps. The absence of two-step behavior could be explained in terms of smaller grain size in the 0.25 wt.% doping samples, so that grains are fully penetrated at a lower field value. We found that in the low-level doped samples, dysprosium enhances the value of the transport critical current densities from 87 A/cm² (0.00 wt.%) to 220 A/cm² (0.25 wt.%). As a result, we could conclude that only a very small amount of Dy atoms can be doped into the Hg-1223 superconducting phase structure.

## ACKNOWLEDGMENTS

This work was supported by Shota Rustaveli National Science Foundation (SRNSF), Grant number: 217524, and project title: Influence of the Polymerization and Various Dopants on the Hg-1223 Superconductive Properties.

## KEYWORDS

- **AC susceptibility**
- **critical currents**
- **dysprosium (III) oxides**
- **Hg-1223**
- **high harmonic**
- **sealed quartz tube technique**

## REFERENCES

1. Putilin, S. N., Antipov, E. V., Chmaissem, O., & Marezio, M., (1993). *Nature, 362*, 226.
2. Schilling, A., Cantoni, M., Guo, J. D., & Ott, H. R., (1993). *Nature, 363*, 56.
3. Cunha, A. G., Orlando, M. T. D., Emmerich, F. G., & Baggio-Saitovitch, E., (2000). *Physica C, 341–348,* 2469.
4. Bastidas, D. M., Pinol, S., Plain, J., Puig, T., Obradors, X., Celotti, G., Sprio, S., & Tampieri, A., (2002). *Physica C, 372–376*, 1171.
5. Mendonca, T. M., Tavares, P. B., Correia, J. G., Lopes, A. M. L., Darie, C., & Araujo, J. P., (2011). *Physica, C, 471*, 1643.
6. Brylewski, T., Przybylski, K., Morawski, A., Gajda, D., Cetner, T., & Chmis, J., (2016). *Journal of Advanced Ceramics, 5*, 185.
7. Li, J. Q., Lam, C. C., Hung, K. C., & Shen, L. J., (1998). *Physica C, 304*, 133.
8. Li, J. Q., Lam, C. C., Peacock, G. B., Hyatt, N. C., Gameson, I., Edwards, P. P., Shields, T. C., & Abell, J. S., (2000). *Supercond. Sci. Technol., 13*, 169.
9. Jasim, K. A., Alwan, T. J., Al-Lamy, H. K., & Mansour, H. L., (2011). *J. Supercond. Nov Magn., 24*, 1963.
10. Jasim, K. A., (2012). *J. Supercond. Nov. Magn., 25*, 1713.

11. Isawa, K., Yamamoto, A. T., Itoh, M., Adachi, S., & Yamauchi, H., (1993). *Physica C, 217*, 11.
12. Isawa, K., Machi, T., Yamamoto, A. T., Adachi, S., Murakami, M., & Yamauchi, H., (1994). *Appl. Phys. Lett., 65*, 2105.
13. Sao, H. M., Lam, C. C., Fung, P. C. W., Wu, X. S., Du, J. H., Shen, G. J., Chow, J. C. L., et al., (1995). *Physica C, 246*, 207.
14. Li, Y., Sastry, P. V. P. S. S., Knoll, D. C., Perterson, S. C., & Schwartz, J., (1999). *IEEE Trans. Appl. Supercon., 9*, 1767.
15. Goto, T., (1997). *Physica C, 282–287*, 891.
16. Sin, A., Odier, P., & Nuez-Regueiro, M., (2000). *Physica C, 330*, 9.
17. Akune, T., Yamada, N., Sakamoto, N., & Matsumoto, Y., (2003). *Physica C, 392–396*, 386.
18. Eleuterio, F. H. S., Amorim, L. S., Belich, H., Orlando, M. T. D., Pasoso, C. A. C., & Epsinoza, O. J. S., (2014). *J. Supercond. Nov. Magn., 27*, 2979.
19. Lee, S., Kiryakov, N. P., Emelyanov, D. A., Kuznetsov, M. S., Tretyakov, Y. D., Petrykin, V. V., et al., (1998). *Physica C, 305*, 57.
20. Michel, C., Hervieu, M., Maignan, A., Pelloquin, D., Badri, V., & Raveau, B., (1995). *Physica. C., 241*, 1.
21. Sastry, P. V. P. S. S., Amm, K. M., Knoll, D. C., Peterson, S. C., & Schwartz, J., (1998). *Advances in Cryogenic Engineering, 44*, 477.
22. Li, J. Q., Lam, C. C., Feng, J., & Hung, K. C., (1997). *Physica C, 292*, 295.
23. Metskhvarishvili, I. R., Dgebuadze, G. N., Bendeliani, B. G., Metskhvarishvili, M. R., Lobzhanidze, T. E., & Mumladze, G. N., (2013). *J. Low Temp. Phys., 170*, 68–74.
24. Metskhvarishvili, I. R., Dgebuadze, G. N., Bendeliani, B. G., Metskhvarishvili, M. R., Lobzhanidze, T. E., & Gugulashvili, L. T., (2015). *J. Supercond. Nov. Magn., 28*, 1491–1494.

CHAPTER 17

# Hot Shock Wave Fabrication of Nanostructured Superconductive $MgB_2$ and $MgB_2$-Fe Composites

AKAKI PEIKRISHVILI,[1] GIORGI TAVADZE,[1] BAGRAT GODIBADZE,[2] GRIGOR MAMNIASHVILI,[3] and ALEXANDER SHENGELAYA[3]

[1]*F. Tavadze Institute of Metallurgy and Materials Science, 10 E. Mindeli St. Tbilisi–0186, Georgia, E-mail: akaki.peikrishvili@yahoo.com (A. Peikrishvili)*

[2]*Iv. Javakhishvili Tbilisi State University, I. Chavchavadze Ave., 1, 0179 Tbilisi, Georgia*

[3]*G. Tsulukidze Mining Institute, 7 E. Mindeli St. Tbilisi–0186, Georgia*

## ABSTRACT

In order to consolidate $MgB_2$ and nanostructured $MgB_2$-F superconductive cylindrical billets near to theoretical density with good integrity of consolidated particles, the original hot explosive consolidation method with the intensity of loading up to 10 GPa was developed. As investigation showed, the application of high temperatures and liquid phase consolidation of Mg-2B and Mg-2B-Fe blend powders above the 900°C provides initiation of a chemical reaction between the Mg and B particles behind and on the shock wave front with the formation of superconductive $MgB_2$ phase. The fabricated billets of pure $MgB_2$ composites are characterized with good integrity and high value of critical temperature with maximal value of 38.5 K. It was found the superconducting characteristics of fabricated $MgB_2$ billets depend not only on the consolidation conditions (temperature and intensity of loading), but from the type of boron precursors and their purity too.

In the case of nanostructured Mg-2B-Fe composites, the high dense billets from $MgB_2$ composites with uniform distribution of nanoscale particles of Fe in whole volume of consolidated billets were obtained. The partial formation of FeB phases in HEC samples were observed too. There were established that the value of critical temperature, magnetic characteristics, and density of consolidated billets depends on the value of temperature, purity, and type of precursors too.

## 17.1 INTRODUCTION

The superconductive properties of $MgB_2$ with C32 structure and critical transformation temperature of $T_c$ = 39 K was discovered in 2001 [1]. Since then, intensive investigation and development of different types of $MgB_2$ superconductive materials in various forms and efforts to increase their $T_c$ above 39 K takes place worldwide [2–5]. The technology of development superconductive materials belongs to traditional powder metallurgy: preparing and densification Mg and B powder blends in static conditions with their further sintering processes [6, 7]. Results described in Ref. [8], where Mg-2B blend powders were first compacted in cylindrical pellets and were after loaded in hot conditions at 2 GPa pressure, also seem interesting. The observation of the clear correlation between the syntheses condition and crystal structure of formed two phases $MgB_2$-MgO composites as well as between their superconductive properties allowed the conclusion that redistribution of oxygen in the $MgB_2$ matrix structure and formation of MgO phase may be considered as a positive effect.

Existing data of the application of shock wave consolidation technology for fabrication of high dense $MgB_2$ billets with higher $T_c$ temperature practically gave the same results, and limit of $T_c$ = 40 K is still considered maximal. Additional sintering processes after the shock wave compression are highly recommended for providing full transformation of consolidating blend phases into the $MgB_2$ composites.

The goals of the current investigation are:

- The development of high dense superconductive billets based on $MgB_2$ using the hot shock wave consolidation technique without any further sintering processes.
- The investigation of the role of temperature in the process of consolidation and sintering of $MgB_2$.
- To investigate the role of nanoscale Fe additives on the superconductive and magnetic characteristics of $MgB_2$ and to fabricate $MgB_2$-Fe nanostructured composites.
- The evaluation and investigation of structure property relationship.

## 17.2 EXPERIMENTAL PROCEDURES

The novelty of the proposed nonconventional approach relies on the fact that the consolidation of the coarse (under 10–15 μ) Mg-2B blend powders was performed in two stages [9]. The explosive pre-densification of the powders was made at room temperatures. In some cases, before dynamic pre-densification, the loading of precursors into the containers were performed by static means. In all cases, the second stage was done by the hot shock wave compaction but at temperatures under 1000°C with an intensity of loading around 5 GPa. The cylindrical geometry of loading was used in all experiments. At the first stage, the Mg-2B powders were placed inside a copper-tube container. The container was sealed at both ends with threaded steel plugs. A concentric cardboard box was filled with the powdered explosive materials and was placed around the cylindrical sample container (Figure 17.1).

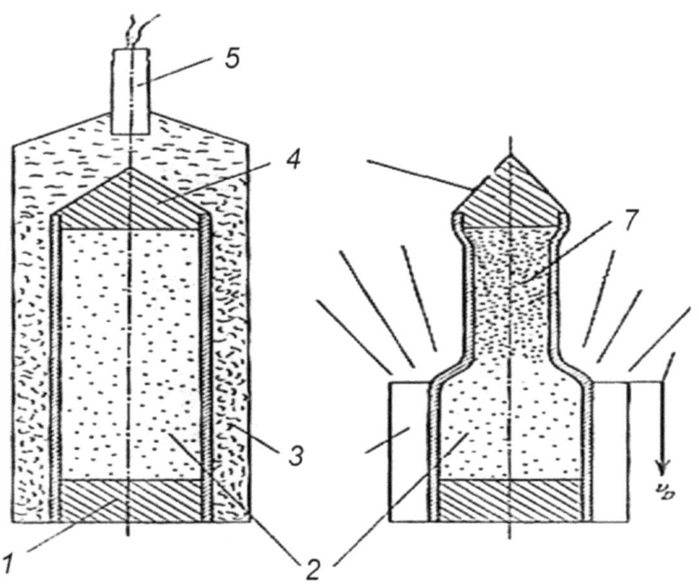


**FIGURE 17.1** The procedure of preliminary densification of Mg-2B blend powders: (1) bottom plug of copper tube; (2) precursor powders; (3) explosive powder; (4) upper plug of steel tube; (5) electric detonator; (6) products of detonation; (7) consolidated powders.

The key operational component of the hot shock wave experiments with vertical configuration of explosive charge that allows to consolidate-syntheses of all type of powders at elevated temperatures is presented on

Figure 17.2. The application of vertical configuration of the explosive charge allows us to increase dimensions and mass of explosive charges without limitations. As a result, the pulse duration on the shock wavefront during the loading increases providing of obtaining billets with higher densities. On the other hand increasing of pulse duration in some cases will allows to decrease of consolidating temperature and the compressing of samples at lower temperatures and as a result the cost reduce of obtained billets too.

The hot shock wave consolidation device (Figure 17.2) consists of three main parts: the heating system (a cylindrical heating furnace), the cylindrical feeding system and the explosive charge set-up. The preliminary pre-densified cylindrical billet (1) is located in the central hole of heating furnace (4). The heating billet is fixed in the furnace by an opening and closing mechanism (6). After reaching the necessary temperature the opening (6) sheet opens the furnace and billet moves through the feeding cylindrical system (9–11) to the explosive charge set-up (17). After receiving the signal that the billet passed through the feeding system and is located in final position (13), the detonation takes place and explosive compression of heated billets occurs. Determined by the volume, type, and density of the sample composition, the heating lasts about 60 min. At the reaching of the desired temperature the furnace is switched off by remotely and feeding mechanism opens. Billets pass through the feeding tube inside of cylindrical charge. As soon as the billet reaches the bottom of explosive charge in requested position, the detonation circuit switches on automatically and the explosive is detonated. The corresponding pressure at the wall of the steel container is around 5 GPa. Figure 17.3 represents the billets after the first stage of pre-densification at room temperatures.

(1) Consolidating powder material; (2) cylindrical steel container; (3) steel container plugs; (4) furnace heating wires; (5) furnace opening and closing; (6) furnace opening sheet; (7) furnace closing sheet; (8) basic construction of HSWC device; (9) feeding steel tube for samples (10) movement tube for heated container; (11) connecting tube from rub; (12) accessory for the fixing of explosive charge; (13) circle fixing passing of steel container; (14) detonator; (15) detonating cord; (16) flying tube for HEC; (17) explosive charge; (18) lowest level of steel container; (19) bottom fixing and stopping steel container; (20) Sand.

## 17.3 MATERIALS

Superconductive Mg-2B and Magnetic nano-composite Mg-2B-1 wt.% Fe ($MgB_2$:Fe1%) were consolidated above 900°C using coarse Mg and B

(5–10 μ) and nano-Fe powder with a particle size of 25 nm [iron powder (carbon coated, av. size 25 nm, Sun Innovations Inc, USA]. Its magnetic and electric properties were studied by using vibrating sample magnetometer (VSM) (Cryogenic Limited, UK). The temperature dependence of magnetization in a zero magnetic field cooling (ZFC) mode was measured in magnetic field B = 20 G and resistance by 4-contact method, correspondingly.

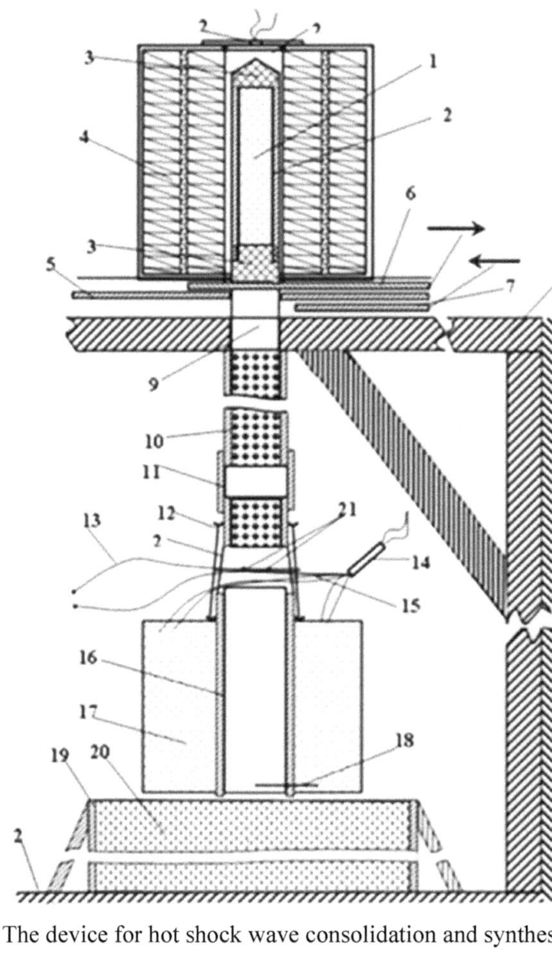

**FIGURE 17.2** The device for hot shock wave consolidation and syntheses of composites.

As explosive materials, there were applied Ammonite with detonation velocity 3.1 km/sec and its mixture with sultpaters providing consolidation/syntheses of precursors near to 5 GPa intensity of loading.

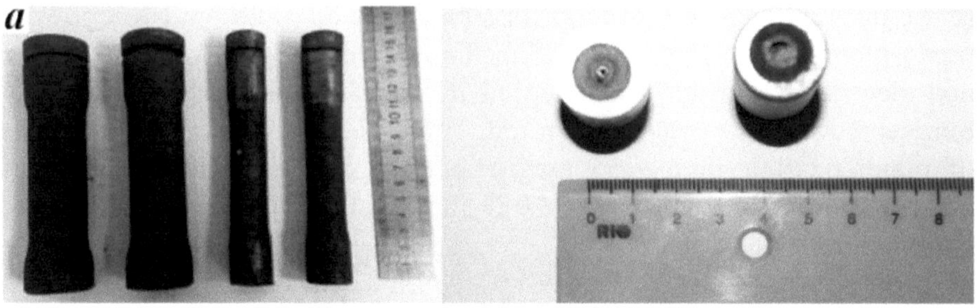


(a) (b)

**FIGURE 17.3** The billets from Mg-2B blend powders after first stage consolidation at room temperature with intensity of loading under 10 GPa: (a) general view; (b) cross-section of after densification: overloading (left) and shortages of loading (compression).

## 17.4 RESULTS

The cylindrical tubes from steel and copper were used in order to consolidate high dense superconductive $MgB_2$ billets. As further investigation showed, the high temperature consolidation of Mg-2B precursors in steel containers have positive effect and allows to fabricate of two phase $MgB_2$-MgO near to theoretical density with critical temperatures around $T_c$ = 38 K. In contrast to steel containers, the shock wave fabrication of Mg-2B powder blends in Cu containers leads to formation of undesirable phases such as $MgCu_2$. This occurs because of the diffusion of copper atoms under the shock wave front towards of center causing further chemical reactions with Mg and the formation of $MgCu_2$. The reaction was so intense and exothermic that the surface between the container's wall and Mg-2B precursors fully melted. The effect connected with the formation of $MgCu_2$ behind the shock wavefront is described in Ref. [3]. In order to prevent the movement of Cu atoms by the shock wavefront, all experiments were implemented using only cylindrical containers from steel.

Figure 17.4 represents the different sections of microstructures of two-stage shock wave consolidated $MgB_2$ samples obtained at 940°C temperature in a steel container. As it's seen from microstructures, there are observed the traces of oxidation that can be explained by its existence in precursors before consolidation and maintained after hot shock wave fabrication too. As for the total, the synthesized composites have high density and have good integrity between the consisting grains.

In order to reduce the existence of oxides in shock fabricated $MgB_2$ samples, there was an applied washing procedure of boron by standard method before preparing Mg-2B powder blends. The application of boron precursors after washing provided obtaining of $MgB_2$ samples after hot shock wave consolidation without traces of oxides and high value of critical temperature ($T_c$) too.

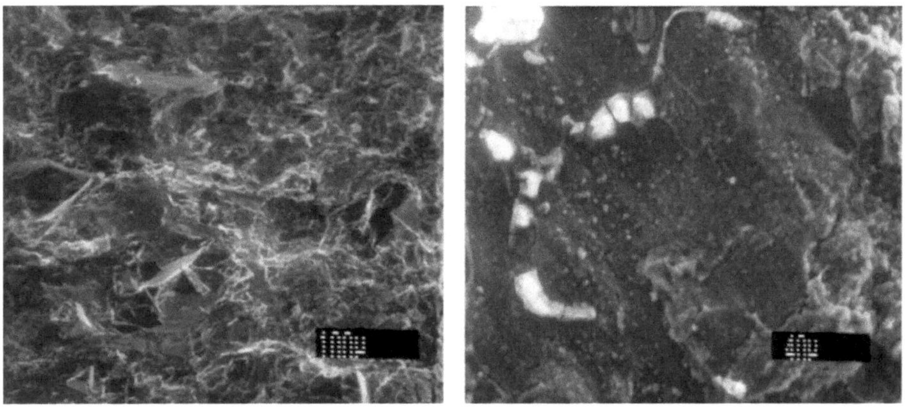

(a) (b)

**FIGURE 17.4** The microstructures of consolidated and synthesized $MgB_2$ after two-stage shock wave loading at 940°C with intensity around of 5GP. (a) The central part; (b) traces of oxidation.

Figure 17.5 represents the microstructure and value of the critical temperature of $MgB_2$ superconductive composites obtained in steel container after washing of amorphous boron (B) powder and synthesized by two-stage hot shock wave consolidation processing at 940°C temperature and intensity of loading under 5GPa.

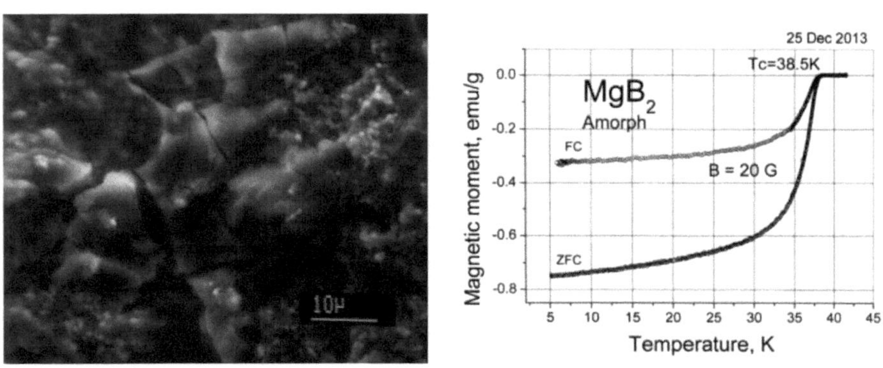


(a) (b)

**FIGURE 17.5** The microstructures of $MgB_2$ after washing of B and two-stage shock loading at 940°C with intensity around of 5GP.

As it's seen from microstructures (Figure 17.5a), there no traces of oxidation are observed. The structure has a high density and shows good integrity between the consisting of grains. The traces of melting-crystallization are observed too. The measuring of critical temperature (Figure 17.5b) shows a high value of $T_c$ too and is equal to 38.5 K.

The experiments of hot shock wave consolidation/syntheses of Mg-2B based blend composites in steel containers were performed below and above the melting point of Mg at temperatures 500, 700, 950, and 1000°C with loading intensity around 5 GPa. It was established that the comparatively low-temperature consolidations at 500°C and 700°C give no results and obtained compacts have no superconducting properties. Only loading by shock waves above the 900°C creates favorable conditions and allows to obtain cylindrical samples near to theoretical densities with a high value of critical temperature under 40 K. The offered approach allows also producing the multilayer cylindrical tubes (pipes) with the steel/$MgB_2$/steel Cu-/$MgB_2$/Cu structure which could find important applications for the production of superconducting cables for simultaneous transport of hydrogen and electrical power in hybrid $MgB_2$-based electric power transmission lines filled with liquid hydrogen [10].

It was found earlier that doping of $MgB_2$ by C, Li, Be, Zn, Al, Ti, Ni, Fe, Co, Mn, and Si leads only to a decrease of $T_c$. The positive effect was found only for the doping of $MgB_2$ with heavier alkali and alkali-earth metals, which are capable of strong carrier donation to electron system and therefore essentially enhanced superconductivity properties of $MgB_2$ up to 45–58 K [13].

In order to improve the magnetic moment of fabricated superconductive compounds, there was carried out a comparative study of effects caused by added to Mg+2B mixture magnetic nanoparticles on the magnetic and electric properties of shock wave consolidated composites. These particles could introduce multiple pinning centers resulting in increased flux pinning and critical current density. The nanoscale Fe with a particle size of 25 nm was added to Mg+2B mixture precursors and were consolidated by 2 stage shock wave fabrication method described above [11, 12]. The results of structural and magnetic investigations are presented in Figures 17.6–17.8.

Figure 17.6 represents the results of SEM investigation of nanostructured $MgB_2$-1%Fe composited consolidated/synthesized at 940°C temperature with the intensity of shock wave loading around 5 GPa. Figure 17.7 represents the diffraction picture of nanostructured $MgB_2$-1%Fe composited consolidated/synthesized at 940°C temperature with intensity of shock wave loading around 5 GPa.

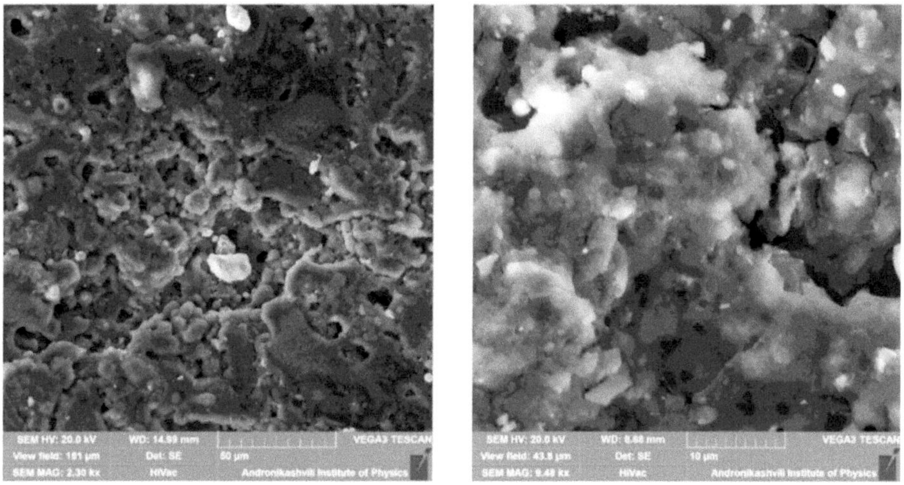

(a) (b)

**FIGURE 17.6** The microstructures of $MgB_2$-1%Fe composites were obtained after two-stage hot shock wave loading at 940°C with an intensity of loading around 5 GPa. The structure has a high density and shows good integrity between the consisting grains. The traces of melting-crystallization are observed too.

As it's seen from diffraction picture of $MgB_2$-1%Fe composite after consolidating/syntheses the existence of magnesium oxides and increased volume of Fe in whole volume are observed. The formation of magnesium oxides takes place due to existence of oxygen in preliminary densified billets before the second stage hot shock wave loading. As for increased amount of Fe, it can be easily explained by the existence in diffraction picture traces of steel containers or by partial diffusion of Fe from steel containers wall towards of edges of fabricated samples. After that we can be sure that after the chemical reactions under the shock wave front full transformation of starting elements into the two-phase composition from $MgB_2$ and MgO occurs. The observation of eutectic colonies on the microstructure confirms the fact of melting/crystallization processes behind the shock wave front.

In order to evaluate the superconductive characteristics of obtained billets the magnetic moment temperature dependence in zero-field-cooled (ZFC) and field-cooled (FC) modes depending on experimental conditions and type of boron precursors were investigated. As reported in Refs. [9, 12, 13], the application of low temperatures up to 900°C and HSWC of Mg-2B precursors in steel containers did not give results. In spite of the high density and uniform distribution of phases, they did not obtain superconductive characteristics. The investigation of HSWC processes for Mg-2B precursors

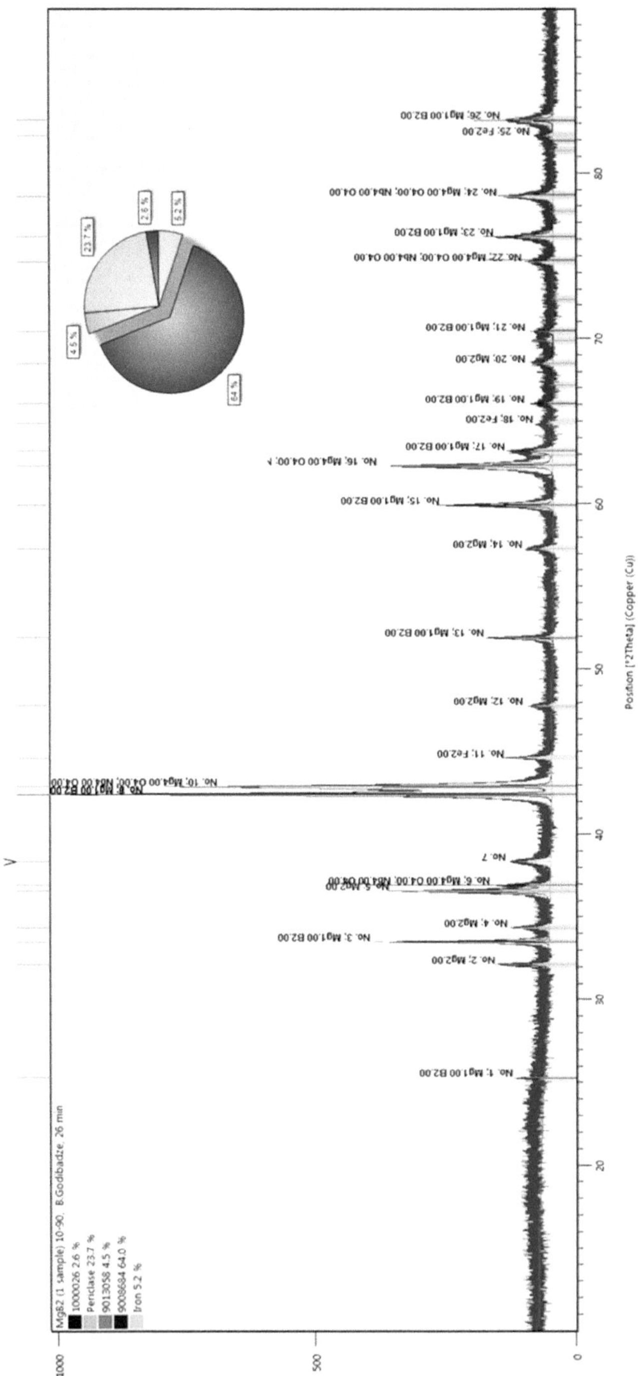


**FIGURE 17.7** The diffraction picture of $MgB_2$-1%Fe composites obtained after two-stage hot shock wave loading at 940°C. The existence of magnesium oxides and increased volume of Fe in whole volume are observed.

in copper containers gave the same results and no superconductive characteristics below 900°C.

Figure 17.8 represents the results of magnetic and electrical characteristics of nanostructured $MgB_2$-1%Fe composited consolidated/synthesized at 940°C temperature with the intensity of shock wave loading around 5 GPa.

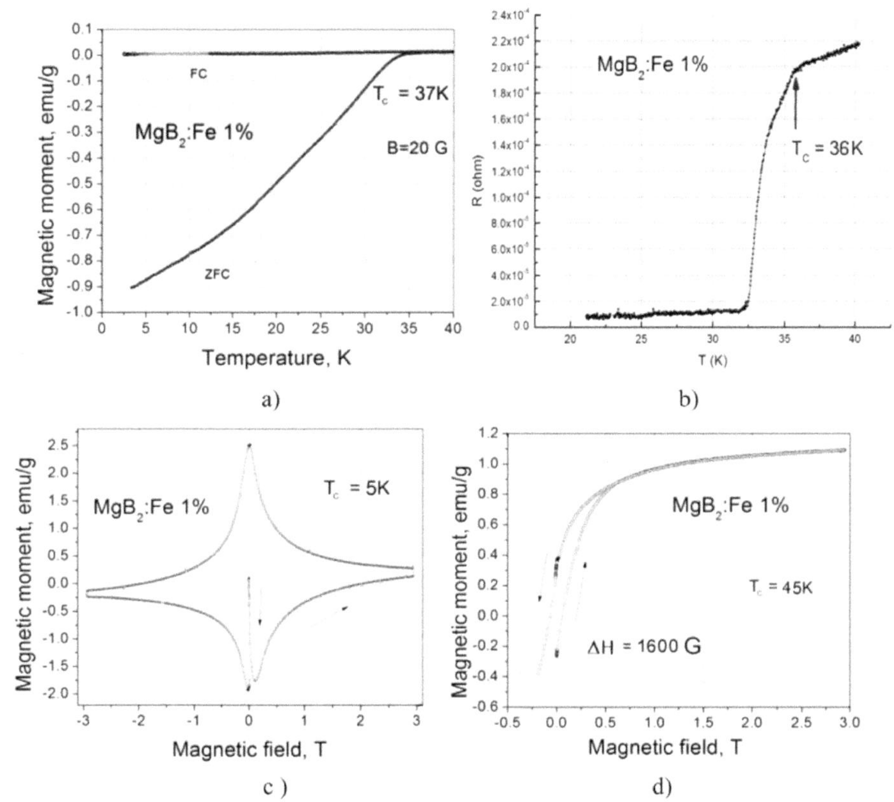


**FIGURE 17.8** The results of investigation magnetic characteristics for $MgB_2$-1%Fe composites fabricated at 940°C temperature with the intensity of shock wave loading around 5 GPa: (a) temperature dependence of magnetization in ZFC and FC modes under a magnetic field of 20 G; (b) temperature dependence of resistance; (c) magnetization hysteresis loop at 5 K; (d) magnetization hysteresis loop at 20 K.

## 17.5 DISCUSSION

The hot shock wave consolidation-syntheses of Mg-2B composite powders were performed in steel containers below and above of Mg melting point. In order to determine the role of temperature, the consolidations were carried out at 500, 700, and 940°C. At 500°C and 700°C, the consolidation gives no results and obtained compacts have no superconductive characteristics. The application of higher temperatures provides the formation of $MgB_2$ composition in the whole volume of fabricated billets with a maximal value of $T_c$ = 38.5 K without any post sintering process. This confirms the important role of temperature in the formation of superconductive $MgB_2$ and corresponds with the literature data where only after sintering processes above 900°C the formation of $MgB_2$ phase with $T_c$ = 40 K takes place. The difference of $T_c$ between the HEC and sintered $MgB_2$ composites may be explained due to unreacted Mg and B phases or due to the existence of oxides in the starting materials. This could be checked by increasing of consolidation-syntheses temperature or by the application of further sintering processes. The careful selection of initial Mg and B phases is important too, and in the case of consolidation Mg-2B precursors with corrections mentioned above, the chance to increase critical temperature ($T_c$) of shock wave synthesized samples increases essentially. The next experimental stage is the fabrication of $MgB_2$ superconductive materials.

As it is seen from the results of magnetic and electric measurements, $MgB_2$-1% Fe composites retain its superconducting properties and reveal a magnetic hysteresis in the normal state. The critical temperature is about 36 K as compared with T = 38.5 for Hot shock wave fabricated $MgB_2$ compounds.

## 17.6 CONCLUSION

The two-stage hot shock wave consolidated-synthesized Mg-B precursors under the 1000°C temperature provides formation $MgB_2$ phase in the whole volume of billets with a maximal value of critical temperature Tc = 38.5 K.

The purity of precursors is an important factor, and existing of oxygen in the form-oxidized phases in precursors leads to reducing Tc and uniformity of HEC billets.

The doping of Mg-2B precursors with 1% nanoscale Fe particles with dimensions 25 nm and fabrication $MgB_2$-1% Fe samples shows that composites retain its superconducting properties and reveals a magnetic hysteresis in the normal state. The critical temperature is about 36 K as compared with T = 38.5 for Hot shock wave fabricated $MgB_2$ compounds.

## KEYWORDS

- **consolidation**
- **shock waves magnetic properties**
- **superconductivity**
- **vibrating sample magnetometer**
- **zero magnetic field cooling**
- **zero-field-cooled**

## REFERENCES

1. Nagamatsu, J., Nakagawa, N., Muranaka, T., Zenitani, Y., & Akimitsu, J., (2001). Superconductivity at 39 K in magnesium diboride. *Nature, 410*(6824), 63–64.
2. Jiang, C. H., Nakane, T., Hatakeyama, H., & Kumakura, H., (2005). Enhanced $J_c$ property in nano-SiC doped thin $MgB_2$/Fe wires by a modified in situ PIT process. *Physica, 422*(3/4), 127–131.
3. Mali, V. I., Neronov, V. A., Perminov, V. P., Korchagin, M. A., & Teslenko, T. S., (2005). Explosive incited magnesium diboride synthesis. *Chemistry for Sustainable Development, 13*(3), 449–451.
4. Orlinska, N., Zaleski, A., Wokulski, Z., & Dercz, G., (2008). Characterization of heat treatment $MgB_2$ rods obtained by PIT technique with explosive consolidation method. *Archives of Metallurgy and Materials, 33*(3), 927–932.
5. Holcomb, M. J., (2005). Supercurrents in magnesium diboride/metal composite wire. *Physica C: Superconductivity and its Applications, 423*(3/4), 103–108.
6. Priknha, T. A., Gawalek, W., Savchuk, Y. M., Moshchil, V. E., Sergienko, N. V., Surzhenko, A. B., Wendt, M., et al., (2003). High-pressure synthesis of a bulk superconductive $MgB_2$-based material. *Physica C: Superconductivity, 386*, 565–568.
7. Mamalis, A. G., Vottea, I. N., & Manolakos, D. E., (2004). Explosive compaction/cladding of metal sheathed/superconducting grooved plates: FE modeling and validation. *Physica C: Superconductivity, 410*, 881–883.
8. Shapovalov, A. P., (2013). High-pressure syntheses of nanostructured superconducting materials based on magnesium diboride. *High Pressure Physics and Engineering, 23*(4), 35–45.
9. Mamniashvili, G., Daraselia, D., Japaridze, D., Peikrishvili, A., & Godibadze, B., (2015). Liquid-phase shock-assisted consolidation of superconducting $MgB_2$ composites. *J. Supercond. Nov. Magn., 28*(7), 1926–1929.
10. Kostyuk, V. V., Antyukhov, I. V., Blagov, E. V., Vysotsky, V. S., Katorgin, B. I., Nosov, A. A., Fetisov, S. S., & Firsov, V. P., (2012). Experimental hybrid power transmission line with liquid hydrogen and $MgB_2$ based superconducting cable. *Technical Physics Letters, 38*(3), 279–282.

11. Gegechkori, T., Godibadze, B., Peikrishvili, V., Mamniashvili, G., & Peikrishvili, A., (2017). One stage production of superconducting $MgB_2$ and hybrid power transmission lines by the hot shock wave consolidation technology. *International Journal of Applied Engineering Research (IJAER), 12*(14), 4729–4734.
12. Gegechkori, T., Mamniashvili, G., Peikrishvili, A., Godibadze, B., & Peikrishvili, V., (2018). Using fast hot shock wave consolidation technology to produce superconducting $MgB_2$. *Engineering, Technology and Applied Science Research, 8*(1), 2374–2378.
13. Sidorov, N. S., Palnichenko, A. V., Shakhrai, D. V., Avdonin, V. V., Vyaselev, O. M., & Khasanov, S. S., (2013). Superconductivity of Mg/MgO interface formed by shockwave pressure. *Physica C: Superconductivity, 488*, 18–24.

# PART II

# Polymer Synthesis and Application

# CHAPTER 18

# Condensed Phosphates: New Inorganic Polymers with a Variety of Applications and Improvement of Their Gravimetric Determination Methods

M. AVALIANI,[1] E. SHAPAKIDZE,[2] V. CHAGELISHVILI,[1] N. BARNOVI,[1] and N. ESAKIA[3]

[1]*Iv. Javahishvili Tbilisi State University, Raphiel Agladze Institute of Inorganic Chemistry and Electrochemistry, Mindeli Street 11, Tbilisi–0186, Georgia, E-mails: avaliani21@hotmail.com; marine.avaliani@tsu.ge (M. Avaliani)*

[2]*Iv. Javakhishvili Tbilisi State University, Alexander Tvalchrelidze Caucasian Institute of Mineral Resources, Mindeli Street 11, Tbilisi–0186, Georgia*

[3]*Iv. Javahishvili Tbilisi State University, Institute of Exact and Natural Sciences, Department of Chemistry, 0179, I. Chavchavadze Ave 3, Tbilisi, Georgia*

## ABSTRACT

We report on the synthesis, properties, and applications of a large number of condensed phosphate polymers, obtained in multicomponent systems $M^{I}_2O$–$M^{III}_2O_3$–$P_2O_5$–$H_2O$, where $M^{I}$ is a monovalent metal and $M^{III}$ a trivalent metal. As an example, we have synthesized a cyclic octa-phosphate $K_2Ga_2P_8O_{24}$ and found that it is an outstanding catalyst for the preparation of low molecular weight digenic olefines. According to the experiments, we established that synthesized acidic triphosphates of gallium and/or indium $M^{III}H_2P_3O_{10}\cdot(1–2)H_2O$ are the best ion exchange material. Earlier, we have synthesized about 80 new formerly unknown double condensed phosphates, otherwise called inorganic polymers, from solution-melts of polyphosphoric acids during the investigation of multicomponent systems $M^I_2O$–$M^{III}_2O_3$–$P_2O_5$–$H_2O$. The studied mixtures were investigated at the temperature range 370–850 K. This work is focused on the synthesis of polymeric phosphates and on gravimetric determination methods' improvement in the cases of double condensed compounds. In the review paper, the opportunity of the development of the gravimetric oxyquinoline method for quantitative analysis of gallium, indium, scandium, and aluminum in samples of inorganic polymers, notably in the double condensed compounds, was considered. A number of variations of proposed gravimetric methods were reflected and detailed analyzed. Due to our numerous experiments, the influence of numerous factors such as: temperature, the quantity of buffer solutions and reagent-precipitant, initial pH of the solution, mixing time of the initial components, and other causes on the accuracy of the results were detailed and carefully studied and our experimental data were compared with the published literature documents.

## 18.1 INTRODUCTION

There is a significant growth in the development of new polymeric materials [1–5]. The reason seems to be the growth of the number of applications of such materials [6–11]. While most of the effort in new polymers development pertains to organic polymers, we are focusing on inorganic ones: condensed compounds of phosphorus. By numerous authors, the oxygen compounds of P, Si, Ge, Se, and other elements may be assumed as inorganic polymers, but they are not always stable, depending on how they are formed, for example, through the synthesis in aqueous solutions at different pH values or by other means and other various methods [1, 5, 8]. The chemistry of inorganic compounds of phosphorous, namely phosphates, has highly evolved in last year caused by the fact that condensed compounds of phosphorus are greatest applicable, useful, and convenient contributing thus to the development of the chemistry of inorganic polymers, and as a final point, they are reasonably presumed as the best materials used in the high-tech areas [12–16]. Multilateral spheres of use of condensed phosphates are very diverse: ion-exchange materials, nanomaterials, efficient applying fertilizers, detergents, cement substances, catalytic agents, and raw materials for phosphates glasses, thermally resistant substances and also as food additive composites, besides, the phosphate's binding agents, phosphate-binders, and laser materials. The configuration, structure, and thermal properties, as well as the vibrational and luminescent features of condensed compounds, determine their use in quantum electronics.

The biomaterials appear on the base of hydroxyapatite and polyphosphates. Sometimes habitual laser materials are replaced by biomaterials on the base of polyphosphates and hydroxyl-apatite [16–19]. Inorganic phosphates are used in the engineering sector and in the construction field, insofar as the industry requires the supplying in necessary phosphate materials. The main component of these composites is the so-called phosphate adhesive used in the form of acid phosphates of various metals, phosphoric acid, or condensed forms, all mixed with polyvalent metal oxides forming autonomous masses. At present, there are many metallophosphate acidic adhesives, but on the outside, among them, aluminum dihydrophosphate is widely used, aluminophosphate is more stable under storage conditions. Potential areas of application of these products have been reported [15–18, 20–22]. Based on the above noteworthy prerequisites, we have focused our efforts on the researches and development to new inorganic polymers, notably on the double condensed acidic and normal di- and triphosphates, cyclic tetraphosphates, cyclic octaphosphates, cyclic dodecaphosphates, and ultra-phosphates [13–15].

## 18.2 EXPERIMENTAL

### *18.2.1 METHODS*

Our group of chemists has been working on inorganic synthesis that represents an environment-friendly method – the technology for used chemicals which are decreased wastes with minimum harm and not producing damaging outputs. Last decades we reported about our studies in the open systems $M^{I}_2O$–$M^{III}_2O_3$-$P_2O_5$-$H_2O$ between temperature range 200–400 K, where $M^I$-alkali metals and $M^{III}$-Ga, In, and Sc. Our various experiments revealed the existence of the many double condensed compounds; in fact, a series of a formerly new class of inorganic polymers. The method of synthesis of double phosphates from solution-melts of phosphoric acids was applied in the course of our experiments. We have established the optimal crystallization regions of various condensed phosphates of mono- and polyvalent metals. This work was also focused on the synthesis of polymeric phosphates and on gravimetric determination methods' improvement in the cases of double condensed compounds. The main objective was the determination of polyvalent metals by 8-oxyquinoline. In the presented review paper, the opportunity of improvement of the gravimetric oxyquinoline method for quantitative analysis of gallium, indium, scandium, and aluminum in samples of inorganic polymers, notably in the double condensed compounds, was considered.

A number of variations of proposed gravimetric methods were reflected and detailed analyzed. Due to our numerous experiments the influence of a numerous factors, such as: temperature, quantity of buffer solutions and reagent-precipitant, initial pH of the solution, mixing time of the initial components, etc., on the accuracy of the results were detailed and carefully studied and our experimental data were compared with the published literature documents [6, 25].

Methodical recommendations for the deposition of trivalent metals of gallium, indium, and scandium from condensed phosphate samples using the oxyquinoline method are proposed.

### 18.2.2 MATERIALS

During investigation of the multicomponent systems at temperature range 400–850 K initial materials were as follows: ortho-phosphoric acid (85%), $AgNO_3$, in some cases-carbonates of alkali metals and oxides of trivalent metals $M^{III}{}_2O_3$ (where $M^{III}$ was Ga, In, Sc, and occasionally-Al). The molar ratio of initial compounds was following: $P_2O_5$: $M^I{}_2O$: $M^{III}{}_2O_3$ = 15.0:2.5:1.0; 15.0:5.0:1.0; 15.0:7.5:1. 0; 15.0:10.0:1.0 and at times 15.0:3.5:1.5; 15.0:5.0:1.5; 15.0:6.0:1.5; 15.0:7.5:1.5; 15.0:8.5:1.5; 15.0:12:1.5. Every now and then, more volume of phosphoric acid was taken during some experiments (for example, sometimes-molar ratio 18.0:8.5:1.5; 18.0:10.0:1.5 was taken). In glassy carbon crucible there were mixed gallium oxide, or scandium oxide, or indium oxide, ortho-phosphoric acid (percentage: 85%), nitrate of silver and/or carbonates of alkali metals in various molar ratio.

### 18.2.3 MEASUREMENT

At the above-mentioned temperature range 400–850 K we have prepared several double condensed phosphates. It is acknowledged that sufficient stability of polymeric phosphates makes it able to identify and classify them by the method of paper chromatography. This method together with the chemical analysis, IR spectroscopy, thermo gravimetric analysis, x-ray diffraction (XRD) analysis, structural analysis was used by us to study the process of formation and composition of many normal, basic, and/or acid of both simple and double di-, tri-, tetra-, octa-, and dodecaphosphates or ultra-phosphates of polyvalent metals [13–15, 19–20, 23–24]. Condensed phosphates so obtained were examined by thermogravimetric analysis

(TGA). Methods of polymer characterization, such as TGA are discussed in Ref. [25]. We have used a Q1500-D Derivatograph with a heating rate of 10 degrees/min, in an air atmosphere and a maximum temperature of 1000 K (occasionally to 1400 K). Some compounds were also examined by paper chromatography. In addition to the above were carried out researches using a scanning electronic microscope of the company JEOL (Japan), equipped with a scanning electronic microscope JSM-6510LV, which in turn was well-appointed by energy-disperse micro-roentgen spectral analyzer produced by Oxford Instruments, the analyzer's type is following: X-Max N 20. Electronic micrographs were carried out by means of reflected (BES) and as well as secondary (SEI) electrons at an accelerating voltage (at 20 kV). The working distance was approximately 15 mm, micrographs have been taken at the diverse enlargements. The micro-spectroscopic analysis was performed from the sampling point zones and its surface area.

## 18.3 RESULTS AND DISCUSSIONS

### *18.3.1 MAIN CLASSES OF SYNTHESIZED PHOSPHATES DURING CONDENSATION PROCESS*

During the investigation of the multicomponent systems, $M^IO–Sc_2O_3–P_2O_5–H_2O$ at temperature range 400–850 K more remarkable condensed phosphates synthesized by us are presented in Table 18.1.

**TABLE 18.1** Leading Classes of Synthesized Condensed Phosphates

| **Doubles Acidic Triphosphates $M^IM^{III}HP_3O_{10H}$ $M^{III}H_2P_3O_{10}$** | **Double Triphosphates $M^I_2M^{III}P_3O_{10}$** | **Double Acidic Diphosphates (Hydrated) $M^IM^{III}(H_2P_2O_7)_2{\cdot}2H_2O$** |
|---|---|---|
| Complex Diphosphates $M^I_2M^{III}H_3(P_2O_7)_2$ | Double Dihydrophosphates $M^IM^{III}(H_2P_2O_7)_2$ | Double Diphosphates $M^IM^{III}P_2O_7$ |
| Double Cyclotetraphosphates $M^IM^{III}(PO_3)_4]_4$ | Double Cyclooctaphosphates $M^I_2M_2^{III}P_80_{24}$ | Double Cyclododecaphophates $M^I_3M^{III}_3P_{12}0_{36}$ |
| Ultraphosphates $M^I_3M^{III}P_8O_{23}$ | Long Chain Double Polyphosphates $[M^IM^{III}(PO_3)_4]_x$ | Various Polyphosphates $M^{III}(PO_3)_3$-(A, C & other forms) |

As we have already noted one of the main aims of our work was to obtain the new inorganic polymers-many double oligomers, polymeric, and cyclic phosphates of monovalent and polyvalent metals from solution-melts of poly-phosphoric acids at various temperature range.

Consequently, presented proceedings are the outcome of chemical synthesis, analysis, and detailed examination of the experimental records. We have determined and evaluated the various properties of synthesized by us compounds and we have examined the correlation with the achievements and improvements in the area of inorganic polymer's chemistry. During last year, we have obtained more than 80 new, formerly unknown double condensed phosphates.

### 18.3.2 GRAVIMETRIC DETERMINATION METHODS EXPEDIENT FOR GALLIUM, INDIUM, AND SCANDIUM

The scientific publications [6–7, 10, 12–13, 16–17, 22] describe the methods of gravimetric determination of metals from inorganic compounds, but there are not detailed and/or sufficient data concerning the identification and definition of metals in samples of condensed phosphates. For this reason, one of the focal targets of our work was the improvement of gravimetric determination methods of Ga, In, and Sc in the samples of inorganic polymers, containing polyvalent and monovalent metals. Table 18.2 shows the correlation between different gravimetric methods/practices and some of their negative aspects.

**TABLE 18.2** The Correlation between Various Gravimetric Practices and Some Revealed Negative Aspects

| Method | Negative Aspects |
|---|---|
| Deposition of hydroxide. | The sludge must be incinerated to oxide |
| Basic thiosulphate method. | It's difficult to isolate scandium, because there's still some sulfur left. |
| Silicon-fluorine hydrogen method. | It's difficult to isolate scandium, remains $SiF_6^{-2}$ |
| Basic tartrate method. | Sludge must be hardened to $Sc_2O_3$; it has a high weighting factor. |
| Oxalate deposition. | Not quantitative method. |
| Pyridine buffer deposition. | The sludge must be incinerated to oxide $Sc_2O_3$. |

During our experiments, we chose the oxychinoline method based on the following arguments:

1. The possibility of using a relatively small weight of condensed samples;

2. High-precision analysis;
3. Simplicity of deposition and definition.

Table 18.3 shows the experimental data regarding the influence of the volumes of buffer and precipitator solutions, pH of solution before deposition, time of sedimentation, pH of solution after deposition and other factors on the sedimentation process. Experimental swatches were samples of condensed mono- and trivalent metal phosphates, synthesized by us from polyphosphoric acid melts.

Sludge composition is the following: $Sc(C_9H_6NO)_3 \cdot C_9H_7NO$.

### 18.3.3 INVESTIGATION OF SYSTEMS CONTAINING LITHIUM, POTASSIUM, AND RUBIDIUM

Following our experiments, we have synthesized one of the first representatives of the long-chain polyphosphates such as lithium-gallium $LiGa[(PO_3)_4]_x$ and $LiIn[(PO_3)_4]_x$, one of the first representatives of double cyclic octaphosphate class, such as the cyclic octaphosphate of potassium-gallium and rubidium-gallium: $K_2Ga_2P_8O_{24}$ and $Rb_2Ga_2P_8O_{24}$. Those compounds were obtained by one of us and the crystal structures were examined and described by Palkina, Chudinova, and all (see Figures 18.1(a), 18.1(b), and 18.2 [1, 4, 12–15].

These our reported accomplishments consisting in the successful synthesis of the first representatives of double cyclic octaphosphates classes, namely $K_2Ga_2P_8O_{24}$ and $Rb_2Ga_2P_8O_{24}$ were noted and marked as excellent achievements by diverse authors in their published works [1, 5, 13, 23]. We have as well synthesized the similar compounds of potassium-scandium, potassium-indium, rubidium-gallium, rubidium-indium, and rubidium-scandium.

### 18.3.4 INVESTIGATION OF SYSTEMS CONTAINING CAESIUM

One of the first representatives of the cyclic dodecaphosphates class was synthesized by us as well, with the following formulas: $Cs_3Ga_3P_{12}0_{36}$, $Cs_3Sc_3P_{12}0_{36}$, and $Cs_3In_3P_{12}0_{36}$. The molar ratio for $P/M^I/M^{III}$ was very variable: 15/2.5/1; 15/3.5/1; 15/5.0/1; 15/6.0/1; 15/7.5/1; 15/10.0/1; And moreover: 15/2.5/1.5; 15/3.5/1.5; 15/5.0/1.5; 15/6.0/1.5; 15/7.5/1.5; 15/10.0/1.5 as well as, 18/2.5/1.5; 18/3.5/1.5; 18/5/0/1.5; 18/6.0/1.5; 18/7.5/1.5; 18/10.0/1.5 [4, 13]. In an exhaustive and remarkable publication [5], it has been described the similar compound, obtained by our method.

**TABLE 18.3** The Influence of Various Factors on Deposition of Sludge

| № of Tests/ Experiments | Sample Quantity (g) | Volume of Oxyquinoline (ml) | Vol. of Buffer $NH_4OH$ $CH_3COOH$ (ml) | Buffer Quantity (g) | pH Before Sedimentation | pH After Sedimentation | Theoretical Quantity of Sc (g) or $Sc_2O_3$ (%) | Practical Yield of the Reaction Products: Sc (g) or $Sc_2O_3$ (%) | Notes (Impact of Other Factors) |
|---|---|---|---|---|---|---|---|---|---|
| 1 | $Sc(NO_3)_3$ 0.2018 | 5.0 | $NH_4OH$ $CH_3COOH$ (ml)<br>15<br>30 | 0.0585 | 6.0 | 9.0 | 0.0044 г<br>Sc | 0.0042 g | Hot solution filtration |
| 2 | « | 6.0 | 15<br>30 | 0.0609 | 6.0 | 8.0 | « | 0.0044 g | Filtering after 30 min. after sedimentation |
| 3 | « | 8.0 | 15<br>30 | 0.0567 | 6.0 | 8.5 | « | 0.0041 g | Filtration immediately after sedimentation. In hot water, the solubility of the complex is greater, so the results are slightly reduced |
| 4 | « | 8.2 | 15<br>30 | 0.0757 | 6.0 | 8.0 | « | 0.0054 g | Filtration about 25 hours after deposition. Results are overestimated due to free oxyquinoline crystallization |
| 5 | « | 9.0 | 15<br>30 | 0.0646 | 6.0 | 8.0 | « | 0.0046 g | Filtering after 15–20 minutes after sedimentation: |
| 6 | « | 10.0 | 20<br>30 | 0.0649 | 6.0 | 8.2 | « | 0.0046 g | |
| 7 | « | 10.0 | 20<br>30 | 0.0649 | 6.0 | 8.2 | « | 0.0046 g | |

**TABLE 18.3** *(Continued)*

| № of Tests/ Experiments | Sample Quantity (g) | Volume of Oxyquinoline (ml) | Vol. of Buffer $NH_4OH$ $CH_3COOH$ (ml) | Buffer Quantity (g) | pH Before Sedimentation | pH After Sedimentation | Theoretical Quantity of Sc (g) or $Sc_2O_3$ (%) | Practical Yield of the Reaction Products: Sc (g) or $Sc_2O_3$ (%) | Notes (Impact of Other Factors) |
|---|---|---|---|---|---|---|---|---|---|
| 8 | $Sc(NO_3)_3$ 0.2960 g | 10.0 | 26<br>30 | 0.0985 | 1.2 | 8.5 | 24.56% $Sc_2O_3$ | 24.6% $Sc_2O_3$ | Many quantity of buffer in the initial solution, pH about 1.0–1.2 |
| 9 | « | 8.0 | 25<br>30 | 0.1000 | 1.2 | 8.5 | 24.56% $Sc_2O_3$ | 25.0% $Sc_2O_3$ | |
| 10 | $Na_3ScP_8O_{23}$ | 3.5 | 15<br>30 | 0.0142 | 6.0 | 8.5 | 5.20% $Sc_2O_3$ | 3.6% $Sc_2O_3$ | Incomplete sedimentation; In fact it was added insufficient volume of oxyquinoline |
| 11 | 0.1613 g | 4.5 | 15<br>30 | 0.0180 | 6.0 | 8.5 | « | 4.10% $Sc_2O_3$ | Filtering after 3 min. after sedimentation |
| 12 | 0.2189 g | 5.0 | « | 0.0132 | 6.2 | 8.5 | « | 4.03% $Sc_2O_3$ | Filtering after 3 min. after sedimentation |
| 13 | 0.2187 g | 6.0 | « | 0.0119 | 6.2 | 8.5 | « | 4.03% $Sc_2O_3$ | Filtering after 4 min. after sedimentation |
| 14 | 0.2189 g | 6.0 | « | 0.0129 | 6.2 | 8.5 | « | 4.17% $Sc_2O_3$ | Filtering after 5 min. after sedimentation |
| 15 | 0.1613 g | 8.0 | « | 0.1118 | 6.2 | 8.5 | « | 5.01% $Sc_2O_3$ | Filtering after 10–13 min. after sedimentation |
| 16 | 0.1613 g | 8.0 | « | 0.1121 | 6.0 | 8.5 | « | 5.02% $Sc_2O_3$ | Filtering after 15–20 min. after sedimentation |

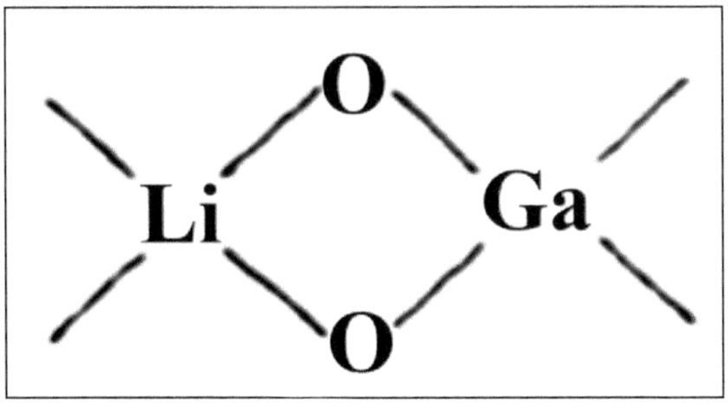


**FIGURE 18.1(a)** One fragment of a long chain polyphosphate $LiGa[(PO_3)_4]_x$.

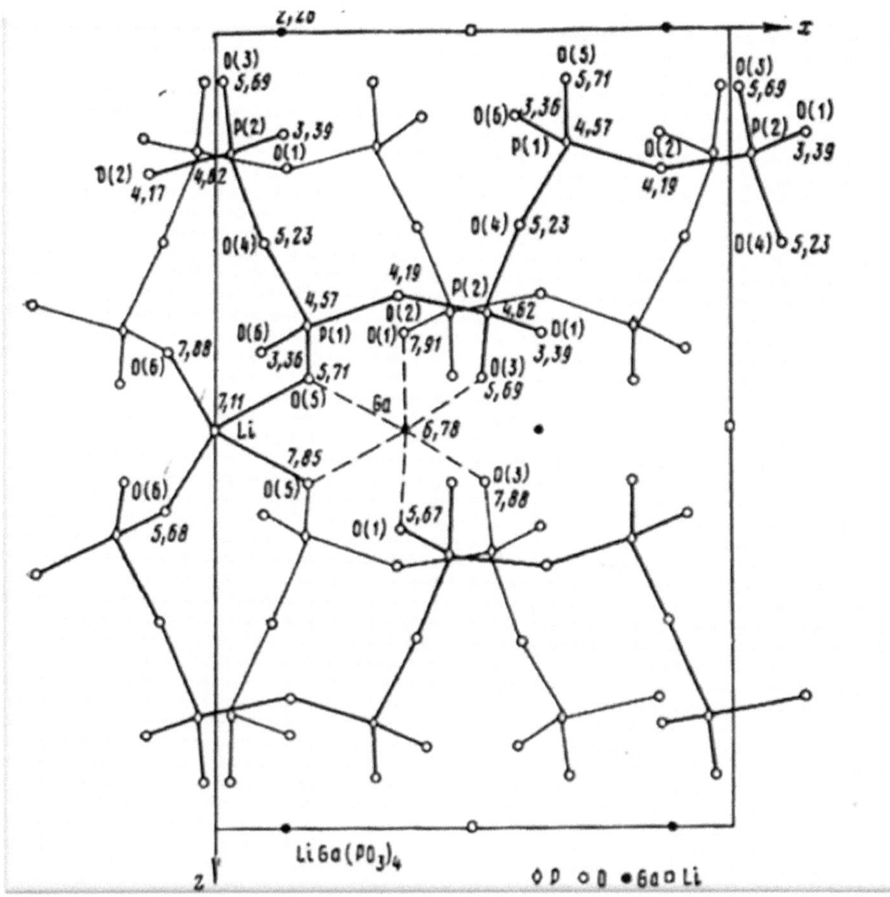

**FIGURE 18.1(b)** The long chain polyphosphate $LiGa[(PO_3)_4]_x$ structure projection along the x-axis.

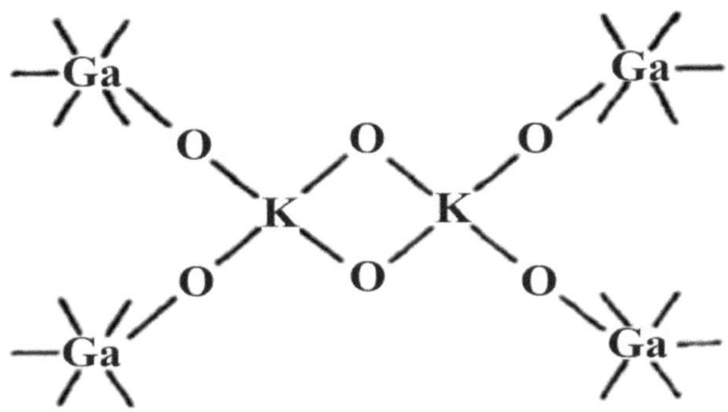

**FIGURE 18.2** The schematic fragment of the structure of cyclooctaphosphate of potassium-gallium.

Numerous synthesized condensed compounds were wholly studied by chemical analysis, also are observed by X-ray structural techniques. The formulas of basic main compounds, so-called inorganic polymers, are presented below: double acidic triphosphates $M^IM^{III}HP_3O_{10}$, (forms I, II, III), $M^IM^{III}HP_3O_{10}\cdot H_2O$, double normal triphosphates $M^I_2M^{III}P_3O_{10}$, double acidic diphosphates $M^I_2M^{III}H_3(P_2O_7)_2$, $M^IM^{III}(H_2P_2O_7)_2$, crystalline hydrate of double acidic diphosphate $M^IM^{III}(H_2P_2O_7)_2\cdot 2H_2O$, double diphosphate $M^IM^{III}P_2O_7$, long-chain double polyphosphates with the general formula $[M^IM^{III}(PO_3)_4]_x$, ultra-phosphate $M^I_3M^{III}P_8O_{23}$, cyclic octaphosphates $M^I_2M_2^{III}P_80_{24}$, cyclic dodecaphosphates $M^I_3M^{III}_3P_{12}0_{36}$, long-chain polyphosphates (forms A and/or C)-$M^{III}$ $(P0_3)_3$ [4, 10–15].

### *18.3.5 INVESTIGATION OF SYSTEMS CONTAINING ARGENTUM*

In order to examine the impact of trivalent and monovalent cations on the formation of inorganic polymers' anionic radicals and the level of condensation, as well as the influence of the molar ratio of initial components we have studied multicomponent systems containing as monovalent metal not only alkaline metals but also Ag and as trivalent metals the following elements: Ga, In, and Sc. The double oligo- and cyclic phosphates have been the first time synthesized and examined by our group. General reliance of structural composition and stability of the double condensed phosphates from ion radius of $M^1$-$M^{III}$ have also been examined. Table 18.4 has shown some of the various oligomeric and polymeric compounds so-called inorganic polymers that we have obtained in recent years.

**TABLE 18.4** Various Newly Synthesized Condensed Phosphates

| $M^{I}M^{III}(H_2P_2O_7)_2$ Double Acidic Diphosphates | $M^{I}M^{III}P_2O_7$ Double Diphosphates | $M^{I}M^{III}HP_3O_{10}$ Double Acidic Triphosphates | $M^{I}_2M^{III}P_3O_{10}$ Double Triphosphates | $M^{I}M^{III}(PO_3)_4$ Long Chain Polyphosphate (a), Ultra-phosphate (b), cycloocta-(c), Cyclotetraphosphate (e) | $M^{III}(PO_3)_3$ Polyphosphate |
|---|---|---|---|---|---|
| $LiSc(H_2P_2O_7)_2$ | $LiScP_2O_7$ | $LiScHP_3O_{10}$ | $Li_2ScP_3O_{10}$ | $\{LiSc(PO_3)_4\}_x$ (a) | $Sc(PO_3)_3$-A |
| $NaSc(H_2P_2O_7)_2$ | $NaScP_2O_7$ | $NaScHP_3O_{10}$ | $Na_2ScP_3O_{10}$ | $Na_3ScP_8O_{23}$ (b) | $Sc(PO_3)_3$-C |
| $KSc(H_2P_2O_7)_2$ | $KScP_2O_7$ | $KScHP_3O_{10}$ | $K_2ScP_3O_{10}$ | $K_2Sc_2P_8O_{24}$ (c) | |
| $RbSc(H_2P_2O_7)_2$ | $RbScP_2O_7$ | $RbScHP_3O_{10}$ | $Rb_2ScP_3O_{10}$ | $Rb_2Sc_2P_8O_{24}$ (c) | |
| $CsSc(H_2P_2O_7)_2$ | $CsScP_2O_7$ | $CsScHP_3O_{10}$ | $Cs_2ScP_3O_{10}$ | $Cs_3Sc_3P_{12}O_{36}$ (d) | |
| $Ag(H_2P_2O_7)_2$ + $AgHScP_3O_{10}$ *Mix phases* | $AgScP_2O_7$ | $AgHScP_3O_{10}$ | | $Ag_3Sc_3P_{12}O_{36}$ (d) | |
| $AgSc(H_2P_2O_7)_2\,H_2O$ | | | | $AgScP_4O_{12}$ (e) | |
| | | | | $AgGaP_4O_{12}$ (e) | |
| | | | | $AgInP_4O_{12}$ (e) | |

### 18.3.6 DETAILED INVESTIGATION OF SYSTEMS CONTAINING SIMULTANEOUSLY SCANDIUM, ARGENTUM AND/OR CAESIUM

We conducted researches and studied in detail the systems $M^I_2O$-$M^{III}_2O_3$-$P_2O_5$-$H_2O$ not only on molar ratio P: $M^I$: $M^{III}$ = 15:2.5:1.5; 15:5:1.5; 15:7.5:1.5; 15:10:1.5 but as well as 15:3:1.5; 15:6:1.5; 15:8:1.5 and at different temperature interval 390–950 K (where $M^I$ = Cs and/or Ag). An earlier unknown compounds-double condensed di- and triphosphates of scandium and silver were obtained by us at the above-mentioned molar ratio. The greatest stable phase at relatively low temperatures from 400–420 K, and even to the 438–450 K are $Ag(H_2P_2O_7)_2$ and $AgHScP_3O_{10}$. Taking into account the circumstance that the arrangement of cations Ag-Sc for double di- and triphosphates (in other words, typical X-ray pictures from single crystalline areas of double oligophosphates of Scandium-Silver samples) have not been nowhere reported in detail. Hence, the mentioned characteristics are not given in the database of file index for testing and materials, and our roentgenograms' data were compared to similar compounds of Ag-P and our standard data models for similar double condensed phosphates of Gallium, Indium, and Scandium with alkali metals. In analyzed samples are not revealed any initial components: $K_2CO_3$, $Sc_2O_3$, $AgNO_3$, and/or their hydrates; as such, they are already completely irreversibly inter-reacted. Based on the founded and pursuant literary data [1–2, 4–11, 19–23] also on our synthesis conditions, on experiences in this field, the structural composition of the compounds is set up. Phases' identification was given in accordance with standard data of the International Center for Diffraction Database of the American Society for Testing and Materials (ASTM).

### 18.3.7 EXAMINATION OF SYSTEMS AT TEMPERATURE RANGE 505–530 K, CONTAINING TRIVALENT METALS, AND ARGENTUM

We conducted researches and studied in detail the systems $M^I_2O$-$M^{III}_2O_3$-$P_2O_5$-$H_2O$ on following molar ratio P: $M^I$: $M^{III}$ = 15:2.5:1.5; 15:5:1.5; 15:7.5:1.5; 15:10:1.5 and 15:3:1.5; 15:6:1.5; 15:8:1.5 and at different temperature interval 390–950 K ($M^I$ = Cs and/or Ag). An earlier unknown compounds-double condensed tetraphosphates of argentums-gallium, argentum-indium, and argentum-scandium were obtained. The formulas of synthesized condensed phosphates named above are the following: double cyclic tetraphosphates $AgGaP_4O_{12}$, $AgScP_4O_{12}$, and $AgInP_4O_{12}$. Phase identification was given in accordance with standard data of the International Center for Diffraction Database of the American Society for Testing and Materials-ASTM. On the basis of the existing scientific publications and conforming to our anterior studies, the structural composition of the compounds has been defined. Comparing the results of double phosphates of scandium, gallium, and indium with literary data, we conclude that condensed compounds of scandium, according to their composition and structure, coincide with phosphates of light trivalent metals (Ga, Fe, Cr, Al) and are not similar to corresponding compounds of rare earth elements.

Discussing about the range of $M^{I}M^{III}(PO_3)_4$ compounds structures where $M^{I}$ is constantly alkali or any other monovalent metal and where $M^{III}$ is any of trivalent metals such as gallium, indium, scandium, and others, even rare earth elements, we made some conclusions. Firstly, we discovered that while the radius of $M^{3+}$ decreases, the polyphosphate chain identity period increases, due to the complication of its form-factors; The cycles slowly appear, the number of structural types increases caused by the correlation of average distances between ($M^{III}$-O) and ($M^{I}$ -O). Secondly, it appears that less is the above-said correlation (molar ratio), the probability of big cycle formation increases.

The analyze of the investigational data has shown that we obtained the following condensed phosphates: acidic di- and triphosphates of Ga, In, and Sc with Ag, such as: $AgSc\ (H_2P_2O_7)_2$, $AgScH\ P_3O_{10}$ $AgGaH\ P_3O_{10}$ and double cyclic tetra-phosphates $AgGaP_4O_{12}$, $AgScP_4O_{12}$, and $AgInP_4O_{12}$. All these compounds have been unknown till now and we obtained them the first time. We have defined the optimal crystallization conditions for those phosphates [19–20, 24].

We have found that condensed compounds of scandium-silver, according to their composition and structure, correspond with phosphates of sodium-gallium and sodium-scandium and sometimes with Cs-Sc. Another additional conclusion: obtained phosphates are not analogous to corresponding compounds of rare earth elements. Likewise, it has been revealed that at moderately low temperatures it's more feasible to produce double acidic phosphates, at relatively high temperature appears double tetra-phosphates of gallium-silver, indium-silver, and scandium-silver [14–15, 19, 24]. It must be emphasized that the influence of molar ratio is nevertheless very important. For example, in the case of molar ratio n = 5 at temperature 610 K and duration of experiment 12–15 days is obtained cyclic dodecaphosphate of Ag-Ga, and at molar ratio 7.5 on the same temperature and similar synthesis conditions the basic phase is tetraphosphate.

Some results of X-ray examination of crystals by a scanning electronic microscope, equipped with energy-disperse micro-roentgen spectral analyzer are presented in Figures 18.3 and 18.4. As we already mentioned above electronic photomicrographs were made by means of reflected (BES) and as well as secondary (SEI) electrons at voltage 20 kV, electron images have been taken at the diverse enlargements (see Figures 18.3(a)–18.3(c)). Micro-spectroscopic analysis was performed from the sampling point zones and their surface area.

**FIGURE 18.3(a)** Electron image 1 for condensed compound, obtained on molar ratio 7.5 (x270).

**FIGURE 18.3(b)** Electron image 2 for condensed compound, obtained on molar ratio 7.5 (x550).

**FIGURE 18.3(c)** Electron image 3, for cyclic condensed compound, obtained on molar ratio 7.5 (x2,700).

Presented Figure 18.4 characterize the spectrum for cyclic phosphate sample, obtained in system, containing Ag and Sc, that we have recently synthesized at 420–425 K.

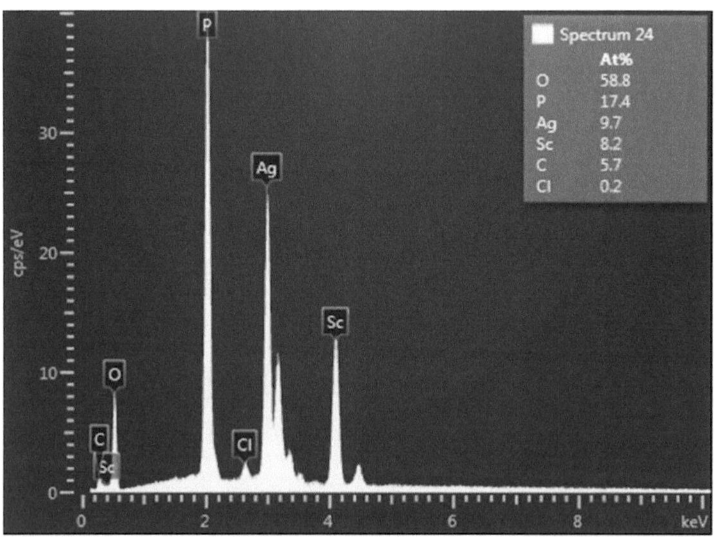


**FIGURE 18.4** Spectrum for condensed compound: cyclic phosphate obtained in system, containing Ag and Sc, synthesis at 420 K.

The tetraphosphates named above are isomorphic among themselves and are iso-structural with the sodium-gallium double condensed tetraphosphates. Performing the comparison of the obtained double condensed compounds data with appropriate phosphates we decided that while the radius of trivalent metal decreases, the polyphosphate chain identity period rises [19, 20].

## 18.4 POTENTIAL AREAS OF APPLICATION OF SYNTHESIZED INORGANIC POLYMERIC MATERIALS

The various possibilities of application of condensed cyclic octaphosphate of potassium-gallium have been studied in detail. We would like to underline that diameter of the channel in the case of gallium cyclic octaphosphate is 5.2 angstrom, which is objectively very good sign. The $K_2Ga_2P_8O_{24}$ synthesized by us was studied for the catalytic activity. The results showed that it has unique properties as an inorganic polymer and can be used as a best catalyst for organic synthesis reactions, specifically for getting low molecular weight dienic olefins. As we have already indicated the properties of this compound in the model dehydration reaction of n-butyl alcohol by the impulsion method have been studied. Based on the piloted experiments, it was founded that the overall conversion was from 52% to 65%. Studied sample is comparable to the activity of the compound $BPO_3$ (50%) achieved by mixing the initial constituents and exceeded the activity of the zeolite catalyst sample NaZr-A (32%), tested under similar and comparable conditions. The properties of the named octaphosphate that we obtained newly have been investigated by the impulsion method at the model dehydration reaction of n-butyl alcohol. It has also been founded that the $C_4$ olefin hydrocarbons were definitely and firmly retained by our catalyst under the experimental conditions.

We have also defined that synthesized by us the acidic triphosphates of gallium, indium, and scandium $M^{III}H_2P_3O_{10}$ (1–2) $H_2O$ are finest ion exchange agents. We have developed an appropriate method for obtaining the ion-exchange form of gallium acid triphosphate – a new ion-exchange material for subsequent use.

Generally, the application of phosphates is wide and it cannot be exposed in one article. In the present work, we would communicate shortly about the use of the mentioned phosphate's systems of hardening which are used in fine arts also. As noted earlier, the procedure of thermo phosphate painting is very simple [6]. The base material (asbestos–cement sheets, fibers, cardboard, glass, ceramics, metals) is first coated with phosphate which is fixed by heating, then with phosphate colors (by a brush or by spraying). Thereafter, the color layer is treated (for example, pulverized) with a fixative containing phosphate adhesive. According to Ivan Tananaev, the material should be heated for 2–3 minutes at 50–150 K using an electric shield-for large size paintings, or an ordinary gas burner or other modes of heating. The painting so obtained is moisture and heatproof and does not call for special conditions necessary for prolonged storage. Certain details concerning thermographic painting-in fact a new trend in the art avowed as "Thermophosphate Pictorial Art," developed by Oleg B. Pavlov, are mentioned in Refs. [24, 26]. He has developed minerals as well as phosphate paints in three forms: thermo phosphate paints, powder colors, pastel, and artistic colors. The additional research in this branch of inorganic chemistry must be motivating for scientists in this sphere, especially in purpose to enlarge the domain of utilization of condensed phosphates.

As a final point, we would like to emphasize that inorganic polymers' chemistry and technology are one of the main fields of new materials science following a number of recent scientific interdisciplinary researches and utilization of advanced technologies [27–36].

Inorganic polymers' chemistry domain touches almost every aspect of the high-technology area and its various applications affect the modern life, from electronics to construction, food, and medicine. They are more and more substituted for metal parts in aviation and automotive industries because of low densities. Polymers are used in electronics and microelectronics, acting as sealants because of low electric conductivity. They also have many other kinds of uses including in cosmetics.

Polymeric materials, due to its thermal stability and flexibility, are used as elastomers, fibers, films, membranes, to name but a few [28, 30, 37]. It is necessary to stress the importance of the compounds with structural diversity including promising systems with a variety of application—the hybrid inorganic-organic materials [31].

## 18.5 CONCLUSIONS

Our study of poly-component systems containing mono- and trivalent metals and phosphoric acid has revealed the dependency of composition versus temperature and molar ratio of the initial components. The reliance of structure from duration of synthesis and radius of the interrelated ions has been discovered. We found that at relatively low temperatures is more probable to produce double acidic phosphates, with increasing temperature the tetraphosphates of gallium-silver and scandium-silver are formed. Tetraphosphates of silver are isomorphs among themselves and are iso-structural with the sodium-gallium and sodium-indium tetraphosphates. Optimum performance for the realization of big cyclic anions is the correlation of the big monovalent cations versus trivalent metals with a small ionic radius.

We note that, for range of structures $M^IM^{III}(PO_3)_4$, while the radius of $M^{3+}$ declines, the polyphosphate chain identity period increases, owing to complication of its form-factors. The cycles gradually appear, the number of structural types increases caused by correlation of average distances between the trivalent metal-oxygen ($M^{III}$ –O) and the monovalent metal-oxygen ($M^I$ – O). The lesser is the correlation (molar ratio), the likelihood of a big cycle formation increases.

We have studied in detail the possibility of improvement of gravimetric determination method of polyvalent metals and the influence of buffer and precipitator solutions, pH of solution before deposition, time of sedimentation, pH of solution after deposition and of other factors on the sedimentation process. Experimental swatches were samples of condensed mono- and trivalent metal phosphates, synthesized by us from polyphosphoric acid melts.

Synthesized cyclooctaphosphate $K_2Ga_2P_8O_{24}$ was studied for the catalytic activity, the results showed that it has exceptional properties as an inorganic polymer and can be used as a best catalyst during chemical synthesis reactions, precisely for the preparation of low molecular weight olefins during organic synthesis process. We established that the overall conversion is from 52% to 65% and this sample doubly exceeds the activity of the zeolite catalyst, experienced under comparable conditions. We have also defined that synthesized by us the acidic triphosphates of gallium and/or indium $M^{III}H_2P_3O_{10}\cdot$ (1–2) $H_2O$ are finest ion exchange agents. We have developed an appropriate method for obtaining the ion-exchange form of gallium acid triphosphate-a new ion-exchange material for subsequent use.

## ACKNOWLEDGMENTS

The important pieces of advice of our colleagues: Eteri Shoshiashvili (Iv. Javakhishvili Tbilisi State University Raphiel Agladze Institute of Inorganic Chemistry and Electrochemistry) and Atinati Mamatsashvili (Ilia State University) is gratefully acknowledged. Our special thanks to Professor Omar Mukbaniani (Iv. Javakhishvili Tbilisi State University) and Professor Witold Brostow (University of North Texas) for their significant consultations.

## KEYWORDS


- **condensed phosphate**
- **cyclophosphate**
- **inorganic polymer**
- **ion exchange agents**
- **oligophosphate**
- **thermogravimetric analysis**


## REFERENCES


1. Durif, A., (2014). *Crystal Chemistry of Condensed Phosphates* (p. 408). Springer Science and Business Media edition.
2. Ribero, D., & Kriven, W. M., (2015). Synthesis of $LiFePO_4$ powder by the organic-inorganic steric entrapment method. *J. Mater. Res., 30*(14), 2133. https://doi.org/10.1557/jmr.2015.181.
3. Castano, V. M., Quezada, J., & De La Borbolla, A. K., (2018). *Polychar., 26*, 1, World Forum on Advanced Materials, Georgia.
4. Avaliani, M. A., (1982). *Synthesis and Characterization of Gallium Indium Condensed Phosphates* (p. 185). Extended abstract of PhD Dissertation.
5. Murashova, E. V., & Chudinova, N. N., (2001). Double condensed phosphates of cesium-indium. *J. Inorgan. Mater., 37*(12), 1521.
6. Komissarova, L. N., (2006). *Inorganic and Analytical Chemistry of Scandium* (p. 512). Editorial URSS.
7. Vassel, N. P., Vassel, S. S., Pavlova, I. V., & Kaklyugin, A. V., (2011). Thermal study of systems involving the metaphosphates of trivalent metals and silver. *Proc. Inst. High. Educ., Phys., 1, 2,* 121.
8. Strutinska, N. Y., Zatovsky, I., Ogorodnyk, O. V., & Slobodyannik, N., (2013). Rietveld refinement of $AgCa_{10}(PO_4)_7$ from X-ray powder data. *Acta Crystallogr. Sect. E. Struct. Rep. Online., 69,* 23.
9. Suthanthiraraj, S. A., & Sarumathi, R., (2013). A new silver ion conducting $SbI_3$-$Ag_4P_2O_7$ nano-composite solid electrolyte. *J. Appl. Nanosci., 3*, 501.
10. Grunze, I., (2009). *Energy Citations Database; Inorg. Mater., 23*(4), 539.
11. Avaliani, M. A., Tananaev, I. V., & Gvelesiani, M. K., (2003). Synthesis and investigation of double condensed phosphates of scandium and alkali metals. *Abstracts for Synthesis Chemists, FIZ CHEMIE Berlin, J. Phosphorus, Sulfur Silicon Relat. Elem.,* 51.
12. Palkina, K. K., Maksimova, S. L., & Chibiskova, N. T., (1981). Structure of crystals $LiGa(PO_3)_4$. Izv. AN SSSR. *J. Neorg. Mater., 17,* 95.
13. Tananaev, I. V., Grunze, X., & Chudinova, N. N., (1984). Prior directions and results in the domain on condensed phosphates' chemistry. *J. Neorg. Mater., 20*(6), 887.

14. Avaliani, M., Gvelesiani, M., Barnovi, N., Purtseladze, B., & Dzanashvili, D., (2016). New investigations of poly-component systems. *J. Proc. Georgian Acad. Sci. Chem. Ser.*, *42,* 308.
15. Avaliani, M., (2016). Main Types of condensed phosphates synthesized in open systems from solution-melts of phosphoric acids. *In Materials of 4th Int. Conf. NANO-2016, GTU,* Georgia, *1,* 51.
16. Marsh, T. P., (2011). *Studies into the Ion Exchange and Intercalation Properties of $AlH_2P_3O_{10}\cdot 2H_2O$.* PhD Thesis, etheses.bham.ac.uk/1599/1/University of Birmingham.
17. Yokoi, T., Kawashita, Kawachi, G., Kikuta, K., & Ohtsuki, C., (2011). Synthesis of calcium phosphate crystals in a silica hydrogel containing phosphate ions. *J. Mater. Res., 24*(6), 2154. https://doi.org/10.1557/jmr.2009.0242.
18. Avaliani, M., (2015). *General Overview of Synthesis and Properties of a New Group of Inorganic Polymers-Double Condensed Phosphates* (p. 244). In book International Conference on Advanced Materials and Technologies ICAMT, GTU, Georgia.
19. Avaliani, M., Dzanashvili, D., Gvelesiani, M., Barnovi, N., & Shapakidze, E., (2017). About new inorganic polymers-double condensed phosphates of silver and trivalent metals. *J. Chem. Chem. Eng. USA, 11,* 60.
20. Avaliani, M., (2018). Investigation and thermal behavior of double condensed compounds of gallium, scandium, and silver. *J. Nano Studies, 17, 18,* 21.
21. Dersch, R., Steinhart, M., Boudriot, U., Greiner, A., & Wendorff, J. H., (2005). *Polymers for Advanced Technologies, 16*(2/3), 276.
22. Harris, D. C., (2010). *Quantitative Chemical Analysis* (8th edn., p. 673). Gravimetric analysis, precipitation titrations and combustion analysis, W. H. Freeman & Co.
23. Zanello, P., (2012). *Chains, Clusters, Inclusion Compounds, Paramagnetic Labels and Organic Rings* (2nd edn.). University of Milan, (Elsevier).
24. Avaliani, M., Purtseladze, B., Shohiashvili, E., & Barnovi, N., (2015). Apropos of inorganic polymers-condensed phosphates and spheres of their applications. *J. Proc. Georgian Acad. Sci. Chem. Ser.*, *41*, 227.
25. Brostow, W., & Hagg, L. H. E., (2017). *Materials: Introduction and Applications* (p. 463), John Wiley & Sons.
26. Kusnetsov, N. T., Chudinova, N. N., & Rozanov, I. A., (2004). Analysis and synthesis, harmony and counterpoint. *J. Proc. Russ. Acad. Sci., 74*, 460.
27. Avaliani, M., Chagelishvili, V., Shapakidze, E., Gvelesiani, M., Barnovi, N., Kveselava, V., & Esakia, N., (2019). Crystallization fields of condensed scandium-silver and gallium-silver phosphates. *J. Eur. Chem. Bull., 8*(5), 164.
28. Mark, J. E., Allcock, H. R., Allcock, H. R., & West, R., (2005). *Inorganic Polymers* (2nd edn.). Oxford University Press.
29. Lampila, L. E., (2013). Applications and functions of food-grade phosphates. *Ann. N.Y. Acad. Sci.* (p. 1301). Wiley Online Library.
30. Brostow, W., (2018). *Flexibility in Relation to Other Properties of Polymers* (Vol. 1, p. 5). "Polychar 26" in Materials of World Forum on Advanced Materials, Georgia.
31. Kickelbick, G., (2003). Concepts for the incorporation of inorganic building blocks into organic polymers on a nanoscale. *Progress in Polymer Science, 28*(1), 83. Elsevier.
32. Trobajo, C., Rodríguez, M. L., Suárez, M., García, J. R., Rodríguez, J., Parra, M. A., Salvad6, J. B., et al., (2011). Layered mixed tin-titanium phosphates. *J. Mater. Res., 13*(3), 754. https://doi.org/10.1557/JMR.1998.0095.

33. Yang, D., & Frindt, R. F., (2000). Structure of polymer intercalated $MnPS_3$ and $CdPS_3$. *J. Mater. Res., 15*(11), 2408.
34. Cheng, H., (2016). Inorganic dissolvable electronics: Materials and devices for biomedicine and environment. *J. Mater. Res., 31*(17), 2549. https://doi.org/10.1557/jmr.2016.289.
35. Viani, A., Sotiriadis, K., & Lanzafa, G., (2019). 3D microstructure of magnesium potassium phosphate ceramics from X-ray tomography: New insights into the reaction mechanisms. *J. Mater Sci., 54*, 3748.
36. Avaliani, M., & Shapakidze, E., (2018). Areas of crystallization of double condensed phosphates of Ag and trivalent metals and regularities of their formation. *J. Chem. Sci., 9*, 63. doi: 10.4172/2150-3494-C4-027.
37. Mukbaniani, O., Aneli, J., Markarashvili, E., & Tatrishvili, T., (2019). Fluorine-containing solvent-free polymer electrolite membranes. In: *6th International Caucasian Symposium on Polymers and Advanced Materials* (p. 77). Batumi, Georgia.

## CHAPTER 19

# New Cationic Polymers Composed of Non-Proteinogenic α-Amino Acids

NINO ZAVRADASHVILI,[1] GIULI OTINASHVILI,[1] TEMUR KANTARIA,[1] NINO KUPATADZE,[1] DAVID TUGUSHI,[1] ASHOT SAGHYAN,[2] ANNA MKRTCHYAN,[2] SERGEY POGHOSYAN,[2] and RAMAZ KATSARAVA[1*]

[1]*Institute of Chemistry and Molecular Engineering, Agricultural University of Georgia, Tbilisi, Georgia*

[2]*Institute of Pharmacy, Yerevan State University, Yerevan, Republic of Armenia*

[*]*Corresponding author. E-mail: r.katsarava@agruni.edu.ge (R. Katsarava)*

## ABSTRACT

The importance of cationic polymers (CPs) is universally recognized because they exhibit unique biological properties; promote cell penetration of various molecules and nano-constructs, etc. Biodegradable CPs, which can be cleared from the body following executing their function, look especially valuable. One of the most convenient approaches for constructing biodegradable CPs is the incorporation of hydrolyzable ester bonds in the polymeric backbones. This could be achieved, e.g., by the application of diamine-diester monomers made of cationic amino acid arginine (R) and diols-bis-(arginine)-alkylene diesters. Promising building blocks for constructing biologically active polymers are also non-proteinogenic amino acids (NPAAs), including those ones containing unsaturated bonds in the lateral chains that revealed a wide range of biological activities. The goal of the present paper is to combine these two classes of biologically active building blocks (i.e., arginine, and NPAAs) for constructing new CPs with an expanded range of potential biological activity. Such kind of CPs is of interest also for fabricating biologically active nanoparticles and as precursors of cationic hydrogels which could be obtained *via* crosslinking the linear CPs by hydrophilic cross-linkers.

## 19.1 INTRODUCTION

The importance of cationic polymers (CPs) is universally recognized since they exhibit unique biological properties. CPs can form electrostatic complexes with anionic biomolecules including nucleic acids and proteins and are of interest as active biological compound carriers to be used in both gene therapy and biotechnology. In addition, inherent bioactive properties such as antimicrobial, antioxidant, antitumor, and anti-inflammatory stimuli responsiveness, make CPs more promising for enhanced therapeutic potential [1, 2]. CPs is also of interest for imparting a positive charge to various nano-particulates that promote their penetration through cell membranes (i.e., intracellular delivery). CPs which can be cleared from the body following executing their function, i.e., biodegradable, look especially valuable.

One of the most attractive CPs is those made of naturally occurring α-amino acid (AA) L-arginine (R). R-based CPs have attracted an attention during recent years owing to findings of Ryser and Hancock [3] that only R-rich, and not lysine-rich, histones were able to transport albumin into tumor cells. Such kind of findings stimulated the search of synthetic (artificial) R-based polymers, which at the same time are biodegradable, i.e., are able to be purged from the body after their function is fulfilled. One of the first representatives of artificial R-based biodegradable CPs promising for intracellular gene delivery were cationic poly(ester amide)s, PEAs, showing at the same time good cell compatibility [4]. A variety of R-based CPs of various classes (PEAs, poly(ester urethane)s, and poly(ester urea)s) showing an excellent cell-compatibility was reported in Ref. [5]. In another approach, biodegradable R-containing CPs were obtained according to the one-pot procedure by *in situ* covalent attachment of arginine methyl ester to intermediary poly(epoxy-succinamides) [5–7]; The CPs showed an original selective transfection [6] and physiological (inhibiting nicotinic acetylcholine receptors (nAChRs)) [7] activities.

Starting from the 1980s, organic cations have also become the most important as bactericidal agents [8]. Antimicrobial peptides, composed of 12–50 amino acids, are enriched by cationic AAs such as arginine (contains guanidine group). A huge group of organic bactericides also contain guanidine group (iminourea) $H_2N\text{-}(C=NH)\text{-}NH_2$, among which one of the most known and widely used is chlorhexidine (as bis-gluconate salt) [9–11].

CPs are also of interest as precursors for obtaining cationic hydrogels after crosslinking with hydrophilic cross-linkers, e.g., with bi-functional derivatives of PEGs [12, 13].

AAs playing a central role in all processes of a live cell are widely used in the biosynthesis of molecules that have specific biological activity. As a rule, in the synthesis of peptides and other amino acid-based biopreparations only natural amino acids were previously used [14–16]. However, in the last 20 years in production of peptides and other medications enantiomerically enriched non-proteinogenic amino acids (NPAAs) containing different functional groups (aliphatic-alkenyl or alkynyl, aromatic, heterocyclic) in the lateral chain have been used [17]. Such amino acids and their derivatives were used as selective inhibitors of endothelin-converting enzymes, inhibitors of thrombin and cathepsin B, inactivators of pyridoxalphosphate-dependent $\gamma$-cystathionase, growth inhibitors of *B. subtillis* B-50, for a protein bioconjugation, derivatives of NPAAs similar to R-based compounds show antibacterial activity, etc. [18–26]. Many of the NPAAs represent very interesting building blocks from chemical point of view. Indeed, alkenyl or alkynyl side chains of the NPAAs can be functionalized by various chemical reactions resulting in a wide range of new products. For example, the NPAAs containing alkynyl side chains can be useful tools for "click" chemistry in peptidomimetic drug design or covalent modification of proteins. Polymers made of NPAAs containing alkenyl side chains are of interest for making hydrogels by crosslinking with hydrophilic cross-agents (e.g., bi-functional derivatives of PEGs) [12, 13].

The present work deals with the synthesis and characterization of new co-polymers (we call "hybrid-polymers," HPs) which can bear properties of „parent" polymers-CPs made of amino acid R and unsaturated polymers made of NPAAs. The new HPs could be of interest for new biological activities as well as platforms for fabricating bioactive nanoparticles and hydrogels for numerous biomedical applications.

## 19.2 EXPERIMENTAL METHODS AND MATERIALS

### *19.2.1 SYNTHESIS OF BIS-NUCLEOPHILIC MONOMERS (R6, ALG6, AND ALG12)*

#### *19.2.1.1 PREPARATION OF TETRA-P-TOLUENESULFONIC ACID SALT OF BIS-(ARGININE)-HEXYLENE DIESTER (R6)*

Typically, for the synthesis of R6 (Scheme 19.1) a mixture of 5.23 g (0.03 mol) of L-arginine, 1.77 g (0.015 mol) of 1,6-hexandiol and 12.55 g (0.066 mol–a slight excess) of *p*-toluenesulfonic acid monohydrate TosOH·$H_2O$ was refluxed in 50 mL of cyclohexane for 24 h. 1.73 mL of water (0.096 mol) liberated after the reaction was collected in a Dean-Stark apparatus. The precipitated white sticky product was filtered of, dried in a vacuum at r.t. After three times recrystallization from the isobutanol/hexane mixture the yield was 71%, m.p. 73–76°C [4].

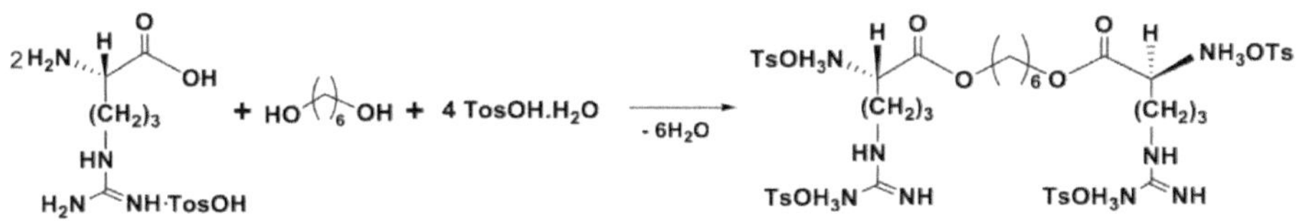


**SCHEME 19.1** The synthesis of bis-nucleophilic co-monomer-tetra-*p*-toluenesulfonic acid salt of bis-(arginine)-hexylene diester (R6).

### *19.2.1.2 PREPARATION OF DI-P-TOLUENESULFONIC ACID SALT OF BIS-(ALLYL-GLYCINE)-1,6-HEXYLEN DIESTER (ALG6)*

For the synthesis of AlG6 (Scheme 19.2) a mixture of 1.84 g (0.016 mol) of allyl-glycine, 0.95 g (0.008 mol) of 1,6-hexandiol and 3.35 g (0.0176 mol-a slight excess) of *p*-toluenesulfonic acid monohydrate TosOH·$H_2O$ was refluxed in 175 mL of cyclohexane for 24 h. 0.6 mL of water (0.0336 mol) liberated after the reaction was collected in a Dean-Stark apparatus. The precipitated white sticky product was filtered of, dried in a vacuum at 60°C. The goal product synthesized for the first time was recrystallized from acetone/ethanol mixture two times and had m.p. 154–155°C. The yield was 71% (Table 19.1).

### *19.2.1.3 PREPARATION OF DI-P-TOLUENESULFONIC ACID SALT OF BIS-(ALLYL-GLYCINE)-1,12-DODECYLEN DIESTER (ALG12)*

For the synthesis of AlG12 (Scheme 19.2) a mixture of 1.84 g (0.016 mol) of allyl-glycine, 1.62 g (0.008 mol) of 1,12-dodecanediol and 3.35 g (0.0176 mol-a slight excess) of *p*-toluenesulfonic acid monohydrate TosOH·$H_2O$ was refluxed in 175 mL of cyclohexane for 24 h. 0.6 mL of water (0.0336 mol) liberated after the reaction was collected in a Dean-Stark apparatus. The precipitated white sticky product was filtered of, dried in a vacuum at 60°C. The goal product synthesized for the first time was recrystallized from acetone/ethanol mixture two times and had m.p. 160–162°C. The yield was 75% (Table 19.1).

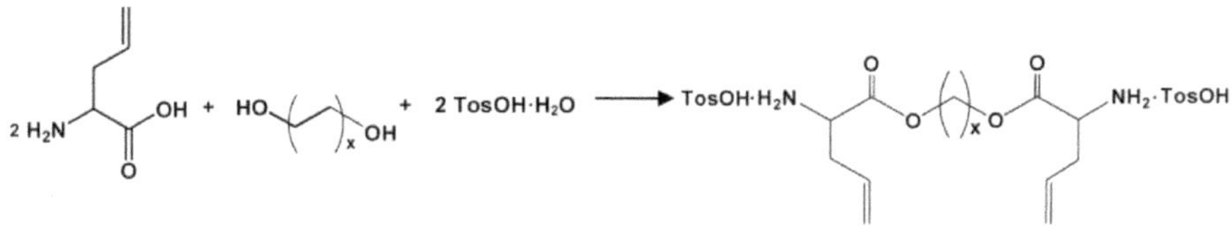

**SCHEME 19.2** The synthesis of bis-nucleophilic co-monomers-AlG6 (when x = 6) and AlG12 (when *x* = 12) based on non-proteinogenic amino acid-AlG.

### 19.2.2 SYNTHESIS OF BIS-ELECTROPHILIC MONOMER (8-N)

Activated di-p-nitrophenyl ester of sebacic acid (8-N) was synthesized by the interaction of 2.78 g (0.02 mol) p-nitrophenol with 2.39 g (0.01 mol) of sebacoyl dichloride in 40 mL of organic solvent (acetone) in the presence of 1.62 mL (0.0205 mol) of dry pyridine as described previously (Scheme 19.3). Di-*p*-nitrophenyl sebacate (8-N) was recrystallized from acetone and had m.p. 182–184°C, the yield was 88% [27, 28].

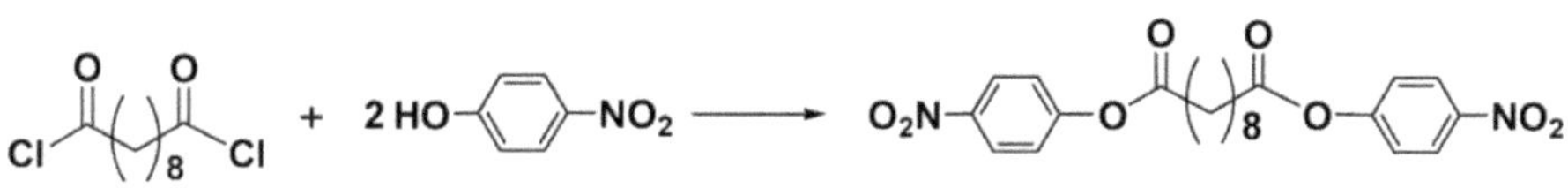

**SCHEME 19.3** The synthesis of bis-electrophilic monomer-Di-*p*-nitrophenyl sebacate (8-N).

### 19.2.3 POLYMER SYNTHESIS

The new HPs-co-polyesteramides (co-PEAs) were synthesized using solution active polycondensation method (SAcP) [4–6, 28] according to the general Scheme 19.4. In particular, 0.003 mole of a mixture of bis-nucleophilic monomers (AlG6 *or* AlG12) and (R6) at a predetermined mole ratio *k*/*l* (Scheme 19.1) and 0.003 mole of bis-electrophilic monomer such as activated di-p-nitrophenyl ester (8-N) were placed into flat-bottom flask, dry DMA (1.58 mL) and dry $NEt_3$ (0.92 mL, 0.0066 mol) (to be used as TosOH acceptor) with a total volume of 2.5 mL corresponding to a monomer concentration 1.2 mol/L, were added and stirred on the magnetic stirrer at 60°C for 24 h, the reactions were performed under an inert atmosphere of argon. Polycondensation reactions proceeded without gelation (i.e., without crosslinking) and resulted in soluble polymers. The obtained viscous solutions were cooled to r.t, and poured drop-wise into dried (using molecular sieves 4 A) acetone where the polymers were precipitated as gum-like sticky mass. The polymers obtained were washed with fresh portions of chilled acetone to remove low-molecular-weight fractions and impurities, and then were dried at 40–50°C in a vacuum up to constant weight yielding a yellowish-white powder form.

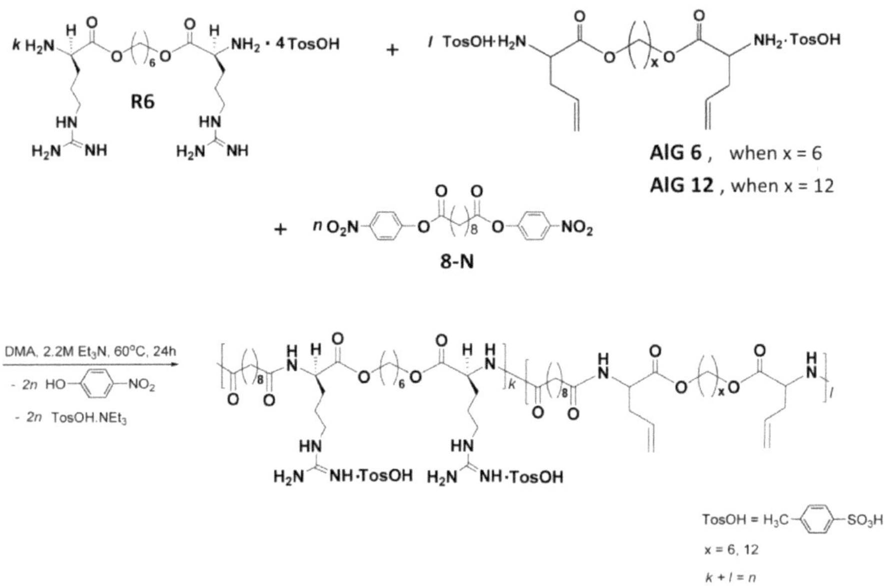


**SCHEME 19.4** Synthesis of AlG and/or R based co-PEAs *via* SAcP.
When; $k = 0.5$, $l = 0.5$, x = 6; co-PEA $[8R6]_{0.5}[8AlG6]_{0.5}$.
When; $k = 0.5$, $l = 0.5$, x = 12; co-PEA $[8R6]_{0.5}[8AlG12]_{0.5}$.

## 19.3 RESULTS AND DISCUSSION

### *19.3.1 MONOMER SYNTHESIS*

The syntheses of new nucleophilic monomers on the basis of non-proteinogenic α-amino acids (NPAAs) such as allylglycine (AlG) have been carried out. In particular, the synthesis of new bis-nucleophilic monomers-di-p-toluenesulfonic acid salts of bis-(allyl-glycine)-1,6-hexylen diester (AlG6) and bis-(allyl-glycine)-1,12-dodecylen diester (AlG12) have been done by direct condensation of AlG (2 moles) with 1,6-hexanediol/1,12-dodecandiol (1 mol) in the presence of p-toluenesulfonic acid monohydrate (2.2 mol) which was used as both amino groups protector and the condensation reaction catalyst

(Scheme 19.2). The synthesis of di-p-toluenesulfonic acid salts monomers has been modified by replacing toxic benzene and toluene (which normally are used in reported papers for their synthesis [27, 28]) with by far less toxic cyclohexane. The characteristics of newly synthesized bis-nucleophilic monomers (AlG6 and AlG12) are summarized in Table 19.1.

**TABLE 19.1** Characteristics of New Non-Proteinogenic Bis-Nucleophilic Monomers

| SL. No. | Monomer | Empirical Formula (MM) | Yield, % (After Recryst.) | m.p. °C | Solvent for Recrystalization |
|---|---|---|---|---|---|
| 1. | AlG6 | $C_{30}H_{44}N_2O_{10}S_2$ (656.81) | 71 | 153–155 | Acetone/ethanol |
| 2. | AlG12 | $C_{36}H_{56}N_2O_{10}S_2$ (891.15) | 75 | 162–164 | Acetone/ethanol |

For the constructing the goal cationic co-polymers (HPs), another bis-nucleophilic co-monomer tetra-p-toluenesulfonic acid salt of bis-(arginine)-1,6-hexylen diesters (R6) was also obtained by direct condensation of amino acid L-arginine, R, (2 moles) with 1,6-hexanediol (1 mole) in the presence of p-toluenesulfonic acid monohydrate (4 moles) according to the Scheme 19.1 [4]. The obtained bis-nucleophilic co-monomer R6 (together with AlG6 and AlG12) was applied for synthesizing the goal cationic HPs.

Activated di-p-nitrophenyl ester of long-chain sebacic acid (8-N) as a bis-electrophilic counter partner was selected rather to impart the hydrophobicity to the goal co-polymers (HPs) required for biological purposes, e.g., for the nanoparticles (NPs) formation based on them. Bis-electrophilic monomer 8-N was obtained according to Scheme 19.2 [27, 28].

### 19.3.2 POLYMER SYNTHESIS

Two new cationic co-PEAs (HPs) ($[8R6]_{0.5}[8AlG6]_{0.5}$ and $[8R6]_{0.5}[8AlG12]_{0.5}$) composed of non-proteinogenic monomers AlG6 and AlG12, and cationic arginine-based monomer R6 have been synthesized using SAcP [4–6, 28]. The aim to construct such co-polymers (HPs) was to combine two classes of biologically active polymers and to receive new HPs bearing the biological activity of both, parent polymers-CPs made of amino acid R and unsaturated polymers made of NPAAs. In other words, to construct new polymers expected to have a new biological activities.

The goal HPs were synthesized using SAcP based on bis-nucleophilic co-monomers-AlG6, AlG12, and R6 (described above). Activated di-p-nitrophenyl ester, di-*p*-nitrophenyl sebacate (8-N) was selected as a bis-electrophilic counter partner in SacP rather to impart the hydrophobicity to the goal HPs synthesized according to the general Scheme 19.3. In brief, one mole of a mixture of bis-nucleophilic monomers (R6) and (AlG6/AlG12) at a predetermined mole ratio *k/l* were interacted with one mole of bis-electrophilic monomer (8-N) in polar organic solvent (DMA) in the presence of 2.2 moles of triethylamine to be used as an acid (TosOH) acceptor. The polycondensation reactions were carried out at 60°C, under this condition reaction mixtures were homogenized and proceeded for 24 h. We have found that the polycondensation reactions (at a k/l = 50/50 ratios) proceeded without gelation (i.e., without crosslinking) and resulted in soluble polymers. This means that the synthetic strategy we have developed for the design of new HPs works and results in soluble polymers. The obtained HPs were precipitated and purified in acetone to remove low-molecular-weight fractions and impurities, and then were dried. HPs received were characterized by standard physical-chemical methods. Monomer ratios in the co-polymers were estimated from the 1H NMR spectra. The characteristics of the HPs are summarized in Table 19.2.

**TABLE 19.2** Properties of New Cationic HPs

| SL. No. | Structure | Yield, % | Monomer Ratio, Arg/AlG | | $\eta_{red}$*, dL/g |
|---|---|---|---|---|---|
| | | | Feed Ratio | Found Ratio** | |
| 1. | $[8R6]_{0.5}[8AlG6]_{0.5}$ | 85 | 0.50: 0.50 | 0.45: 0.55 | 0.25 |
| 2. | $[8R6]_{0.5}[8AlG12]_{0.5}$ | 78 | 0.50: 0.50 | 0.53: 0.47 | 0.18 |

*The reduced viscosity ($\eta_{red}$) of the polymers was determined in DMF solutions at concentrations of 5 g/L at 25°C using an automatic viscometer (Lauda PVS 1, Königshofen, Germany).

**Calculated on the bases of $^1H$ NMR spectra of CPs ($^1H$ NMR (400 MHz) spectra were recorded on a JEOL JNM-ESC400 spectrometer using DMSO-*d6* as a solvent calibrated using tetramethylsilane (TMS) as the internal standard).

## 19.4 CONCLUSION

New cationic co-polymers (HPs) composed of non-proteinogenic bis-nucleophilic monomers AlG6 and AlG12, and cationic arginine-based co-monomer R6 have been synthesized using the SAcP. For this, modified method of the synthesis of bis-nucleophilic monomers-di-p-toluenesulfonic acid salts of bis-(NPAA)-alkylene diesters (AlG6 and AlG12)-has been elaborated using by far less toxic cyclohexane as a reaction media. The goal co-polymers were successfully synthesized under the condition of SAcP which means the synthetic strategy we have developed for constructing HPs works and results in soluble polymers. Although the polycondensation experiments were done at k/l = 50/50 ratio of bis-nucleophilic monomers so far it is also possible to vary the mole ratio (k/l) of R/AlG rather to expend the new family of HPs with increased biological activities. The incorporation of NPAAs into the polymeric backbones can result in multiple positive effects such as: enhanced stability compared to NPAAs based peptides, better bioavailability (e.g., *via* the preparation of nanoparticles), increased membrane penetration (e.g., *via* imparting cationic nature to macromolecules), prolonged action and decreased immunogenicity. The incorporation of R-based cationic fragments into the new polymeric backbones could enhance the cellular uptake of the nanoparticles hence increasing their biological activity. The new CPs containing double-bond moieties of AlG residues are also of interest for preparing hydrogels via crosslinking using, e.g., bis-acryloyl polyethyleng-lycols (PEGs) as hydrophilic cross-linkers.

## ACKNOWLEDGMENT

This work was supported by ISTC [Grant # A-2289 "Synthesis and screening of a new generation of optically active non-proteinogenic α-amino acids, peptides, and polymers containing unsaturated groups in the side chain"].

## KEYWORDS

- **amino acid**
- **arginine**
- **cationic polymer**
- **glycine**
- **hybrid polymer**
- **non-proteinogenic amino acid**

## REFERENCES


1. Samal, S. K., Dash, M., Van, V. S., Kaplan, D. L., Chiellini, E., Van, B. C., Moronid, L., & Dubrue, P., (2012). Cationic polymers and their therapeutic potential. *Chem. Soc. Rev., 41*, 7147–7194.
2. Moroson, H., & Rotman, M., (1975). Biomedical applications of polycations. In: Rembaum, A., & Sélégny, E., (eds.), *Springer Book Series-Charged and Reactive Polymers: Polyelectrolytes and their Applications* (Vol. 2, pp. 187–195).
3. Ryser, H., & Hancock, R., (1965). Histones and basic poly(amino acid)s stimulate the uptake of albumin by tumor cells in culture. *Science, 150*, 501–503.
4. Memanishvili, T., Zavradashvili, N., Kupatadze, N., Tugushi, D., Gverdtsiteli, M., Torchilin, V. P., Wandrey, C., et al., (2014). Arginine-based biodegradable ether-ester polymers with low cytotoxicity as potential gene carriers. *Biomacromolecules, 15*, 2839–2848.
5. Zavradashvili, N., Memanishvili, T., Kupatadze, N., Baldi, L., Shen, X., Tugushi, D., Wandrey, C., & Katsarava, R., (2014). In: Adhikari, R., & Thapa, S., (eds.), *Advances in Experimental Medicine and Biology: Infectious Diseases and Nanomedicine I* (Vol. 807, pp. 59–73). Springer: New Delhi.
6. Zavradashvili, N., Sarisozen, C., Titvinidze, G., Otinashvili, G., Kantaria, T., Tugushi, D., Puiggali, J., et al., (2019). Library of cationic polymers composed of polyamines and arginine as gene transfection agents. *ACS Omega, 4*(1), 2090–2101.
7. Lebedev, D. S., Kryukova, E. V., Ivanov, I. A., Egorova, N. S., Timofeev, N. D., Spirova, E. N., Tufanova, E. Y., et al., (2019). Oligoarginine peptides: A new family of nicotinic acetylcholine receptor inhibitors. *Molecular Pharmacology, 96*(5), 664–673.
8. Kenawy, E. R., Worley, S. D., & Broughton, R., (2007). The chemistry and applications of antimicrobial polymers: A state-of-the-art review. *Biomacromolecules, 8*, 1359–1384.
9. Jenkins, S., Addy, M., & Wade, W., (1988). The mechanism of action of chlorhexidine. A study of plaque growth on enamel inserts *in vivo*. *J. Clin. Periodontol., 15*(7), 415–424.
10. Manetti, F., Castagnolo, D., Raffi, F., Zizzari, A. T., Rajamäki, S., D'Arezzo, S., Visca, P., Cona, A., et al., (2009). Synthesis of new linear guanidines and macrocyclic amidinourea derivatives endowed with high antifungal activity against *Candida* spp. and *Aspergillus* spp. *J. Chem. Med., 52*, 7376–7379.
11. Hensler, M. E., Bernstein, G., Nizet, V., & Nefzi, A., (2006). Pyrrolidine bis-cyclic guanidines with antimicrobial activity against drug-resistant gram-positive pathogens identified from a mixture-based combinatorial library. *Bioorganic and Medicinal Chemistry Letters., 16*, 5073–5079.
12. Pang, X., & Chu, C. C., (2010). Synthesis, characterization and biodegradation of poly(ester amide)s based hydrogels. *Polymer, 51*(18), 4200–4210.
13. Pang, X., Wu, J., Chu, C. C., & Chen, X., (2014). Development of an arginine-based cationic hydrogel platform: Synthesis, characterization and biomedical applications. *Acta Biomaterialia, 10*(7), 3098–3107.
14. Rutjes, F. P., Wolf, L. B., & Schoemaker, H. E., (2000). Applications of aliphatic unsaturated non-proteinogenic α-H-α-amino acids. *J. Chem. Soc., Perkin Trans. I, 24*, 4197–4212.
15. Wiesmuller, K. H., (1996). In: Jung, G., (ed.), *Combinatorial Peptide and Nonpeptide Libraries: A Handbook*. Wiley-VCH, Weinheim.

16. Stevenazzi, A., Marchini, M., Sandrone, G., Vergani, B., & Lattanzio, M., (2014). Amino acidic scaffolds bearing unnatural side chains: An old idea generates new and versatile tools for the life sciences. *Bioorg. Med. Chem. Lett., 24*(23), 5349–5356.
17. Lee, K., Hwang, S. Y., & Park, C. W., (1999). Thrombin inhibitors based on a propargyl-glycine template. *Bioorg. and Med. Chem. Lett., 9*(7), 1013–1018.
18. Kaiser, J., Kinderman, S. S., Van, E. B. C., Van, D. F. L., Schoemaker, H. E., Blaauw, R. H., & Rutjes, F. P., (2005). *Organic and Biomolecular Chemistry, 3*(19), 3435–3467.
19. Wolf, L. B., Sonke, T., Tjen, K. C., Kaptein, B., Broxterman, Q. B., Schoemaker, H. E., & Rutjes, F. P., (2001). *Advanced Synthesis and Catalysis, 343*, 662–674.
20. Washtien, W., & Abeles, R. H., (1977). Mechanism of inactivation of γ-cystathionase by the acetylenic substrate analog propargylglycine. *Biochemistry, 16*(11), 2485–2491.
21. Marcotte, P., & Walsh, C., (1976). Vinylglycine and propargylglycine: Complementary suicide substrates for L-amino acid oxidase and D-amino acid oxidase. *Biochemistry, 15*(14), 3070–3076.
22. Cheung, K. S., Wasserman, S. A., Dudek, E., Lerner, S. A., & Johnston, M., (1983). Chloroalanyl and propargylglycyl dipeptides. Suicide-substrate-containing antibacterials. *J. Med. Chem., 26*(12), 1733–1741.
23. Wallace, E. M., Moliterni, J. A., Moskal, M. A., Neubert, A. D., Marcopulos, N., Stamford, L. B., Trapani, A. J., et al., (1998). Design and synthesis of potent, selective inhibitors of endothelin-converting enzyme. *J. Med. Chem., 41*(9), 1513–1523.
24. Aoyagi, Y., & Sugahara, T., (1985). 2 (S)-Aminohex-5-ynoic acid, an antimetabolite from *Cortinarius* claricolor var. tenuipes. *Phytochemistry, 24*(8), 1835–1836.
25. Maza, J. C., McKenna, J. R., Raliski, B. K., Freedman, M. T., & Young, D. D., (2015). Synthesis and incorporation of unnatural amino acids to probe and optimize protein bioconjugations. *Bioconjugate Chem., 26*(9), 1884–1889.
26. Fanelli, R., Jeanne-Julien, L., René, A., Martinez, J., & Cavelier, F., (2015). Stereoselective synthesis of unsaturated α-amino acids. *Amino Acids, 47*(6), 1107–1115.
27. Katsarava, R., (2003). Active polycondensation: From peptide chemistry to amino acid based biodegradable polymers. In: *Macromolecular Symposia* (Vol. 199, No. 1, pp. 419–430). Weinheim: WILEY-VCH Verlag.
28. Katsarava, R., Beridze, V., Arabuli, N., Kharadze, D., Chu, C. C., & Won, C. Y., (1999). Amino acid-based bioanalogous polymers. Synthesis, and study of regular poly (ester amide)s based on bis (α-amino acid) α, ω-alkylene diesters, and aliphatic dicarboxylic acids. *J. Polym. Sci. Part A: Polymer Chem., 37*(4), 391–407.

## CHAPTER 20

# Investigation of the Causes of Stability Violation of Propagating Polymerization Heat Waves in the Process of Frontal Polymerization

ANAHIT O. TONOYAN, ARAM H. MINASYAN, ANAHIT Z. VARDERESYAN, ARMENUHI G. KETYAN, and SEVAN P. DAVTYAN

*National Polytechnic University of Armenia, Department of General Chemistry, and Chemical Processes, Teryan Str., 105, Yerevan–0009, Armenia, E-mail: atonoyan@mail.ru (A. O. Tonoyan)*

### ABSTRACT

In this chapter, the causes of the stability violation of self-propagating heat waves in the process of frontal polymerization (FP) are studied. FP is an autowave process of the propagation of polymerization heat waves. One of the major factors for the practical implementation of FP is the necessity to establish the causes and boundaries of the stability violation of propagating heat waves in the process of FP. The paper clarifies one of the causes of stability violation during FP of complexes of acrylamide (AAm) with transition metals. Taking into account that, for the specified monomers, the cause of stability violation and appearance of spin modes is shrinkage of polymer, we have investigated nanoparticle additions to the polymerizing media. It is shown that it is possible to regulate the stability violation during FP of the monomers depending on the amount of nanoparticle additions.

### 20.1 INTRODUCTION

In relation with the synthesis of functional gradient materials (FGM), which is one of the main advantages of frontal polymerization (FP) [1, 2], it has become necessary to study the stability of FP heatwaves for the synthesis of high-temperature superconducting polymer composites with an objective of developing a strong linkage between superconducting composite and polymer according to a prescribed program. As it is known [3, 4], superconducting composites were obtained on the basis of cobalt- and nickel-complexes of acrylamide (AAm) using additions of superconducting ceramic, and studied for superconducting properties, transitions, and electrical conductivity. However, in terms of the adhesion of these composites to the required polymers for the synthesis of FGM, we faced with the task of a detailed investigation of the polymerization of metal-complexes of AAm both with and without additions. In Ref. [5], FP of the monomers, as well as stability violation and appearance of spin modes, were observed. It was shown that the main cause of the above-mentioned phenomena is the shrinkage of polymer in the process of FP. To verify this, we specifically carried out FP in strict adiabatic conditions and confirmed that there was no heat loss from the walls of the reaction vessel [5]. A possible mass transfer from a hot polymeric part to a cool monomeric was checked as well. To that end, experiments were carried out where FP was initiated both from the top-down and from the bottom up. The data obtained were consistent, and the possibility of mass transfer was also excluded. Thus, in the work [5], the main causes of stability loss as a result of shrinkage of polymer were established.

Accordingly, the investigations with the additions of nanoparticles make it possible to synthesize compatible components of FGM by the method of FP.

## 20.2 EXPERIMENTAL SECTION

Reagents used: AAm (>99%) (AAm), cobalt nitrate hexahydrate (99.5%) ($Co(NO_3)_2$ $6H_2O$), nickel nitrate hexahydrate (99.5%) ($Ni(NO_3)_2{\cdot}6H_2O$), silicon dioxide nanoparticles (10 nm, 99.9%) (all purchased from Sigma-Aldrich), bentonite from the local origin (63–80 μm, 99.5%).

Synthesis of cobalt, and nickel-containing acrylamide complexes ($Co(AAm)_4(NO_3)_2$, $Ni(AAm)_4(NO_3)_2$) was carried out according to the method described in Ref. [6].

The FP of the monomers was carried out in glass ampoules with a diameter of 8 mm. The mechanical mixtures of the monomer and nanoparticles and bentonite were prepared by mixing the substances for a long time to thoroughly distribute the nanoparticles in a monomeric medium. The prepared mixtures were then loaded into the ampoules. To commence the FP reaction of the monomeric mixtures, an instant heat was locally applied on the upper end of the ampoules.

The measurements were taken using thermocouples located in the recess holes on the reaction ampoules. The samples obtained were examined on microscope MB30.

## 20.3 RESULTS AND DISCUSSION

The FP of the monomers was investigated using thermocouples. The temperature profiles of the processes are given in Figures 20.1 and 20.2 (the thermocouples were located at a distance of 3 mm from each other).

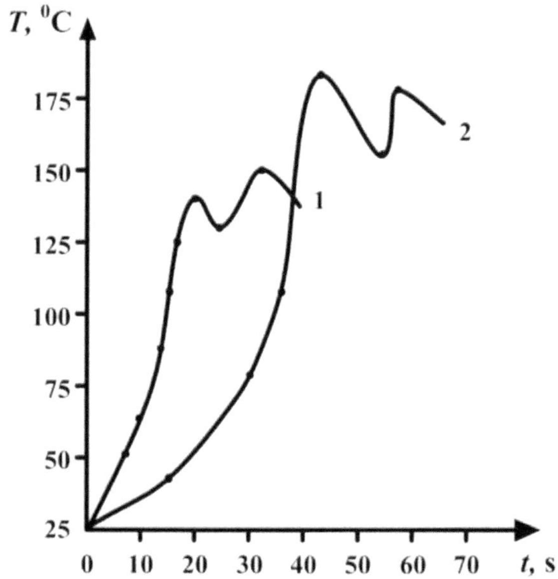


**FIGURE 20.1** Temperature profiles of FP of Ni (curve 1) and Co (curve 2) complexes of AAm.

As can be seen from the figures, the temperature of the waves reach the adiabatic heating temperature for the polymerizing monomer and, at a certain moment, a part of the polymer formed during FP shrinks and gets detached from the monomer. At this detachment zone, the wave temperature declines, which in turn decreases the polymerization rate. When a sufficient amount of heat, released from the exothermic polymerization, transfers to the neighboring layer of the monomer on account of thermal conductivity, the FP process is restored, and the temperature again reaches the adiabatic heating temperature.

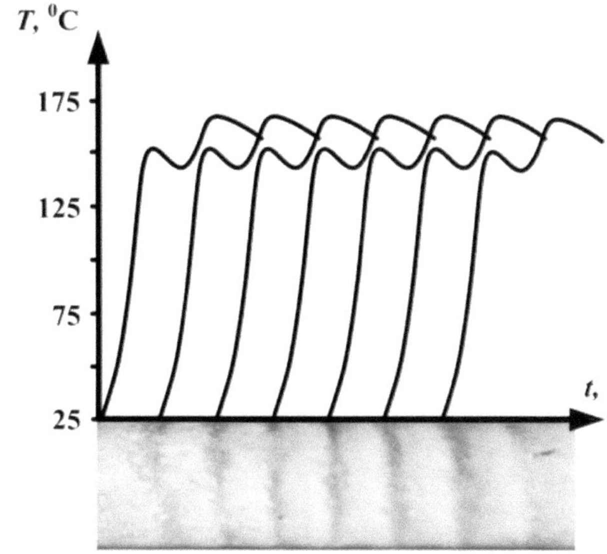


**FIGURE 20.2** Temperature profiles of FP of Ni-complex of AAm in comparison with spin rings formed on the samples as a result of polymer shrinkage.

Figure 20.3 shows the temperature profiles of FP of $Co(AAm)_4(NO_3)_2$ with additions of different amounts of $SiO_2$ nanoparticles. As can be seen from the figure, FP of $Co(AAm)_4(NO_3)_2$ proceeds with formations of instabilities and spin modes (curve 1). The addition of up to 5 wt.% of $SiO_2$ nanoparticles leads to a decrease in the gap between the polymer and monomer (curve 2), which, in turn, may be resulted from intermolecular interactions of the nanoparticles with the polymer chains [6].

Next, the further increase (10–15 wt.%) in the amount of the nanoparticles leads to a uniform propagation of the FP heat waves (curve 3, 4) with a decrease in both the rate and maximal adiabatic heating of the polymerization process. The behavior of the curves in relation to the adiabatic heating of the FP of $Co(AAm)_4(NO_3)_2$ is shown in Figure 20.4.

Figure 20.5 shows the samples of nanocomposites obtained by FP of $Co(AAm)_4(NO_3)_2$ with the additions of 0 (a), 5 (b), 10 (c), 15 (d), and 20 (e) wt.% of $SiO_2$ nanoparticles. It is clearly seen from the figure that addition of the nanoparticles up to 10 wt.% gives the samples a homogeneous structure

(Figures 20.5(b) and 20.5(c)). The fact is that, as mentioned above, the nanoparticles, interacting with the polymer chains, attract them to each other and change the structure of the resulting composite. Nevertheless, with a further increase in the amount of nanoparticles, the structure of the obtained composites becomes inhomogeneous with the formation of uneven ruptures of the polymer matrix of the composite (Figures 20.5(c) and 20.5(d)).

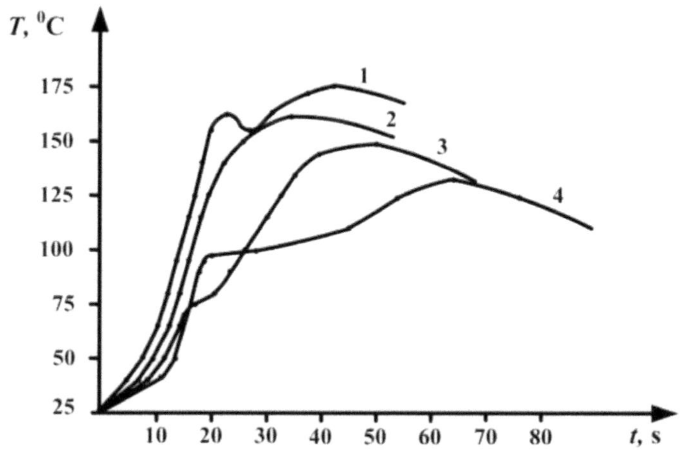


**FIGURE 20.3** Temperature profiles of FP of $Co(AAm)_4(NO_3)_2$ with the additions of 0 (curve 1), 5 (curve 2), 10 (curve 3), 15 (curve 4), and 20 (curve 5) wt.% of $SiO_2$ nanoparticles.

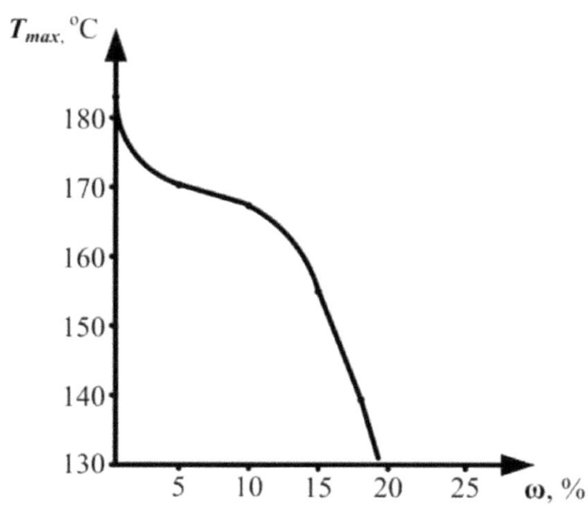


**FIGURE 20.4** Dependence of the maximal adiabatic heating temperature of FP of $Co(AAm)_4(NO_3)_2$ on the concentration of $SiO_2$ nanoparticles.

**FIGURE 20.5** Samples of nanocomposites obtained by FP of $Co(AAm)_4(NO_3)_2$ with the additions of 0 (a), 5 (b), 10 (c), 15 (d), and 20 (e) wt.% of $SiO_2$ nanoparticles.

The samples of the obtained nanocomposites were also examined on the microscope MB30. The microimages of the composites (Figure 20.6) confirm the conclusions made. As can be seen from the microscopy photographs, indeed, with additions of nanoparticles more than 15 wt.%, the structure of the nanocomposites becomes non-uniform and crumbly (Figures 20.6(c) and 20.6(d)).

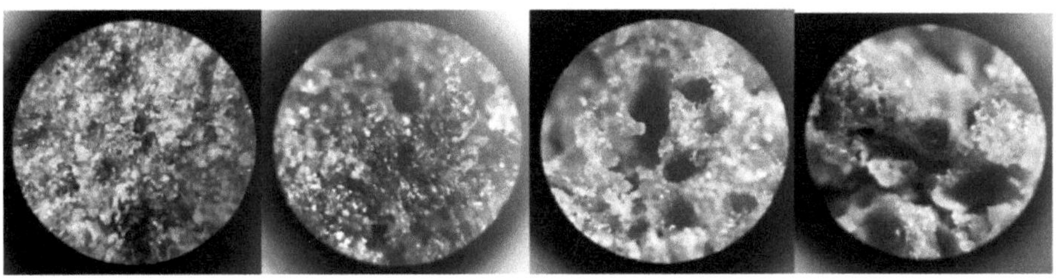


**FIGURE 20.6** Microscopic images of the samples of nanocomposites (at 100 × magnification) containing 5 (a), 10 (b), 15 (c), and 20 (d) wt.% of $SiO_2$ nanoparticles.

FP of $Co(AAm)_4(NO_3)_2$ with the addition of 5, 10, 15, and 20 wt.% of bentonite as an inert filler was investigated as well. The results are demonstrated in Figure 20.7.

As seen from the figure, the shrinkage area decreases as the bentonite concentration increases. This phenomenon is caused by the dilution of the monomer with inert filler that reduces the front velocity, and consequently, the shrinkage of the resulting composite.

## 20.4 CONCLUSION

The work presents investigations of stability loss of self-propagating heat waves in the process of FP. In the chapter, the effects of nanoparticles on the stationarity of FP heatwaves were studied. From the presented data, it can be concluded that the additions of nanoparticles diminish the effect of polymer shrinkage, and therefore instabilities and spin modes. At the same time, however, the denser filling disrupts the smoothness of the propagation of the heat waves, which is due to the agglomeration of the nanoparticles as a result of their very tight packing. From the presented data, it can be concluded that by adjusting the quantity and quality of nanoparticles, it is possible to regulate the properties of the resulting composites. The results indicate that, by means of nanoparticles, it is possible to regulate the kinetics of the FP process, and therefore, obtain polymeric nanocomposites with prescribed properties and synthesize compatible components of FGM by the method of FP.

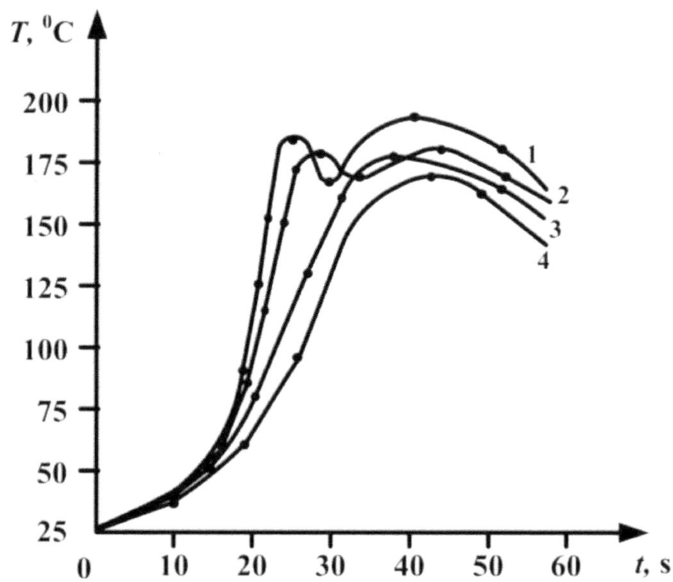


**FIGURE 20.7** Temperature profiles of FP of $Co(AAm)_4(NO_3)_2$ with the addition of 5 (curve 1), 10 (curve 2), 15(curve 3), and 20 (curve 4) wt.% of bentonite.

## ACKNOWLEDGMENT

This work was supported by the Science Committee of the Ministry of Education and Science of Armenia.

## KEYWORDS


- **acrylamide**
- **frontal polymerization**
- **functional gradient materials**
- **nanoparticles**
- **polymer shrinkage**
- **stability loss**


## REFERENCES


1. Davtyan, S. P., & Tonoyan, A. O., (2014). *Theory and Practice of Adiabatic and Frontal Polymerization.* Palmarium Academic Publishing: Germany.
2. Davtyan, S. P., & Tonoyan, A. O., (2018). The frontal polymerization method in high technology applications, *Rev. J. Chem., 8*(4), 432–455.
3. Davtyan, S. P., & Tonoyan, A. O., (2017). Possibilities of current carrying superconducting polymer-ceramic nanocomposites obtainment. *Chemical Engineering of Polymers. Production of Functional and Flexible Materials, Part III: Materials and Properties* (p. 482).
4. Davtyan, S. P., Tonoyan, A. O., Michaelyan, A. R., & Stefan, M., (2017). Charge-carrying superconducting polymer-ceramic nanocomposites. *Chemical Journal of Armenia 70* (1/2), 254–264.
5. Tonoyan, A. O., Minasyan, A. H., Varderesyan, A. Z., Ketyan, A. G., & Davtyan, S. P., (2019). Influence of shrinkage of polymer on the stationarity of frontal polymerization heat waves. *J. Polym. Eng., 39*(8), 769–773.
6. Tonoyan, A. O., Davtyan, D. S., Varderesyan, A. Z., Hamamchyan, M. G., Davtyan, S. P., **(**2016**)**. Synthesis of bentonite and diatomite-containing polymer nanocomposites and their characteristics. In: Mukbaniani, O. V., Mark, J. M. A., & Tatrishvili, T., (eds.), *Performance Polymers for Engineering-Based Composites. Part 2: Engineered-Based Composites and Models* (pp. 203–218). Apple Academic Press; Chapter 19.
7. Pomogailo, A. D., & Savostyanov, V. S., (1988). *Metal-Containing Monomers and Polymers on the Basis of Them.* Moscow.

# CHAPTER 21

# Synthesis and Investigation of Properties of Comb-Type Methylsiloxane Copolymers with Pendant Diphenylsiloxane Groups

TAMARA TATRISHVILI[1,2], KALOIAN KOYNOV[3], and OMARI MUKBANIANI[1,2]

[1]*Department of Macromolecular Chemistry, Ivane Javakhishvili Tbilisi State University, I. Chavchavadze Ave., 1, 0179 Tbilisi, Georgia*

[2]*Institute of Macromolecular Chemistry and Polymeric Materials, Ivane Javakhishvili Tbilisi State University, I. Chavchavadze Ave., 1, 0179 Tbilisi, Georgia*

[3]*Max Planck Institute for Polymer Research, Department of Physics of Interfaces, Ackermannweg 10 D-55128 Mainz, Germany,*
*E-mail: tamar.tatrishvili@tsu.ge*

## ABSTRACT

Dehydrocondensation reactions of linear trimethylsiloxy group terminated methyl-hydro siloxane and methyl-hydro siloxane-dimethylsiloxane copolymers with α-hydroxyl-ω-trimethylsiloxydiphenylsiloxanes in the presence of dry potassium hydroxide have been studied and comb-type copolymers and block-copolymers (BCP) with various lengths of the side diphenylsiloxane fragments have been obtained. Depending on the lengths of the rigid and flexible initial oligomers microdomain structure of BCP were observed. The synthesized BCP were characterized by gel permeation chromatography, differential scanning calorimetric and X-ray methods.

## 21.1 INTRODUCTION

The possibility of realizing novel technical solutions in many branches of industry is always defined by thermal and frost resistance of modern elastic materials. That is why development of methods for obtaining new polymers for composites with improved resistance to destabilizing influence of high and low temperatures is always relevant.

One of the most interesting ways in this direction is synthesis of block-copolymers (BCP), which macromolecules represent a "hybrid" of two or more blocks different by chemical structure or composition.

In the majority of cases, thermodynamic incompatibility of the blocks induces stable microphase splitting, which, finally, allows an original combination of properties of heterogeneous fragments of BCP. Depending on difference in the chemical origin of the blocks, their length, number, and alternation sequence, as well as their ability to crystallize, materials with structures and properties, significantly different from the properties of the initial components, may be prepared. Of special interest are BCP, composed of blocks with different physical properties: glass transition temperature, ability to crystallize, solubility parameters, etc. Typically, soft, flexible blocks are combined with more rigid blocks that impose increased mechanical strength and, in some cases, thermal stability of BCP. Mechanical properties of BCP generally depend on both length of flexible and rigid blocks and their ratio in the macromolecular chain.

The main aim of this proposal is the synthesis of regular comb-type copolymers combining soft poly (dimethylsiloxane) blocks and rigid blocks with diphenylsiloxane fragments in the side chain and investigation of their properties. The poly(dimethylsiloxane) blocks possessing glass transition temperature of- 123°C and dielectric constant of 28 impart high elasticity at low temperatures, good dielectric properties, high resistance to ozone-induced aging, durability, and good gas permeability. On the other hand, the more rigid high-melting diphenylsiloxane blocks imparts high strength and other valuable properties to the BCP. These newly synthesized BCP materials can be applied as stationary phases for gas-liquid chromatography or thermoplastic elastomers.

There are two main routes for synthesis of comb-type polymers. One is polymerization of unsaturated or cyclic monomers, in which side fragments with mesogenic properties already exist in initial monomer molecules. Alternatively, one can attach new units to a preexisting matrix. We have decided to follow the second route and to obtain new comb-type organosilicon oligomers by insertion of various lengths of diphenylsiloxane side fragments into methylhydrosiloxane oligomers. Our approach has several advantages. For example, we will use commercially available polymethylhydrosiloxane (PMHS), which is one of the most interesting products in modification reactions [1–6]. Furthermore, the catalyst that we will use is one of the cheapest available.

It is known [7] that compounds with ≡Si-H bond interact with water with the elimination of hydrogen in the presence of alkali catalysts. The reaction of catalytic dehydrocondensation between ≡Si-H containing compounds and hydroxyl-containing organic compounds proceeds similarly. The reaction capacity of ≡Si-H bond is determined by the nature and concentration of catalyst, temperature, reactant concentration, solvent character, etc., [8, 9].

In the reaction of catalytic dehydrocondensation of hydroxyorganosilanes with hydroorganosilanes such catalysts as colloidal nickel, anhydrous zinc chloride, platinum hydrochloric acid, platinum on the carbon and amines [10, 11] were used and the reactions proceeds according to the following scheme:

$$\equiv Si\text{-}H + HO\text{—}Si\equiv \xrightarrow[-H_2]{Cat.} \equiv Si\text{—}O\text{—}Si\equiv$$

The reactions of catalytic dehydrocondensation of hydride-containing organosiloxanes, with hydroxyl-containing organosilanes and organosiloxanes in the presence of powder-like caustic potassium as nucleophilic catalyst were studied by several groups [12, 13]. It has been shown [14, 15] that under the action of nucleophilic catalysts the competing reactions of dialkylsiloxane elimination does not take place.

By catalytic dehydrocondensation of linear α,ω-dihydridediorganosiloxanes with α,ω-dihydroxydiorganosiloxane in the presence of caustic potassium, depending on the length of the siloxane fragment, both individual organosiloxanes and linear copolymers were synthesized [16]. Dehydrocondensation reaction of trimethylsiloxy-group terminated ethylhydridesiloxane with hydroxyorganocyclosiloxane in the presence of caustic potassium or platinum hydrochloric acid studied and ethylsiloxane oligomers with cyclic fragments in the side chain were obtained [17]. It was established that not all ≡Si-H bonds take place in dehydrocondensation reaction and various-linked oligomers soluble in organic solvents were obtained. Therefore, catalytic amount of caustic potassium in dehydrocondensation reaction did not cause opening of siloxane backbone. In the literature, there is practical information on the reaction of catalytic dehydrocondensation of hydride-containing organocyclosiloxanes with hydroxyl-containing organocyclosiloxanes.

Therefore, the aim of presented work is investigation of dehydrocondensation reactions of linear trimethylsiloxy group terminated methylhydrosiloxane and methylhydrosiloxane-dimethylsiloxane copolymers with α-hydroxyl-ω-trimethylsiloxydiphenylsiloxanes in the presence of dry potassium hydroxide and synthesis of comb-type copolymers and BCP with various lengths of the side diphenylsiloxane fragments and study their properties. The reaction order, rate constants, and activation energies of dehydrocondensation reaction.

## 21.2 EXPERIMENTAL PART

### *21.2.1 MATERIALS*

Dichloro diphenylsilane, dihydroxy diphenylsilane, chloro trimethylsilane was purchased Aldrich and were used as received. Trimethylsiloxy group terminated PMHS with degree of polymerization ($n = 35, \bar{M}_n = 2262$, Si-H% = 1.55) were also purchased from Aldrich. Toluene was dried over and distilled from sodium under an atmosphere of dry nitrogen.

The initial α,ω-dichlorodiphenylsiloxanes and corresponding α,ω-dihydroxydiphenylsiloxanes have been obtained by heterofunctional condensation of dihydroxydiphenylsilane with dichlorodiphenylsilane at various ratio of initial compounds in the presence of dry pyridine and toluene according to the well-known method [18].

- Heterofunctional condensation reaction of 1.5-dihidroxyhexaphenyltrisiloxane with trimethylchlorosilane. To a 10 mL dry toluene, simultaneously was added solution of 2.9866 g (0.0488 mole) 1.5-dihidroxyhexaphenyltrisiloxane and 0.772 g (0.0488 mole) pyridine in 20 mL dry toluene and the solution of 0.52 g (0.0488 mole) trimethylchlorosilane in 20 mL dry toluene at 5 to 0°C temperature. The reaction mixture was stirred for 5 h at room temperature and then heated for 2 h up to the boiling point of toluene. Then, the reaction mixture was filtered, washed, concentrated, and precipitated with methanol and subjected to vacuum up to constant weight. White-like compound I (3.1 g, 92.8%) was obtained. The synthesis of other monohydroxyl-derivatives was carried out analogically with the aforementioned method.

- Copolymerization reaction of α,ω-bis(trimethylsiloxy)methylhydrosiloxane with octamethylcyclotetrasilxoxane have been carried out according to the well-known method [21]. To a 22.62 g (0.01 Mole) of α,ω-bis(trimethylsiloxy)methylhydrosiloxane and 17.76 g (0.06 Mole) of octamethylcyclotetrasiloxane in 100 mL dry toluene 0.01% anhydrous potassium hydroxide was added and heated up to 80–90°C. The polymerization reaction was controlled by gas-liquid chromatography up to the disappearance of the traces of octamethylcyclotetrasiloxane. Then the reaction mixture was washed by water up to the neutral area, dried on the anhydrous Sodium sulfate, filtered, the solvent was removed and dried on the vacuum up to constant mass. The 37.0 g (92%) transparent viscous product have been obtained. Where: Si-H% content = 0.84, m≈35, p≈24; $\eta_{sp}$ ($\bar{M}\omega$≈4038). Calculated values: Si-H % = 0.86; M = 4038.
- Dehydrocondensation reaction of α-hydroxyl-ω-trimethylsiloxydiphenylsiloxane with α,ω-bis(trimethylsiloxy)methylhydrosiloxane. Catalytic dehydrocondensation reaction was carried out in a two-necked flask equipped with a catalyst inlet tube and reflux condenser connected with a gasometer. Between the gasometer and the reflux condenser, a cold trap and wash bottle were installed. The reaction products were placed into the flask and dissolved in absolute toluene and thermostat in an oil bath until the constant temperature was achieved, and then the catalyst was introduced. After which, hydrogen having ceased to be released, the reaction products were washed until neutral area from the used catalyst, dried over anhydrous $Na_2SO_4$, precipitated from toluene solution by n-hexane, and subjected to vacuum up to constant weight.

### 21.2.2 CHARACTERIZATION

FTIR investigation has been carried out on Nicolet Magna 850 FTIR spectrometer. Differential scanning calorimetric (DSC) investigation was performed on a Nietzsche DSC 200 F3 Maia apparatus. The glass-transition temperatures (Tg) were read from endothermic DSC traces, which were approximated to be midpoints between the extrapolated tangents to the baselines above and below the glass-transition region. The heating and cooling scanning rates were 10 K/min.

Gel-permeation chromatography (GPC) was carried out with a Waters Model 6000A chromatograph with an R 401 differential refraction meter detector.

The column set comprised of $10^3$ and $10^4$ Å ultrastyragel columns. Sample concentration was approximately 3% by weight in toluene; typical injection volume for the siloxane was 5 μL, flow rate about 1.0 ml/min. Standardization of the GPC was accomplished by the use of styrene or polydimethylsiloxane standards with the known molecular weight. Determination of ≡Si-H content was calculated according to the method described in [20].

## 21.3 RESULTS AND DISCUSSION

### *21.3.1 SYNTHESIS OF MONOMERS*

In the literature, there are no data about comb-type siloxane-siloxane copolymers with the specified arrangement of linear diphenylsiloxane fragments as a lateral group. For the synthesis of comb-type copolymers and BCP on a first stage, we have obtained the initial compounds-α-hydroxyl-ω-trimethylsiloxydiphenylsiloxanes. We have investigated the heterofunctional condensation reaction of dihydroxydiphenylsilane and dichlorodiphenylsilane at 1:2 ratios of initial compounds, in 20–50% solution of dry toluene, in the presence of dry pyridine, at 5–0°C according to well-known method [18] by the following Scheme 21.1:

$$pPh_2Si(OH)_2 + qPh_2SiCl_2 \xrightarrow[-2Py\ HCl]{2Py} Cl\text{-}(Ph_2SiO)_{n-1}Ph_2SiCl$$

**SCHEME 21.1** Heterofunctional condensation reaction of dihydroxydiphenylsilane and diphenyldichlorosilane.
Where; n ≈ 3 (I) [18, 19], 5 (I) [18, 19], 11 (III).

The obtained compounds I-III are represented as solid white type products. The structure and composition of synthesized compounds were determined via elemental analysis, determination of molecular masses and FTIR spectra. Physical-chemical properties of compound I and II correspond to literature data [18, 19].

We have investigated hydrolytic condensation reaction of synthesized α,ω-dichlorodiphenylsiloxanes in the presence of sodium hydroxide, at 0–5°C temperature corresponding α,ω-dihydroxydiphenylsiloxanes has been obtained, according to the following Scheme 21.2:

$$Cl\text{-}(Ph_2SiO)_{n-1}Ph_2SiCl + 2H_2O \xrightarrow{2NaOH} HO\text{-}(Ph_2SiO)_{n+1}Ph_2SiOH$$

**SCHEME 21.2** Hydrolytic condensation reaction of synthesized α,ω-dichlorodiphenylsiloxanes.

Where; n ≈ 3 (IV) [], 5 (V) [18, 19], 11 (VI).

Some characteristics of compounds IV and VI are presented in Table 21.1. Monofunctional hydroxydiphenylsiloxanes have been obtained by heterofunctional condensation reaction of α,ω-dihydroxydiphenylsiloxane with trimethylchlorosilane in the presence of ammines, according to the Scheme 21.3:

$$HO\text{-}(Ph_2SiO)_{n-1}Ph_2SiOH + Me_3SiCl \xrightarrow[-Py\,HCl]{Py} HO\text{-}(Ph_2SiO)_{n-1}Ph_2SiOSiMe_3$$

**SCHEME 21.3** Heterofunctional condensation reaction of α,ω-dihydroxydiphenylsiloxane with trimethylchlorosilane in the presence of ammines.

Where; n ≈ 3 (VII), 5 (VIII), 11 (IX).

The synthesized monohydroxy derivatives are white color semicrystalline products well soluble in ordinary organic solvents. Structure and composition of compounds VII–IX have been determined by Chugaeve-Tserevetinov method [18] and by FTIR spectra.

In the spectra, one can observe characteristic bands 3400–3600 $cm^{-1}$ for Si-OH bonds and as well as for trimethylsiloxy groups at 840 $cm^{-1}$. Some physical-chemical properties of synthesized oligomers is presented in Table 21.1.

The $Si\text{-}CH_3$ group is easily recognized by a strong, sharp band at about 1260 $cm^{-1}$ together with one or more strong bands in the range 865–750 $cm^{-1}$. Some $(CH_{3)3}$ Si compounds show a 1250 $cm^{-1}$ band split into two components with the weaker component often appearing as a shoulder on the higher frequency side of the band. Blocks of dimethyl D units show a relatively weak band at 860 $cm^{-1}$. In many copolymers containing dimethyl D units (random or alternative, not block), the 860 $cm^{-1}$ band shifts to 845 $cm^{-1}$ and becomes stronger. For Si-Cl bonds in the spectra $SiCl_2$ and $SiCl_3$ compounds usually show two bands in this range 625–425 $cm^{-1}$.

The synthesized oligomers VII-IX have been used in dehydrocoupling reaction PMHS and with PMHS-dimethylsiloxane copolymer for obtaining of comb-type copolymers.

**TABLE 21.1** Some Physical-Chemical Properties of α,ω-Dichloro(Dihydroxy)Diphenylsiloxanes

| SL. No. | $(Ph_2SiO)_n$ | $T_{melt}$ (°C) | Cl/OH% | $M^{a,b}$ | Yield, % |
|---|---|---|---|---|---|
| III | 11 | 45–47 | 3.18<br>2.95 | 2233<br>2100 | 85 |
| IV | 3 | 211–213 | 5.55<br>5.44 | 612<br>600 | 83 |
| V | 5 | 218–224 | 3.37 | 1068<br>1030 | 84 |
| VI | 11 | 205–210 | 1.55<br>1.48 | 2196<br>2100 | 82% |
| VII | 3 | 79–80 | 2.48<br>2.24 | 684<br>620 | 85 |
| VIII | 5 | 80–85 | 1.57<br>1.60 | 1080<br>1100 | 88 |
| IX | 11 | 93–97 | 0.74<br>0.69 | 2268<br>2115 | 86 |

[a]Over the line, there are calculated values and under the line, experimental values.
[b]Molecular masses were determined by hydroxyl group analysis.

### 21.3.2 SYNTHESIS OF COPOLYMERS AND BCP

For the synthesis of comb-type copolymers as an initial product was used trimethylsiloxy group terminated PMHS (Aldrich) with degree of polymerization ($n = 35, \bar{M}_n = 2262$, Si-H% = 1.55) and trimethylsiloxy group terminated PMHS-dimethylsiloxane copolymer which was synthesized by copolymerization reaction of PMHS with octamethylcyclotetrasiloxane in the presence of powder-like anhydrous potassium hydroxide, at 1:6 ratio of initial compounds, by the known method [21]. The polymerization reaction was carried out in a 50% solution of dry toluene, at 80°C temperature up to the constant value of the reaction mixture. According to the following Scheme 21.4:

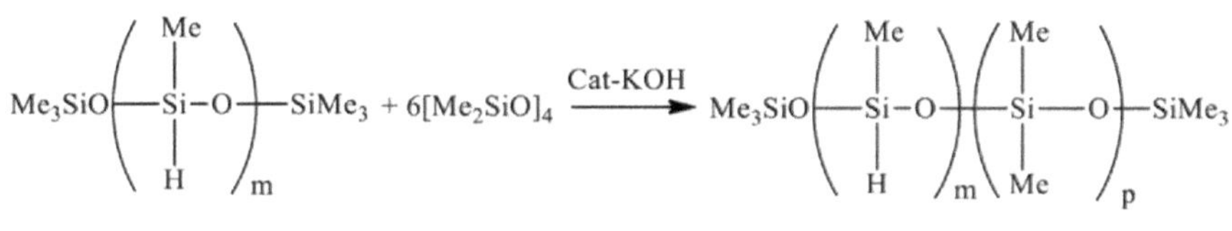


X

**SCHEME 21.4** Copolymerization reaction of PMHS with octamethylcyclotetrasiloxane.

The contend and the structure of obtained oligomer X has been determined by FTIR spectra and determination of Si-H% contend and molecular masses. By dehydrocondensation reactions of PMHS and PMHS-dimethylsiloxane copolymers with monohydroxydiphenylsiloxanes and ethyl alcohol at 1:35:0, or 1:28:7 ratio of initial compounds in the presence of 0.1% (mass) of dry potassium hydroxide, at 30–40°C temperature comb-type copolymers have been obtained, according to the following Scheme 21.5:

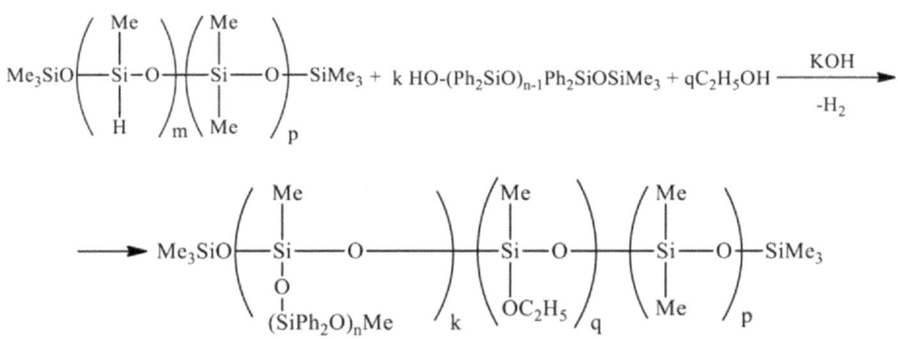


**SCHEME 21.5** Dehydrocondensation reactions of PMHS and polymethylhydrosiloxane-dimethylsiloxane copolymers with monohydroxydiphenylsiloxanes in the presence of dry potassium hydroxide [where; $m \approx 35$, $p = 0$, $k = 35$, $q = 0$: $n = 3$ (XI), 5 (XII), 11 (XIII); $m \approx 35$, $p = 0$, $k = 28$, $q = 7$: $n = 5$ (XIV), 11 (XV); $m \approx 35$, $p \approx 24$: $n = 3$ (XVI), 5 (XVII), 11 (XVIII); $m \approx 35$, $p = 24$, $k = 28$, $q = 7$: $n = 5$ (XIX)].

During the dehydrocoupling reaction, a decrease of active Si-H groups' concentration with time was observed. Dehydrocondensation reaction was performed in dry toluene solution ($C \approx 6.301 \cdot 10^{-2}$ mole/l). As it is evident from Table 21.2, all active ≡Si-H groups participate in dehydrocondensation reaction. From Figure 21.1, it is evident that the depth of dehydrocondensation reaction rises with the increase of temperature and all active ≡Si-H bonds takes participation in dehydrocondensation reaction. After 20 min hydrogen conversion depending of temperature changes from 85% up to 72%.

In the FTIR spectra of oligomers XII-XIX (Figures 21.2 and 21.3) one can observe absorption bands for asymmetric valence oscillation of linear ≡Si-O-Si≡ bonds at 1020 and 1079 $cm^{-1}$, for ≡Si-Me bonds at 1276 $cm^{-1}$ and for ≡C-H bonds at 2800–3100 $cm^{-1}$. In the FTIR spectra there is now absorption band characteristic for un-reacted ≡Si-H bonds at 2160 $cm^{-1}$. Some properties of obtained comb-type copolymers are presented in Table 21.2.

**TABLE 21.2** Some Properties of Obtained Comb-Type Copolymers

| SL. No. | m | p | n | k | q | Yield, % | $\eta_{sp}$ | T g, °C | | $d_1$, Å |
|---|---|---|---|---|---|---|---|---|---|---|
| XI | 35 | 0 | 3 | 35 | 0 | 86.7 | 0.081 | +160 | $7.8 \times 10^3$ (2.6) | 11.0 |
| XII | 35 | 0 | 5 | 35 | 0 | 86.7 | 0.084 | +170 | | - |
| XIII | 35 | 0 | 11 | 35 | 0 | 97.4 | 0.09 | +170 | | - |
| XIV | 35 | 0 | 5 | 28 | 7 | 86.4 | 0.87 | - | | - |
| XV | 35 | 0 | 11 | 28 | 7 | 84.2 | 0.08 | - | | - |
| XVI | 35 | 24 | 3 | 35 | 0 | 94.8 | 0.07 | −125<br>+180 | | 10.60<br>6.81 |
| XVII | 35 | 24 | 5 | 35 | 0 | 83.5 | 0.075 | −125<br>+165 | 42592 (1.3) | 9.70<br>7.09 |
| XVIII | 35 | 24 | 11 | 35 | 0 | 86.9 | 0.08 | −125<br>+170 | | 10.61<br>7.13 |

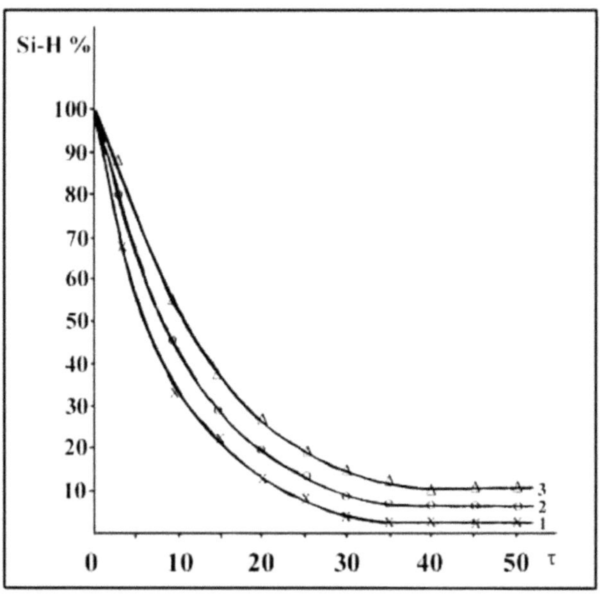


**FIGURE 21.1** Dependence of changes of active ≡Si-H groups concentration on the time during dehydrocondensation reactions of PMHS with monohydroxydiphenylsiloxane (VII) at 40°C (1), 50°C (2) and 60°C (3).

For synthesized copolymers, GPC investigations have been carried out (Figures 21.4 and 21.5). It was found that during the investigation of copolymers XI-XVIII the obtained results of molecular masses are higher compared with theoretical values.

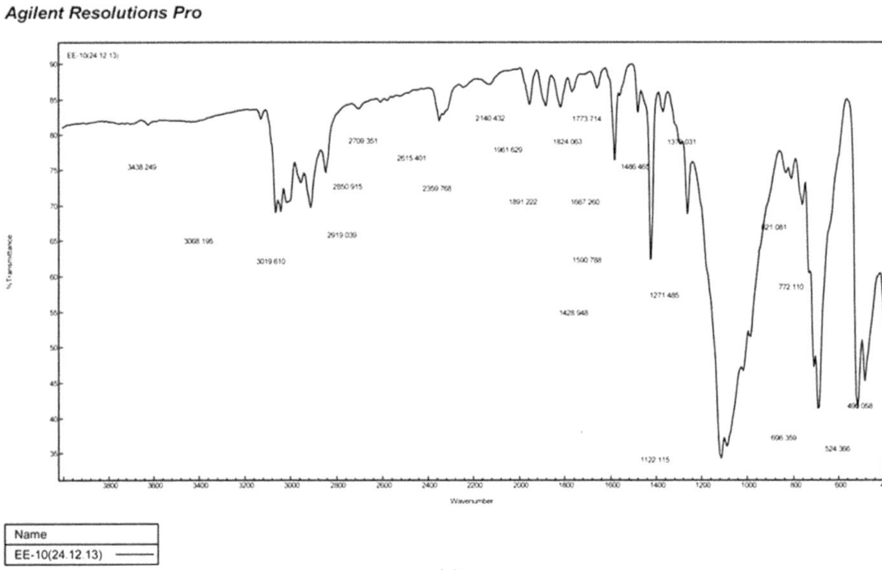


(a)

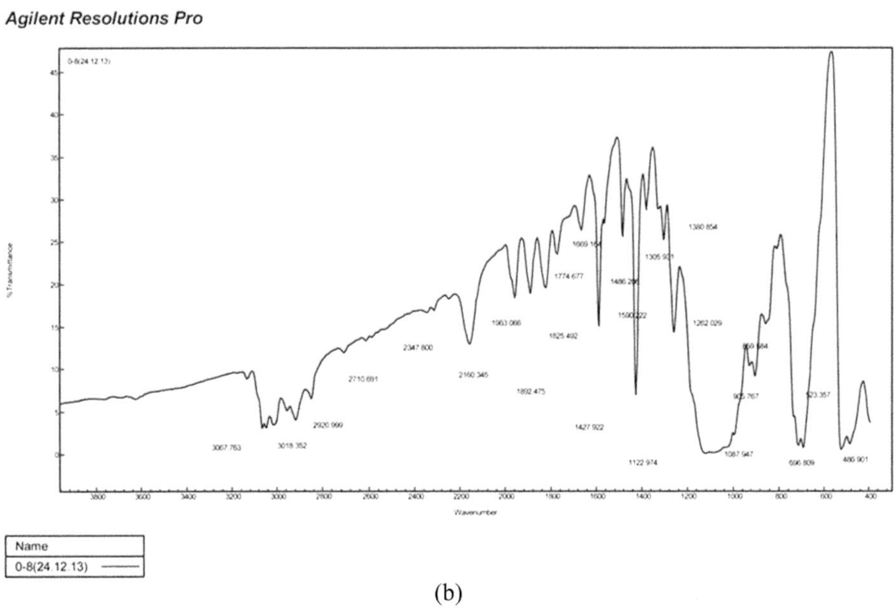


(b)

**FIGURE 21.2** FTIR spectra of copolymer XI (a) and XII (b).

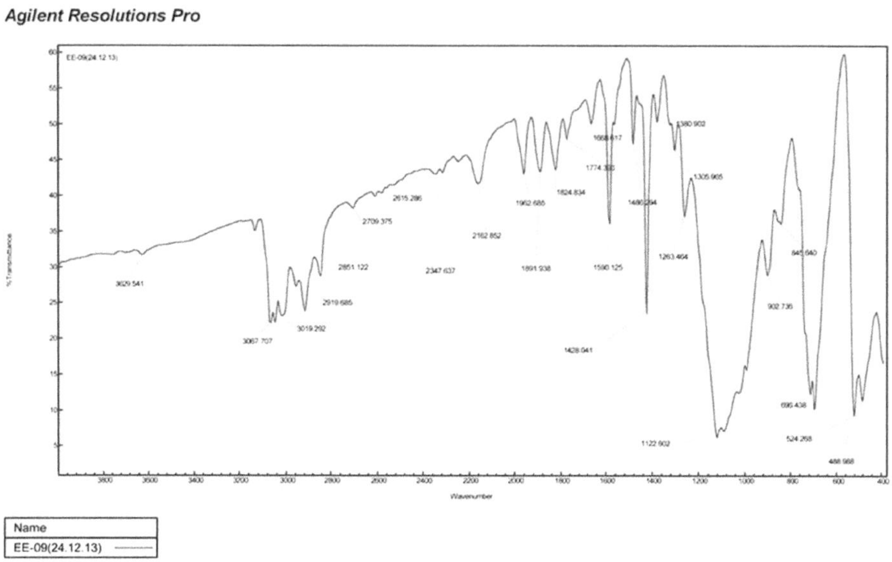


(a)

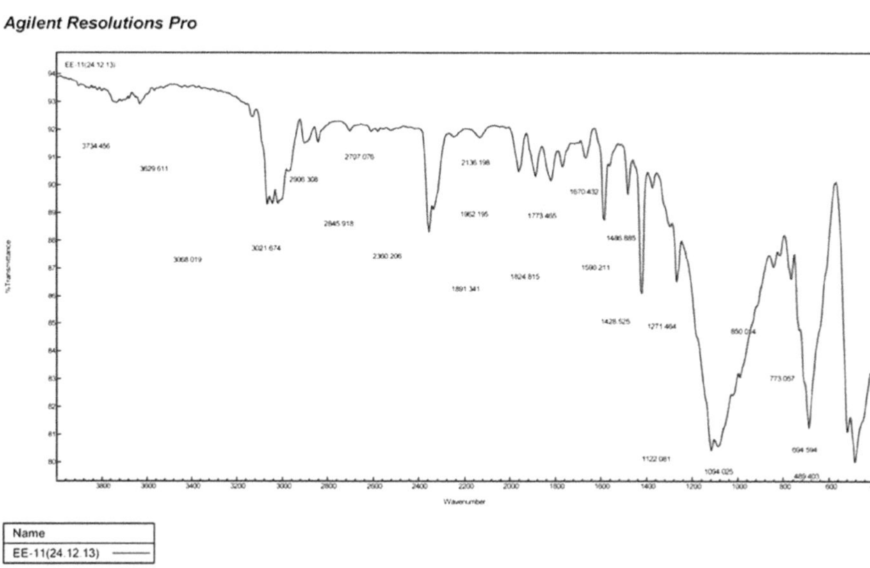


(a)

**FIGURE 21.3** FTIR spectra of copolymer XIII (a) and XIV (b).

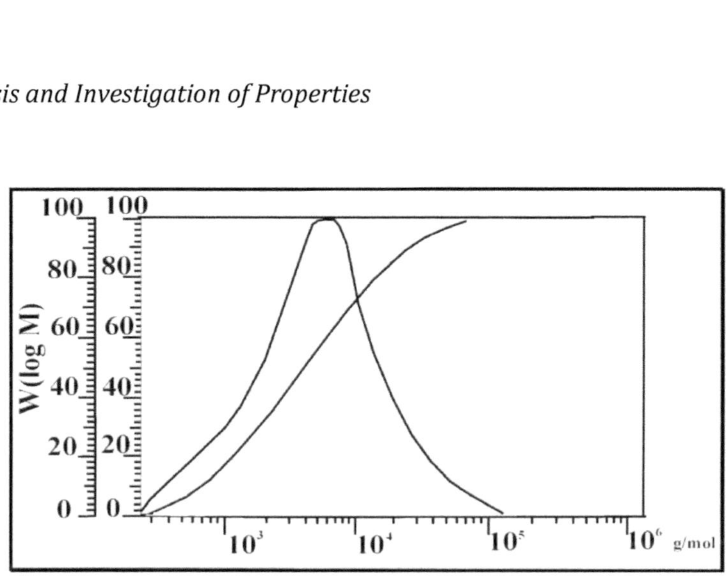


: $\bar{M}_{\omega} \approx 7{,}8 \times 10^{3}$, $p \approx 2{,}6$.

**FIGURE 21.4** GPC curves of copolymer XI.

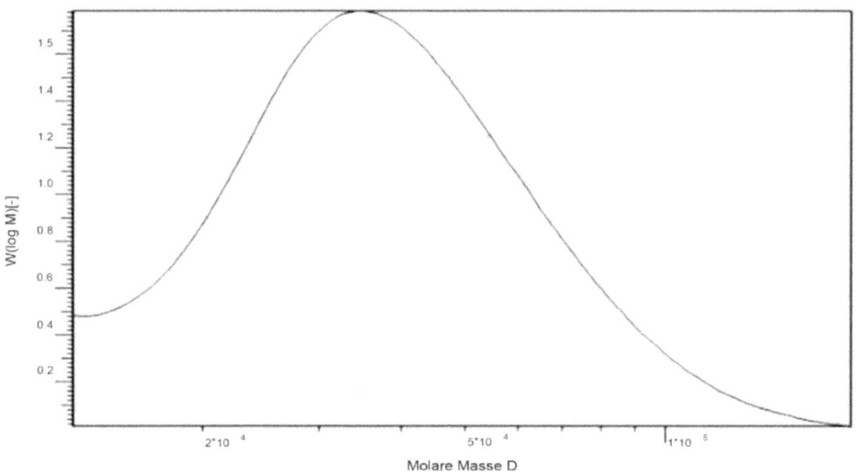


**FIGURE 21.5** GPC curves of copolymer XVII.

| Detector | Mn (g/mol) | Mw (g/mol) | D (Mw/Mn) | Vp (ml) | Mp (g/mol) | Area (ml*V) |
|---|---|---|---|---|---|---|
| RI | 32822,40 | 42592,10 | 1,30 | 23,63 | 34456,50 | 0.01 |

Here it is obvious that the average molecular weights of the synthesized oligomers exceed several times the theoretical values of the molecular weights calculated in case of full dehydrocondensation. It indicates that during inter-molecular dehydrocondensation reaction branching processes on active ≡Si-H groups also takes place, which is in agreement with literature data [17]. Hence, the obtained copolymers are various linked branched systems.

Thermogravimetric investigation of synthesize copolymers have been carried out. In Figure 21.9, the thermogravimetric curves is presented. Thermogravimetric investigation has been studied in open area. The heating rate of the system is 10°C/min.

### 21.3.3 TGA INVESTIGATIONS

TGA investigations have been carried out on Mettler Toledo DSC-822 device. The heating rate was 10°C/min. As it is seen from Figures 21.6 and 21.7 with an increase of the length of diphenylsiloxy group in the side chain the thermal stability of copolymers rises. 5% mass losses for copolymers is observed in temperature range 250–300°C. Therefore, the modification effect of methylsiloxane copolymers have been achieved by the insertion of rigid dipenylsiloxane fragments in the side chain. Because, from the literature it is known that linear polydimethylsiloxane polymers decomposed fully at 300°C [22, 23].

By TGA it was shown that during introduction in the methylsiloxane side chain diphenylsiloxane fragments and the major destruction process begins approximately at 100°C higher than that in linear dimethylsiloxane polymers, which at 300°C completely decays. This may be explained by the presence of aromatic groups that are highly resistant to oxidation and have an inhibiting effect on the oxidation of methyl groups [22, 23] and by the break of the spiral structure of the linear PDMS chain [25]. Moreover, the insertion of rigid diphenylsiloxane fragments in the side chain leads to breakdown of the destruction processes of the depolymerization mechanism, because of the impossibility of the formation of a transitive complex, which is in agreement with literature data [25].

### 21.3.4 DSC INVESTIGATIONS

DSC investigation has been carried out on Mettler Toledo DSC-822 necessary for the DSC studies. For obtaining copolymers differential scanning calorimetric investigation have been carried out. As it is seen from Figures 21.8 and 21.9 copolymers XI-XIII represented as a one-phase systems and they are characterized by one glass transition temperature and their Tg changes

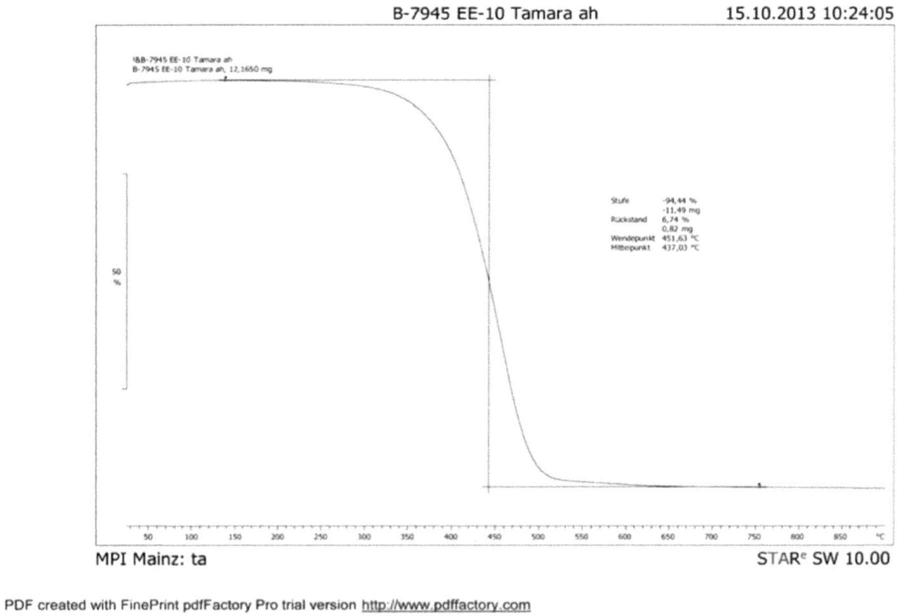


(a)

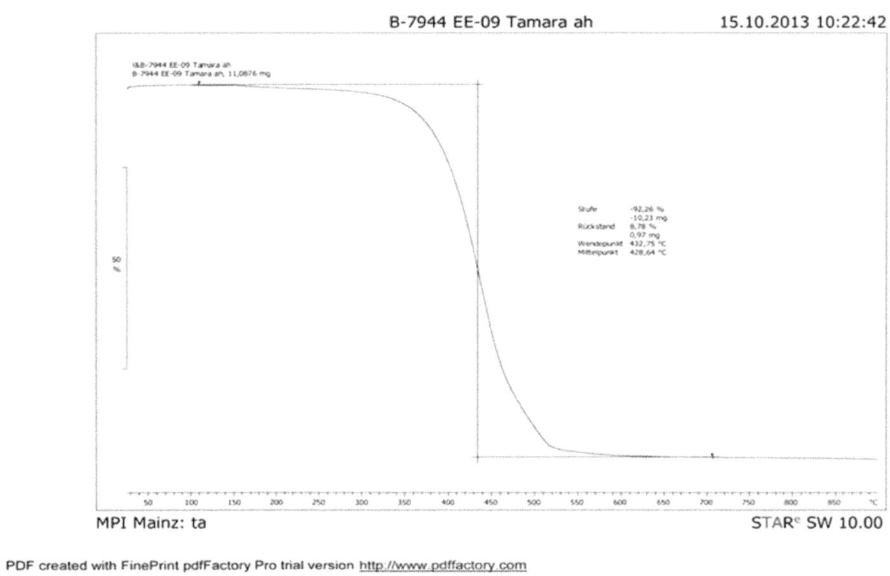


(b)

**FIGURE 21.6** TGA of copolymer a-XI, b-XIII.

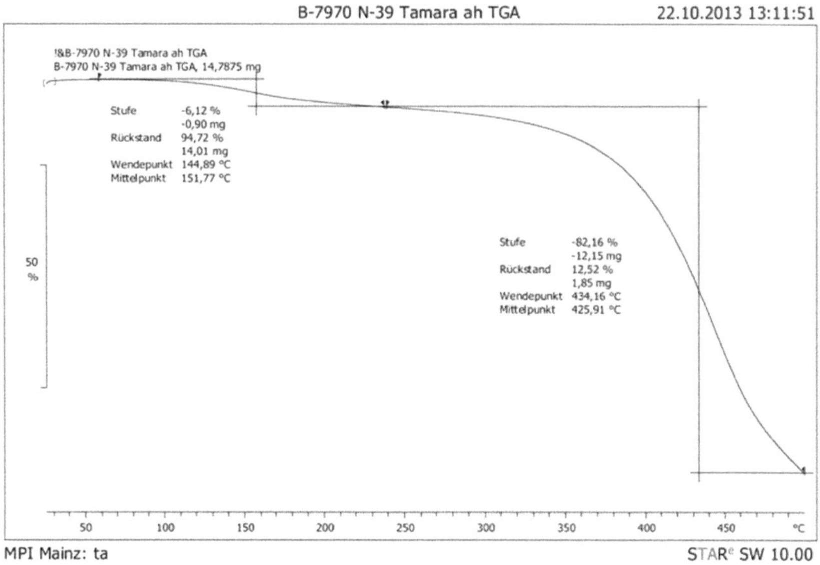




(a)

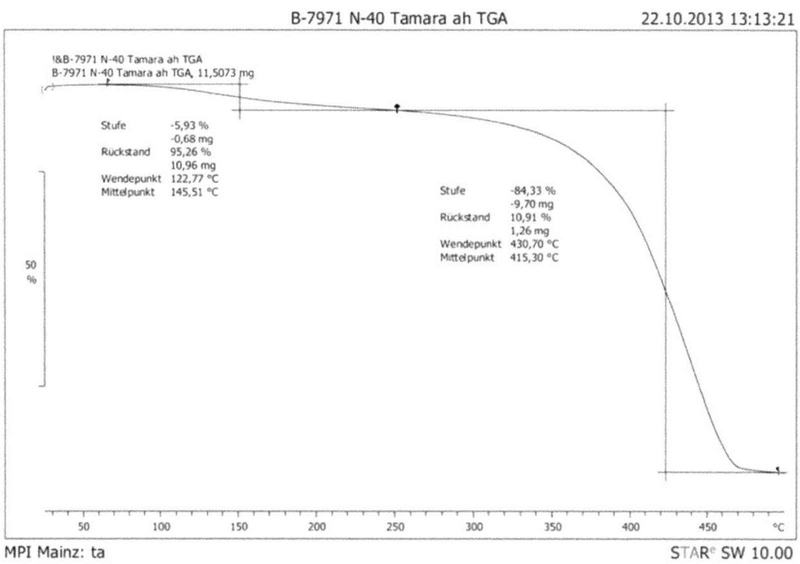




(b)

**FIGURE 21.7** TGA of copolymers a-XIV, b-XVIII.

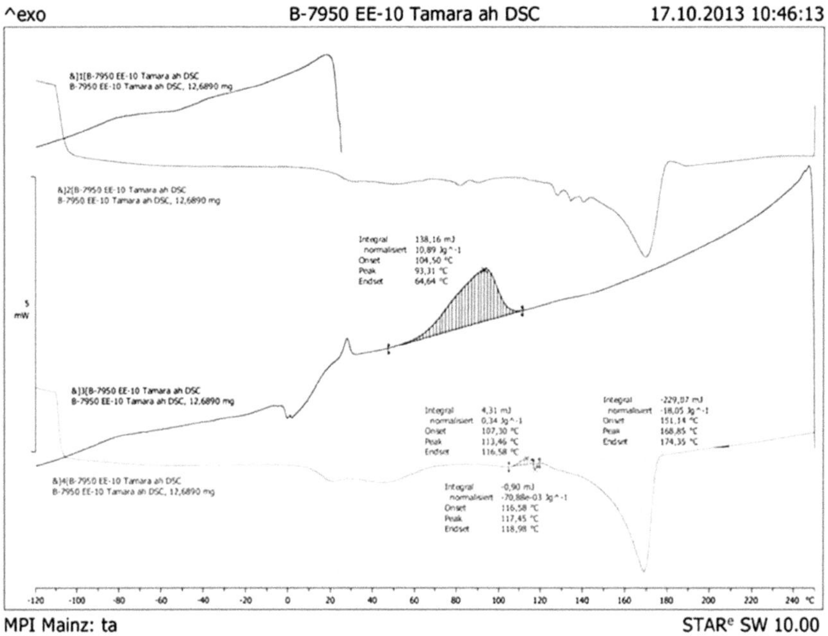


(a)

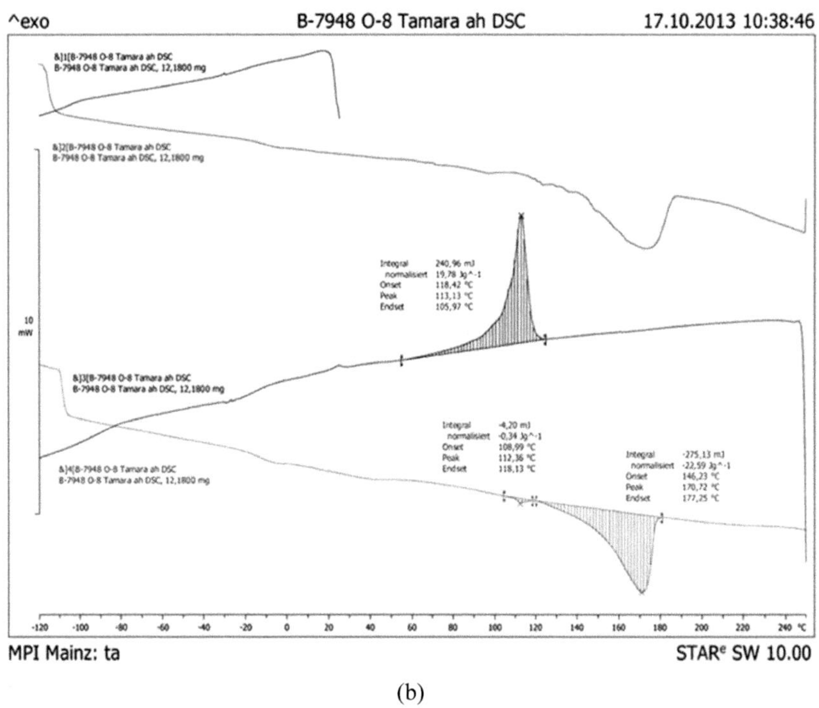


(b)

**FIGURE 21.8** DSC curves of copolymers a-XI, b-XII.

in the interval 80–102°C. For copolymers XIV-XVIII one can observe two Tg characteristic for flexible dimethylsiloxane chain at –125 ÷ –130°C and for rigid diphenylsiloxane chain in the range +160 ÷ +180°C, which proved incompatibility of phases, i.e., obtaining two-phase systems.

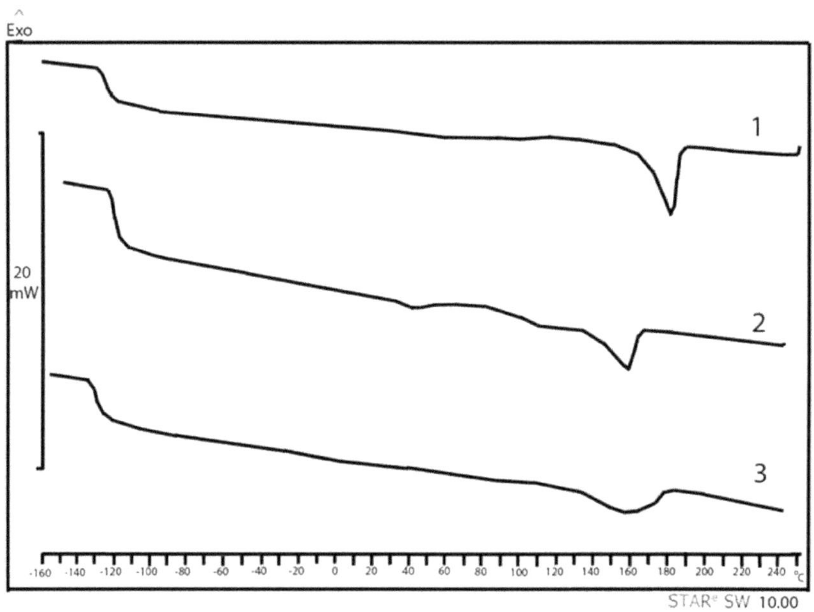


**FIGURE 21.9** DSC curves of copolymer 1-XVI, 2-XVII, and 3-XVIII.

Therefore, during modification of methylhydridesiloxane oligomers with diphenylsiloxane fragments, the copolymers represent a continuous phase with a rigid structure occupying the whole volume and discrete aggregates methylsiloxane fragments distributed in it. Formation of the microdomain structure, two-phase systems, is observed only during modification of methylhydridesiloxane-dimethylsiloxane copolymers by rigid diphenylsiloxane fragments.

In Figure 21.10, DSC curves of copolymer XXII showed that the copolymer is characterized by one Tg at +38°C.

### *21.3.5 WAX INVESTIGATION*

A wide-angle X-ray analysis of synthesized copolymers was carried out. As is seen from Figure 21.10, comb-type copolymers (XI-XIII) are one-phase amorphous systems and are characterized only by one value of interchain distances $d_1$ = 10.10 Å. The given value of interchain distance should determine interchain distance between lateral branches of rigid diphenylsiloxane fragments, because in linear diphenylsiloxanee oligomers, the interchain distances are equal to 10.10–11.0 Å accordingly, and the interchain distance in linear oligomethylhydrosiloxane chain should not exceed 7.40 Å. The second maximum $d_2$ = 4.73 Å characterized both interchain and interatomic interactions.

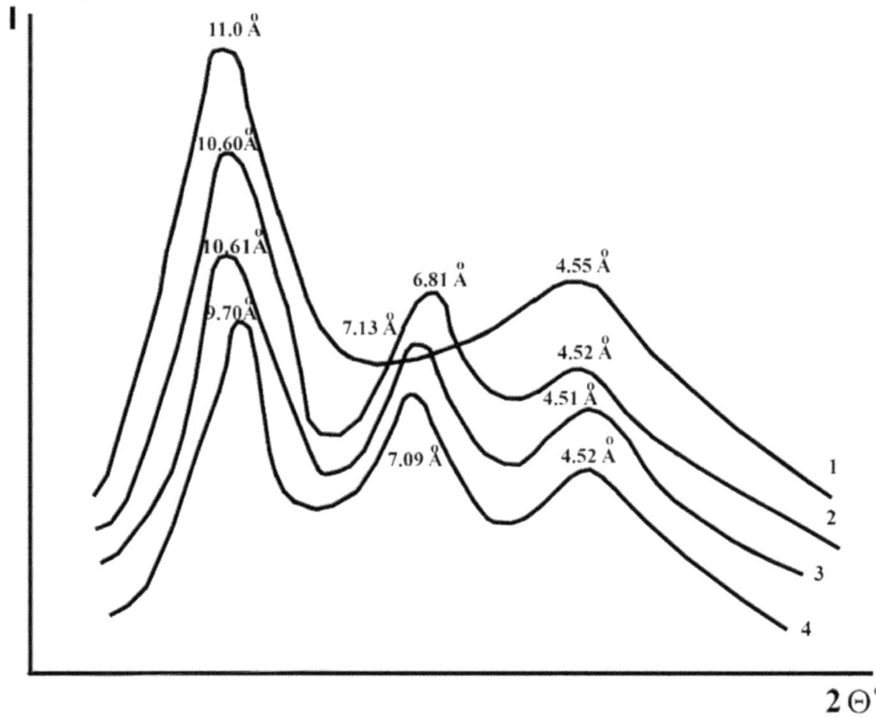


**FIGURE 21.10** Roentgenographic curves of copolymers. Where curve 1 corresponds to copolymer XI, 2-XVI, 3-XVIII, 4-XVII.

By the roentgenographic analysis for copolymers XIV-XVIII, it is shown that on diffractogram curves, the third diffraction maximum in the region 2Θ = 11.4–10.70° is observed, which corresponds to values of interchain distances $d_1$ = 6.81–7.13 Å in flexible linear dimethylsiloxane chain. This means that here segregation processes precede with the formation of independent domains both for the main flexible dimethylsiloxane chain and for lateral rigid side groups in a parallel plane, which is in agreement with early stated literature data [26].

## KEYWORDS


- **block-copolymers**
- **dehydrocondensation reaction**
- **differential scanning calorimetric**
- **gel-permeation chromatography**
- **polysilane**
- **thermal-oxidative stability**


## REFERENCES


1. Mukbaniani, V. O., Tatrishvili, N. T., & Zaikov, E. G., (2007). *The Book, Modification Reactions of Oligo Methyl Hydride Siloxanes* (pp. 1–228). Nova Science Publisher, Inc. Huntington, New York.
2. Mukbaniani, O., Tatrishvili, T., & Mukbaniani, N., (2007). *Journ. Applied Polymer Science, 104*(2), 2161.
3. Mukbaniani, O., Tatrishvili, T., Titvinidze, G., Mukbaniani, N., Brostow, W., & Pietkiewicz, D., (2007). *Journ. Applied Polymer Science, 104*(2), 1176.
4. Mukbaniani, O., Tatrishvili, T., Titvinidze, G., & Mukbaniani, N., (2007). *J. Applied Polymer Science, 104*, 2168.
5. Mukbaniani, O., Aneli, J., Esartia, I., Tatrishvili, T., Markarashvili, E., & Jalagonia, N., (2013). Siloxane oligomers with epoxy pendant groups. *Macromolec. Symposia, 328*(1), 25.
6. Mukbaniani, O., Zaikov, G., Tatrishvili, T., Titvinidze, G., & Mukbaniani, N., (2007). *Macromolecular Symposia, 247,* 364.
7. Khananashvili, L. M., Mukbaniani, O. V., Inaridze, I. A., Porchkhidze, G. V., & Koberidze, K. E., (1998). *Eur. Polym. J., 34*, 581.
8. Dolgov, B. N., Kharitonov, I. P., & Voronkov, M. G., (1954 &1955). *Zhurn. Obshch. Khim., 24*, 861 (in Russian); *Chem. Abstr., 49*, 8094i.
9. Baines, J., & Eaborn, C., (1955). *Chem, J., Soc.,* 4023.
10. Borisov, S. N., Sviridova, I. G., & Orlov, V. S., (1966). *Zhurn. Obshch. Khim., 36*, 687 (in Russian); *Chem. Abstr., 65*, 8946b.
11. Bannister, B. J., Doyle. M., & MacFarlane, D. R., (1985). *Polym, J. Sci. Polym., Lett. Ed., 23*, 465.
12. Andrianov, A. K., Nogaideli, I. A., Khananashvili, M. L., & Nakaidze, I. L., (1968 &1969). *Izv. Akad. Nauk. Ser. Khim.*, 2146 (in Russian); *Chem. Abstr., 70*, 20148c.

13. Andrianov, A. K., Nogaideli, I. A., Khananashvili, M. L., & Nakaidze, I. L., (1968). *Izv. Akad. Nauk. Ser, Khim.*, 828 (in Russian); *Chem. Abstr., 69*, 77318s.
14. Lee, L. C., (1966). *J. Organomet. Chem., 6*, 603.
15. Lichy, L. E., (1965). *J. Organomet, Chem., 4*, 431.
16. Nogaideli, I. A., Tkeshelashvili, S. R., Nakaidze, I. L., & Mukbaniani, V. O., (1976 & 1977). Bull. Tbilisi State University, *167*, 59 (in Russian); *Chem. Abstr., 86*, 17405J.
17. Mukbaniani, V. O., Khananashvili, M. L., Esartia, G. I., & Khaduri, D. S., (1994). *Intern. Polym. J. Mater., 24*, 131.
18. Koyava, A. N., Mukbaniani, V. O., Khananashvili, M. L., & Tsitsishvili, G. V., (1980). *J. of General Chemistry, 50*(8), 197.
19. Mukbaniani, V. O., Achelashvili, A. V., Meladze, M. S., Koyava, A. N., & Khananashvili, L. M., (1986). *Bulletin of the Georgian Academy of Sciences, 122*(1), 105.
20. Iwahara, T., Kusakabe, M., Chiba, M., & Yonezawa, K., (1993). *Polym. J. Sci. A., 31*, 2617.
21. Khananashvili, M. L., Bochorishvili, Z. I., Gvirgvliani, A. D., Markarashvili, G. E., & Vardosanidze, T. N., (1996). *Intern. Polym. J. Mater., 3*, 37.
22. Andrianov, A. K., Papkov, S. V., Slonimski, L. G., Zhdanov, A. A., & Yakushina, E. S., (1969). *Vysokomol. Soed, 11A*, 2030 (in Russian).
23. Papkov, S. V., Ilina, N. M., Kvachev, P. V., Makarova, N. N., Zhdanov, A. A., Slonimski, L. G., & Andrianov, A. K., (1975). *Vysokomol. Soed., 17A*, 2700 (in Russian).
24. Mileshkevich, P. V., (1978). *Kauchuk Rezina, 6*, 4(in Russian).
25. Thomas, H. T., & Kendrick, C. T., (1969). *Polym. J. Sci., 2A*, 537.
26. Mukbaniani, O., Tatrishvili, T., & Mukbaniani, N., (2007). *J. Applied Polymer Science, 104*, 2161.

## CHAPTER 22

# Reaction Hydrosilylation of Allyl-2,3,4-Tri-O-Acetyl-β-D-Ramnopyranose with Methyl- and Phenylcyclodisilazanes

N. N. SIDAMONIDZE, R. O. VARDIASHVILI, and M. O. NUTSUBIDZE

*Iv. Javakhishvili State University, Department of Chemistry,*
*I. Chavchavadze Ave., 1, 0179, Tbilisi, Georgia,*
*E-mail: neli.sidamonidze@tsu.ge (N. N. Sidamonidze)*

### ABSTRACT

By hydrosilylation 1-0-allyl-2,3,4-tri-O-acetyl-β-L-ramnopyranose (1) with 1,3-bis(dimethylsilyl)-2,2,4,4-tetramethylcyclodisilazane (2) and 1,3-bis (diphenylsilyl)-2,2,4,4-tetraphenylcyclodisilazane (3) in the presence of catalyst-$Co_2(CO)_8$, we obtained 1,3-di[3-(2,3,4-tri-O-acetyl-β-L-ramnopyranosyloxy)propyldimethylsilyl]-2,2,4,4-tetramethylcyclodisilazane (4) and 1,3-di[3-(2,3,4-tri-O-acetyl-β-L-ramnopyranosyloxy)propyldiphenylsilyl] -2,2,4,4-tetraphenylcyclodisilazane (5). The structure of obtained compounds was established by physical-chemical methods of analysis.

### 22.1 INTRODUCTION

Recently, the silylation reaction (especially trimethylsilylation) is increasingly used in organic synthesis. Usually, the term "silylation" means the introduction of a silyl group instead of a mobile hydrogen atom or a metal replacing it, and it must be possible to easily remove the group introduced during the silylation process with the restoration of active hydrogen. Silylation makes it possible to widely modify various properties of the starting materials: it changes the reactivity of the products and, in particular, allows you to block some reaction centers; improves the solubility of compounds in non-polar solvents; due to the disappearance of hydrogen bonds increases the volatility of products, which makes it possible use for their analysis.

Silicon compounds play a particularly important role for many living creatures that are at the lowest stage of evolutionary development (silicate bacteria, protozoa, spore plants, etc.), in the body of which they are contained in very large quantities. Substantial silicon is important for many higher plants, even called "flint," due to the high silicon content in them. The silicon compounds perform very important functions in the body of higher animals and humans, despite the fact that in most of their organs, the content of this element is relatively small [1–6].

The simplest organosilicon compounds–serve as the basis for obtaining the most important silicon–contains polymeric materials for various purposes, siloxane liquids, modifying agents, sorbents, and biologically active preparations. Silicone reagents used are often distinguished by increased reactivity in nucleophilic substitution reactions at the silicon atom, which may be due to the formation of intermediates with an extension of its coordination number to five or six. Organic bases, various anions and carbonyl compounds, including amides and lactams, can act as ligands during the formation of hypervalent Si intermediates. The latter, in turn, serve as the basis for the synthesis of many biologically active preparations, polymers, solvents, and intermediates.

Synthesis of low-toxicity compound has become important in biological and pharmacological studies, and so there is interest in using carbohydrates to modify linear and cyclolinear siloxanes, which may lead to a substantial change in the nature of the drug action [7–10].

## 22.2 EXPERIMENTAL PART

FTIR spectra were recorded on a Nicolet Nexus 470 machine with MCTB detector. The KBr pellets of samples were prepared by mixing (1.5–2.0) mg of samples, finely grounded, with 200 mg KBr (FTIR grade) in a vibratory ball mixer for 20 s. The $^{1}H$ NMR spectrum was taken on a Bruker WM-250 spectrometer (250 MHz); the $^{13}C$ NMR spectrum was taken on Bruker AM-300 spectrometer (75 MHz) in $CDCI_3$. The purity of the compounds obtained and the $R_f$ values were determined on Silufol UV-254. The optical rotation was measured on an SU-3 general-purpose saccharimeter at 20±2°C.

1,3-di[3-(2,3,4-tri-O-acetyl-β-L-ramnopyranosyloxy) propyldimethylsilyl]-2,2,4,4-tetrame-thylcyclodisilazane (4). 1.31 g (0.005 moll or 5 mmol) 1,3-bis(dimethylsilyl)-2,2,4,4-tetramethylcyclodisi-lazane (2) in dry chloroform (20 ml) and $Co_2(CO)_8$ (0.15 g) were added drop-wise to a solution of 4.12 g (0.0125 moll or 12.5 mmol) compound (1) in dry chloroform (25 ml). The reaction was carried out under a nitrogen atmosphere with constant stirring for 1.5 h (60–65°C). After cooling and separating on a column (2:1 benzene-chloroform system, silica gel L (50/100), a chromatographically pure product (4) was obtained in a yield of 3.94 g (68.4%); m.p. 159–160°C. $R_f$ 0.82 (2:1 benzene-chloroform system). $[\alpha]_D^{18}$ +77° (c 0.52, chloroform). FTIR spectrum, ν, $cm^{-1}$: 640 (Si-C); 1140, 1025 (C-O-C); 1725 (C = O); 920 (Si-N); 1450 ($CH_3$).

Found, %: C 49.13; H 6.90; Si 12.53; N 2.85. $C_{38}H_{68}Si_4N_2O_{16}$. Calculated, %: C 49.56; H 7.39; Si 12.17; N 3.04;

$^{13}C$ NMR spectrum, δ, ppm. $CDCI_3$: C1-81.20 and 80.68 ($C_{(1)}$ and $C_{(1')}$); C2-71.50 and 70.98 ($C_{(2)}$ and $C_{(2')}$);; C3-72.10 and 74.03 ($C_{(3)}$ and $C_{(3')}$);; C4-70.80 and 72.02 ($C_{(4)}$ and $C_{(4')}$); C5-72.79 and 72.61 ($C_{(5)}$ and $C_{(5')}$);; C6-18.12 and 18.09 ($C_{(6)}$ and $C_{(6')}$); 169.20–175.60 (RO-<u>C</u>O-$CH_3$); 20.60 (RO-CO-<u>C</u>$H_3$); 71.62–72.255 (RO-<u>C</u>$H_2$-$CH_2$-$CH_2$-$SiR_3$); 28.46–30.70 (RO-$CH_2$-<u>C</u>$H_2$-$CH_2$-$SiR_3$); 21.74–23.52 (RO-$CH_2$-$CH_2$-<u>C</u>$H_2$-$SiR_3$); 8.12–12.30 (Si-$CH_3$);

$^1H$ NMR spectrum, δ, ppm (*J*, Hz):4.15 (1H, d, $J_{1,2}$ = 8, H-1); 5.50 (1H, d, $J_{1,2}$ = 4, H-1'); 4.18 (1H, dd, $J_{2,1}$ = 8.1; $J_{2,3}$ = 9.4, H-2); 4.65 (1H, dd, $J_{2,1}$ = 4; $J_{2,3}$ = 10.6, H-2'); 5.63 (1H, dd, $J_{3,2}$ = 9.4; $J_{3,4}$ = 10, H-3); 5.45 (1H, dd, $J_{3,2}$ = 10.6; $J_{3,4}$ = 9.8, H-3'); 4.20–4.22 (dd, $J_{4,3}$ = 10; $J_{4,5}$ = 12.3, H-4); 5.20–5.34 (dd, $J_{4,3}$ = 9.8; $J_{4,5}$ = 9.9, H-4'); 3.30–3.72 (1H, m, H-5); 3.96–4.10 (1H, m, H-5'); 4.70–4.10 and 4.22–4.28 (2H, d, H-6 and H-6' C<u>H</u>$_2$OCO$CH_3$);2.13–2.18 (18H, m, 6CO-C<u>H</u>$_3$); 3.65–3.75 and 3.80–3.84 (2H, 2m, RO-C<u>H</u>$_2$-$CH_2$-$CH_2$-$SiR_3$); 1.90–1.95 and 1.92–1.98 (2H, 2m, RO-$CH_2$-C<u>H</u>$_2$-$CH_2$-$SiR_3$); 1.72–1.80 and 1.60–1.69 (2H, 2m, RO-$CH_2$-$CH_2$-C<u>H</u>$_2$-$SiR_3$); 1.05–1.10 (24H, m, 8 Si-C<u>H</u>$_3$); 1.44–1.47 (24H, s, 8C<u>H</u>$_3$).

1,3-di[3-(2,3,4-tri-O-acetyl-β-L-ramnopyranosyloxy)propyldiphenyl silyl]-2,2,4,4-tetraphenylcyclodisilazane (5). 3.79 g (0.005 moll or 5 mmol) 1,3-bis(diphenyl silyl)-2,2,4,4-tetraphenylcyclodisilazane (3) in dry chloroform (20 ml) and $Co_2(CO)_8$ (0.15 g) were added drop-wise to a solution of 4.12 g (0.0125 moll or 12.5 mmol) compound (1) in dry chloroform (25 ml). The reaction was carried out under a nitrogen atmosphere with constant stirring for 2.5 h (65–70°C). After cooling and separating on a column (2:1

benzene-chloroform system, silica gel L (50/100), a chromatographically pure product (5) was obtained in a yield of 6.33 g (71.5%); p.m. 178–178.5°C. $R_f$ 0.54 (2:1 benzene-chloroform system). $[\alpha]_D^{18}$ +32° (c 1.5, chloroform). IR spectrum, ν, $cm^{-1}$: 690 (Si-C); 1120, 1040, 1030 (C-O-C); 1720 (C = O); 940 (Si-N); 1450, 1460 ($C = C_{arom}$); 732, 839 ($C\text{-}H_{arom}$). Found,%: C 49.13; H 6.90; N 2.85; Si 12.53. $C_{78}H_{84}Si_4N_2O_{16}$. Calculated,%: C 49.56; H 7.39; N 3.04; Si 12.17.

$^{13}C$ NMR spectrum, δ, ppm. $CDCl_3$: C1-80.38 and 81.30 ($C_{(1)}$ and $C_{(1')}$); C2-71.28 and 70.65 ($C_{(2)}$ and $C_{(2')}$);; C3-72.10 and 74.20 ($C_{(3)}$ and $C_{(3')}$);; C4-72.41 and 72.18 ($C_{(4)}$ and $C_{(4')}$); C5-72.12 and 72.47 ($C_{(5)}$ and $C_{(5')}$);; C6-18.08 and 18.20 ($C_{(6)}$ and $C_{(6')}$); 165.35–170.40(RO-$\underline{C}$O-$CH_3$); 20.20 (RO-CO-$\underline{C}H_3$); 71.84–72.45 (RO-$\underline{C}H_2$-$CH_2$-$CH_2$-$SiR_3$); 29.15–31.43 (RO-$CH_2$-$\underline{C}H_2$-$CH_2$-$SiR_3$); 21.14–23.74 (RO-$CH_2$-$CH_2$-$\underline{C}H_2$-$SiR_3$); 123–130 (Si-$\underline{C}_6H_5$).

$^1H$ NMR spectrum, δ, ppm (*J*, Hz):4.15 (1H, d, $J_{1,2}$ = 8, H-1); 5.50 (1H, d, $J_{1,2}$ = 4, H-1'); 4.18 (1H, dd, $J_{2,1}$ = 8.1; $J_{2,3}$ = 9.4, H-2); 4.65 (1H, dd, $J_{2,1}$ = 4; $J_{2,3}$ = 10.6, H-2'); 5.63 (1H, dd, $J_{3,2}$ = 9.4; $J_{3,4}$ = 10, H-3); 5.45 (1H, dd, $J_{3,2}$ = 10.6; $J_{3,4}$ = 9.8, H-3'); 4.20–4.22 (dd, $J_{4,3}$ = 10; $J_{4,5}$ = 12.3, H-4); 5.20–5.34 (dd, $J_{4,3}$ = 9.8; $J_{4,5}$ = 9.9, H-4'); 3.30–3.72 (1H, m, H-5); 3.96–4.10 (1H, m, H-5'); 4.70–4.10 and 4.22–4.28 (2H, d, H-6 and H-6' $C\underline{H}_2OCOCH_3$); 2.13–2.18 (18H, m, 6CO-$C\underline{H}_3$); 3.65–3.75 and 3.80–3.84 (2H, 2m, RO-$C\underline{H}_2$-$CH_2$-$CH_2$-$SiR_3$); 1.90–1.95 and 1.92–1.98 (2H, 2m, RO-$CH_2$-$C\underline{H}_2$-$CH_2$-$SiR_3$); 1.72–1.80 and 1.60–1.69 (2H, 2m, RO-$CH_2$-$CH_2$-$C\underline{H}_2$-$SiR_3$); 7.0–7.5 (8$C_6\underline{H}_5$, m, aromatic protons).

## 22.3 RESULTS AND DISCUSSION

In order to obtain biologically active compounds, we have devised a new method for synthesizing glycosides containing silicon. Insertion of silicon atoms in known medicinal preparations can result in an essential change of character of the action of preparation, and sometimes can give them a number of new properties.

By hydrosilylation 1-0-allyl-2,3,4-tri-O-acetyl-β-L-ramnopyranose (1) with 1,3-bis(dimethylsilyl)-2,2,4,4-tetramethylcyclodisilazane (2) and 1,3-bis (diphenyl silyl)-2,2,4,4-tetraphenylcyclodisilazane (3) in the presence of catalyst-$Co_2(CO)_8$, we obtained 1,3-di[3-(2,3,4-tri-O-acetyl-β-L-ramnopyranosyloxy)propyldimethylsilyl]-2,2,4,4-tetramethylcyclodisilazane (4) and 1,3-di[3-(2,3,4-tri-O-acetyl-β-L-ramnopyranosyloxy)propyldiphenylsilyl]-2,2,4,4-tetraphenylcyclodisi-lazane (5).

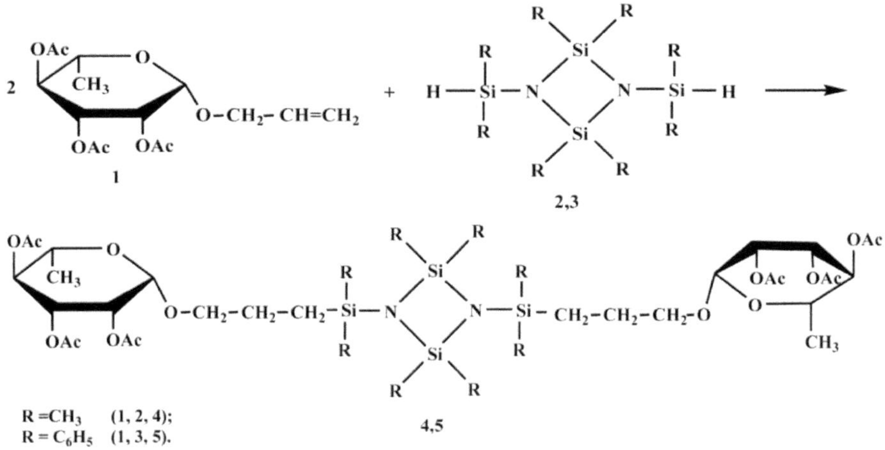


The reaction mainly occurs according to Farmer rule, although a small amount of Markovnikov addition product is also formed. The course of the reaction was monitored from the decrease in active hydrogen on the silicon over time (Figure 22.1). We have established that in 1.5 h, the hydrogen in °Si-H group is completely removed, which is supported by the IR spectrum.

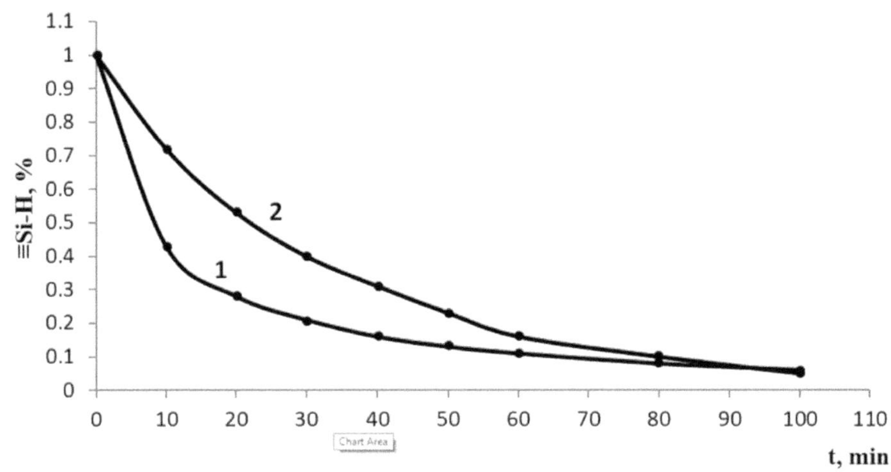


**FIGURE 22.1** Amount of active hydrogen on the silicon atom vs. hydrosilylation reaction time for compound 1 with 1,3-bis(dimethylsilyl)-2,2,4,4-tetramethylcyclodisilazane (curve 1) and 1,3-bis(diphenyl silyl)-2,2,4,4-tetraphenylcyclodisilazane (curve 2).

By deacetylation of compounds 4 and 5 in absolute methanol in the presence of sodium methoxide, we obtained 1,3-di[3-(β-L-ramnopyranosyloxy) propyldimethylsilyl]-2,2,4,4-tetramethylcyclodisilazane (6) and 1,3-di[3-(β-L-ramnopyranosyloxy)propyldiphenylsilyl]-2,2,4,4-tetraphenylcyclodisilazane (7).

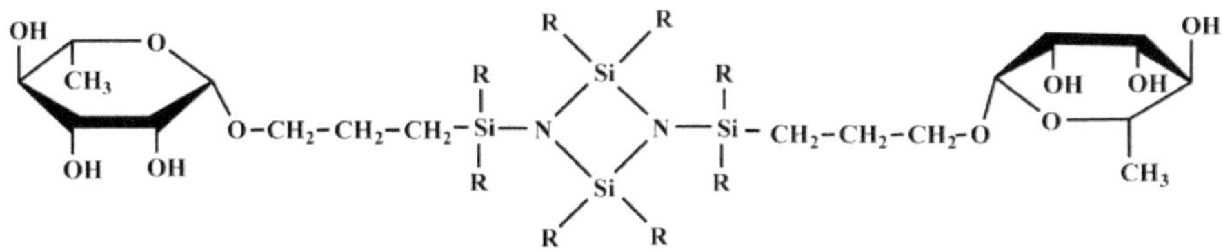

$R = CH_3$ (6)
$R = C_6H_5$ (7)

6,7

The structures of obtained compounds were established by physical-chemical methods of analysis. Addition of 1,3-bis-(dimethylsilyl)-2,2,4,4-tetramethylcyclodisilazane (2) to allyl ramnoside (1) was selected to study the pathway and mechanism of the model reaction. Quantum-chemical computation was performed using CS MOPAC (Chem 3D Ultra version 8.03). Each computation by the AM1 method was preceded by optimization of the compounds, i.e., minimization of the energy by molecular (MM) and quantum-chemical method. This reaction was examined in two directions, according to the Farmer and Markovnikov rules. The calculated heats formations of products: $DH_{form}$ = –844.02 kcal/mole compound 4 and $DH_{form}$ = –801.61 kcal/mole compound 8 showed that production of compound 4 was highly probable.

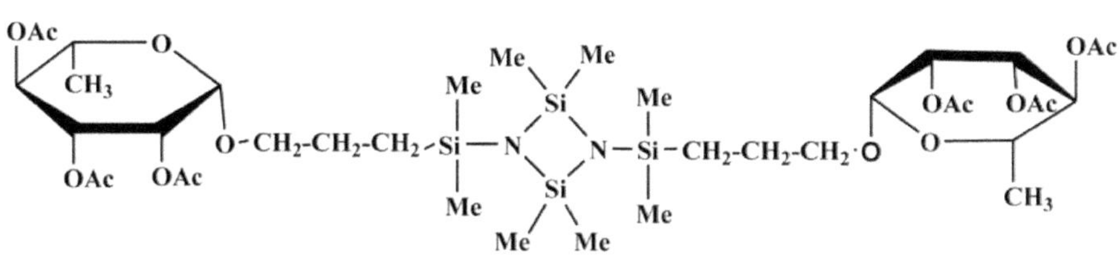

4

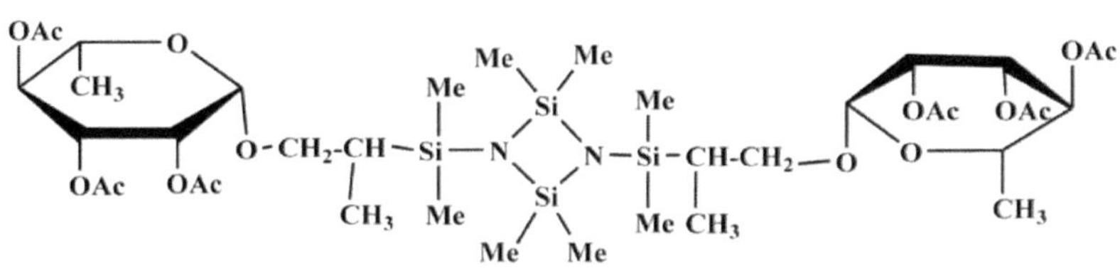

8

The reactions in the presence of $Co_2(CO)_8$ probable occurred according to the following mechanism. The $Co_2(CO)_8$ itself cannot catalyze a hydrosilylation process of allyl glucosides, the first step in the formation of the active catalyst species, $HCo(CO)_4$, is important and results from the reactions of $Co_2(CO)_8$ with compound 3:

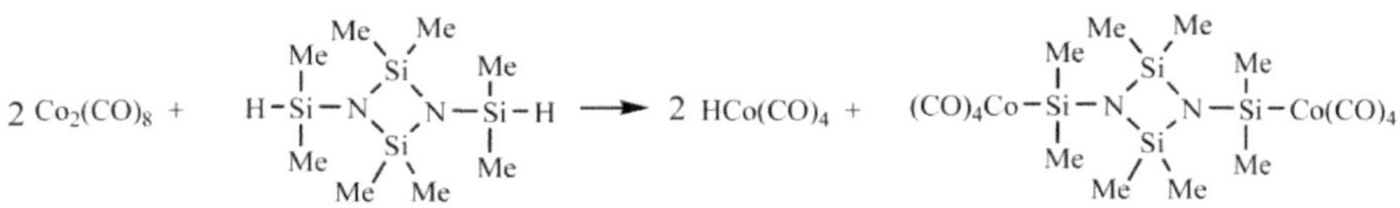


3

It is known that $HCo(CO)_4$ adds readily to olefins to form intermediate **11,** that is, unstable because of the excess of electron density on the metal and tends to convert into the saturated Co complex **11a** by loss of the most labile ligand, in this instance hydride. The last step of the mechanism is the formation of the final product and regeneration of the catalysts:

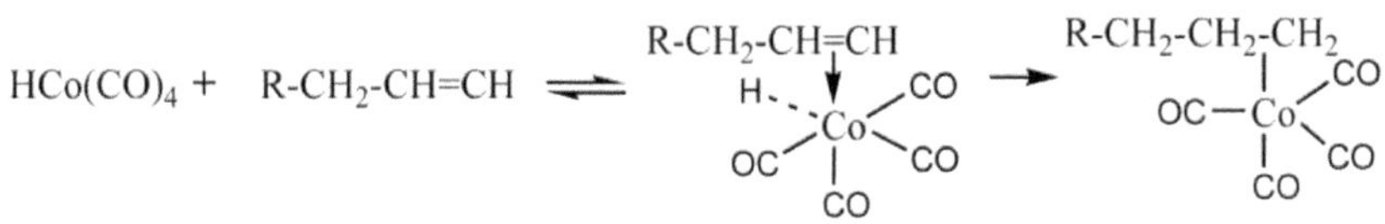


**11** **11a**

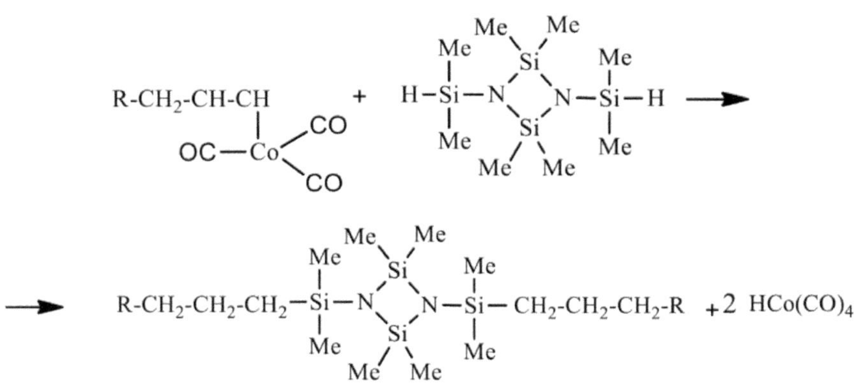


The proposed mechanism of cyclodisilazane addition to allyl glycosides agrees well with the experimental results.

## KEYWORDS

- **allyl glycosides**
- **FTIR spectrum**
- **hydrosilylation**
- **molecular mechanics**
- **NMR spectrum**
- **quantum-chemical calculation**

## REFERENCES

1. Zuev, V., & Kalinin, A. V., (2003). A facile and efficient synthesis of organocyclosiloxanes, *Phosphorus Sulfur and Silicon and the Related Elements, 178*(6), 1289–1294.
2. Andrianov, K. A., Bagratishvili, G. D., Kantariya, M. L., Sidorov, V. I., Khananashvili, L. M., & Tsitsishvili, G. V., (1965). Synthesis of organocyclosiloxanes containing alkenyl groups attached to silicon. *Journal of Organometallic Chemistry, 4*(6), 440–445.
3. Sidamonidze, N. N., Vardiashvili, R. O., Isakadze, M. O., & Djaniashvili, L. K., (2007). Boilogical activity of some aldose-containing compounds. *Pharmaceutical Chemistry Journal, 41*(3), 131–132.
4. Gakhokidze, R., Sidamonidze, N., Tabatadze, L., Tatarishvili, M., & Bedukadze, L., (2005). *Synthesis of Some N-Glycosides Containing the Lactamine Cycle* (No. 13, pp. 69–72). "*Moambe,*" Sukhum branch of Tbilisi State University Iv. Javakhishvili.
5. Ionas, G., & Stadler, R., (1994). Carbohydrate modified polysiloxanes. *Acta Polymer., 45*(11), 14–20.
6. Kramarova, V. P., Shipov, A. B., & Baukov, Y. I., (1992). Synthesis of N-silatranyl-methyllactams. *Journal of General Chemistry, 62*(11), 2559–2567.
7. Kochetkov, N. K., Bochkov, A. F., Dmitriev, B. A., Usov, A. I., Chizhov, O. S., & Shibaev, V. N., (1967). *Ctmistry of Carbohydrate*. Moscow, Publisher "Khimiya."
8. Lonas, G., & Stadler, R., (1994). *Acta Polymer., 45*, 14.
9. Sidamonidze, N. N., Gakhokidze, R. A., Chan, V. T., & Khidzsheli, Z. G., (1987). *Zashchita Rastenii, 7*, 41.
10. Sidamonidze, N. N., Janiashvili, L. K., Isakadze, M. O., & Vardiashvili, R. O., (2005). *Ceorg. Eng. News, 2*, 153.

# CHAPTER 23

# Synthesis and Characterization of Novel Star-Type Triarm Block Copolymers Including Poly(Β-Methyl Β-Alanine) by Raft Polymerization

B. SAVAŞ,[1] E. ÇATIKER,[2] T. ÖZTÜRK,[3] E. MEYVACI,[3] M. ATAKAY,[4] and B. SALIH[4]

[1]*Kafkas University, Kars Vocational School, 36100 Kars, Turkey*

[2]*Ordu University, Department of Chemistry, 52200 Ordu, Turkey, E-mail: ecatiker@gmail.com*

[3]*Giresun University, Department of Chemistry, 28200 Giresun, Turkey*

[4]*Hacettepe University, Department of Chemistry, 06800 Ankara, Turkey*

## ABSTRACT

Synthesis of novel poly(β-methyl β-alanine-b-methyl methacrylate) (poly(MBA-b-MMA)) star-type triarm block copolymers were achieved via reversible addition-fragmentation chain transfer (RAFT) polymerization of methyl methacrylate (MMA) using a novel RAFT macroinitiator (RAFT-macro agent). For this purpose, poly(β-methyl β-alanine) (PMBA-diBr) with olefinic end-group was obtained by hydrogen transfer polymerization (HTP) of crotonamide. Olefinic end-group were brominated by bromine to achieve terminally dibromo PMBA-diBr. RAFT-macro agent was acquired by reaction of PMBA-diBr and potassium ethyl xanthogenate. By reacting RAFT-macro agent and MMA, the star-type triarm block copolymers were obtained. Characterization of the products was performed by using FT-IR, $^1$H-NMR and MALDI-MS analyses. Thermal transitions and degradation features of the star-type triarm block copolymers were investigated by using DSC and TGA methods. Spectroscopic and thermal analyses revealed that both the group modifications and the RAFT polymerizations were successfully achieved.

## 23.1 INTRODUCTION

Reversible addition-fragmentation chain transfer (RAFT) polymerization represents the most recently developed controlled/living radical polymerization method and is a powerful technique for the macromolecular synthesis of a broad range of well-defined polymers [1–11]. The versatility of the technique is proved by its compatibility with a very wide range of monomers and reaction conditions [8]. Reversible chain transfer includes homolytic substitution, addition fragmentation, or some other transfer mechanisms [1]. Block and graft copolymers are used in technological applications and theoretical research [12–23]. Especially, these type polymers are great importance in fields such as drug delivery [24, 25]. Templates for inorganic particles [26–28], molecular actuators [29], and various carbon nanostructures [30, 31]. Poly(epichlorohydrin) (PECH) is present in many syntheses of copolymers [10, 32–34].

This paper demonstrates the synthesis of poly(β-methyl β-alanine-b-methyl methacrylate) (poly(MBA-b-MMA)) star-type triarm block copolymers by using RAFT polymerization. To the best of our knowledge, synthesis of poly(β-methyl β-alanine) [PMBA]/poly(methyl methacrylate) [PMMA] block copolymers by RAFT polymerization has not been reported until now. We synthesized PMBA with olefinic end-group by hydrogen transfer polymerization (HTP) of crotonamide as shown in our previous work [35]. Terminally dibromo poly (β-methyl β-alanine) (PMBA-diBr) was obtained with bromination of PMBA. A novel RAFT macroinitiator (RAFT-macro agent) was acquired by chemical reaction of PMBA-diBr with potassium ethyl xanthogenate. RAFT-macro agent and methyl methacrylate (MMA) were used to obtain novel poly(MBA-b-MMA) star-type triarm block copolymers by RAFT polymerization. Characterizations of the products were discussed in detail.

## 23.2 EXPERIMENTAL

### 23.2.1 MATERIALS

Potassium ethyl xanthogenate, α,α'-azoisobutyronitrile (AIBN), and MMA were supplied by Merck. Formic acid, petroleum ether, diethyl ether, sodium tert-butoxide, bromine, acetone, aluminum oxide, tetrahydrofuran (THF), and methanol were obtained by Sigma-Aldrich. Ammonia was supplied by Linde Group. Crotonyl chloride, and xylene were supplied by Fluka. An alumina column was used to remove the inhibitor from MMA.

### 23.2.2 INSTRUMENTATION

$^1$H-NMR spectra were recorded using Bruker Ultra Shield Plus, ultra-long hold time 400 NMR spectrometers. Trifluoroacetic acid/$CDCl_3$ mixture was used as solvent. FT-IR spectra were detected using Jasco FT/IR 6600 FT-IR spectrometer in the range of 600–4000 $cm^{-1}$. Thermogravimetric analysis (TGA) measurements were conducted using a Seiko II Exstar 6000 model instrument. The samples were heated at a rate of 10°C/min from 25°C to 800°C under nitrogen atmosphere. Differential scanning calorimetry (DSC) measurements were conducted at a rate of 10°C/min from- 80°C to 150°C under nitrogen atmosphere using a Perkin Elmer DSC 8500 series instrument. MALDI-MS analyses were performed using a Bruker Rapiflex MALDI-ToF/ToF mass spectrometer (Bruker Daltonics, Bremen, Germany) equipped with a smartbeam™ 3D laser. The data were acquired in positive ion mode by averaging 10000 laser shots for each sample. 2,5-dihydroxy-benzoic acid (DHB) solution (20 mg/mL in 1:1, ACN:THF containing 1.0% (v/v) formic acid) was prepared and used as the matrix. Samples (1.0 mg/mL) were dissolved in THF containing 10.0% formic acid by volume. The mass calibration was performed using standards at m/z scale of interest. The matrix solution was spotted as the bottom and top layers of the sandwich-type sample spot on MALDI target, whereas the sample solution was placed in the middle of matrix layers. After air-drying of each spot at room temperature, analyses were carried out.

### 23.2.3 SYNTHESIS OF TERMINALLY DIBROMO POLY(β-METHYL β-ALANINE) (PMBA-DIBR)

Crotonamide was obtained through bubbling of ammonia in crotonyl chloride solution. PMBA were prepared by HTP of crotonamide using sodium tert-butoxide (t-NaOBu) as basic catalyst [35]. 10% (wt/wt) solution of PMBA was prepared using formic acid (99%) as solvent. Several drops of bromine were added into the solutions. Bromination of PMBA oligomer with olefinic chain-end was carried out in a 50 mL Schlenk tube with continuous stirring under argon atmosphere at ambient temperature for 12 hours. Excess bromine was evaporated using water jet pump. Products were precipitated in excess anhydrous acetone, separated by filtration, and dried under vacuum at ambient temperature.

### *23.2.4 SYNTHESIS OF A NOVEL RAFT MACROINITIATOR (RAFT-MACRO AGENT)*

0.201 g of PMBA-diBr was reacted with 0.108 g of potassium ethyl xanthogenate in 1 mL formic acid in a Schlenk tube at 30°C for 48 hours. Nitrogen gas was presented by injection. After this time, the tube content was poured in cold diethyl ether/petroleum ether (1/1) (v/v) solution to precipitate novel RAFT-macro agent. The mixture was kept overnight in a fridge. After decantation, RAFT-macro agent was dried under vacuum at 25°C for 1 day. The yield of RAFT-macro agent was defined gravimetrically.

### *23.2.5 SYNTHESIS OF NOVEL POLY(B-METHYL B-ALANINE-B-METHYL METHACRYLATE) [POLY(MBA-B-MMA)] STAR-TYPE TRIARM BLOCK COPOLYMERS BY RAFT POLYMERIZATION*

The amounts of chemicals used in the copolymerization were shown in Table 23.1. Specified amounts of novel RAFT-macro agent, MMA, AIBN, and formic acid were put separately into a Schlenk tube followed by injecting nitrogen gas in the tube for 1 minute. The tube was put in a silicone oil bath on a magnetic stir plate at fixed temperature for fixed time. After polymerization, the mixture was drained into excess methanol to precipitate novel poly(MBA-b-MMA) star-type triarm block copolymer. The mixture was kept overnight in a fridge. After decantation, the star-type triarm block copolymer was dried under vacuum at 25°C for 1 day. The yield of the copolymer was defined gravimetrically.

## 23.3 RESULTS AND DISCUSSION

### *23.3.1 SYNTHESIS OF PMBA-DIBR*

Scheme 23.1 includes the reaction pathway for the synthesis of PMBA and PMBA-diBr (FT-IR: 3282 $cm^{-1}$ for -NH, 2700–3060 $cm^{-1}$ for aliphatic -CH,

-$CH_2$, $CH_3$, 1645 $cm^{-1}$ for -C = O, 1140 $cm^{-1}$ for -OC; $^1$H-NMR: 1.3 ppm for -C<u>H</u>$_3$, 2.0 ppm for -C<u>H</u>$_2$, 2.8 ppm for -NC<u>H</u>, 4.4 ppm for -BrC<u>H</u>, 8.1 ppm for -N<u>H</u>). The gravimetric yields obtained from the weights of PMBA and PMBA-diBr were 78.0% (wt) and 97.0% (wt), respectively.

**TABLE 23.1** Synthesis of Novel Poly(MBA-b-MMA) Star-Type Triarm Block Copolymers by RAFT Polymerization*

| Code | RAFT-Macro Agent (g) | MMA (g) | Temperature (°C) | Time (h) | Yield | |
|---|---|---|---|---|---|---|
| | | | | | (g) | (wt.%) |
| HZM | 0.050 | 0.200 | 65 | 72 | 0.062 | 24.80 |
| HZM-1 | 0.051 | 0.107 | 85 | 24 | 0.048 | 30.38 |
| HZM-2 | 0.051 | 0.211 | 85 | 24 | 0.108 | 41.22 |
| HZM-3 | 0.056 | 0.314 | 85 | 24 | 0.153 | 41.35 |
| HZM-4 | 0.051 | 0.416 | 85 | 24 | 0.193 | 41.33 |

*Formic acid: 0.5 mL, AIBN: 0.001 g.

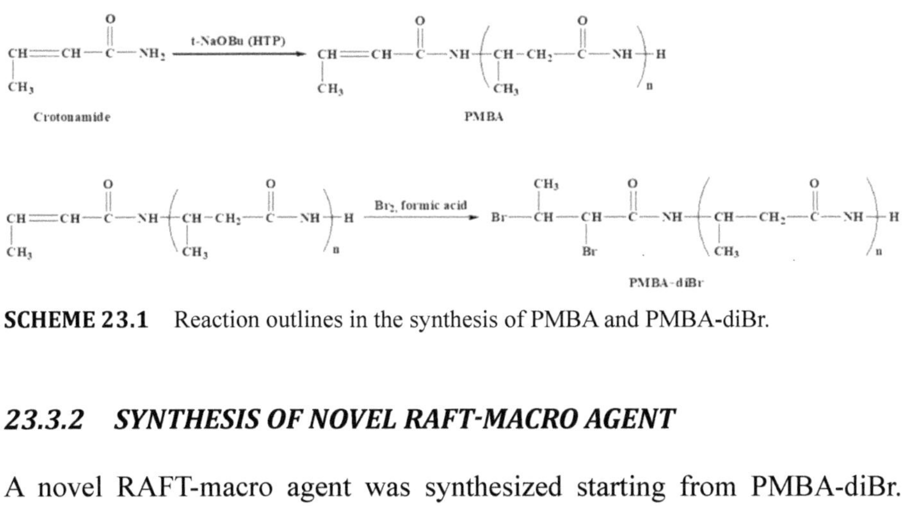


**SCHEME 23.1** Reaction outlines in the synthesis of PMBA and PMBA-diBr.

### *23.3.2 SYNTHESIS OF NOVEL RAFT-MACRO AGENT*

A novel RAFT-macro agent was synthesized starting from PMBA-diBr. Scheme 23.2 includes the reaction outline for the synthesis of RAFT-macro agent. The gravimetric yield of RAFT-macro agent was 60.7% (wt). The $^1$H-NMR spectrum of RAFT-macro agent (Figure 23.1) displayed peaks at 1.4 ppm for -C<u>H</u>$_3$ of PMBA group, 2.0 ppm for -C<u>H</u>$_3$ of xanthogenate group, 2.8 ppm for -C<u>H</u>$_2$ linked carbonyl unit of PMBA group, 4.0 ppm for -OC<u>H</u>$_2$ of xanthogenate group, 4.4 ppm for -NC<u>H</u> of PMBA group, 6.0 ppm for -SC<u>H</u>, and 8.1 ppm for -N<u>H</u>. Furthermore, the peak at 2.0 ppm for -$CH_3$ protons and 4.0 ppm for -$OCH_2$ protons of xanthogenate group in the $^1$H-NMR spectrum was further evidences that RAFT-macro agent was successfully obtained.

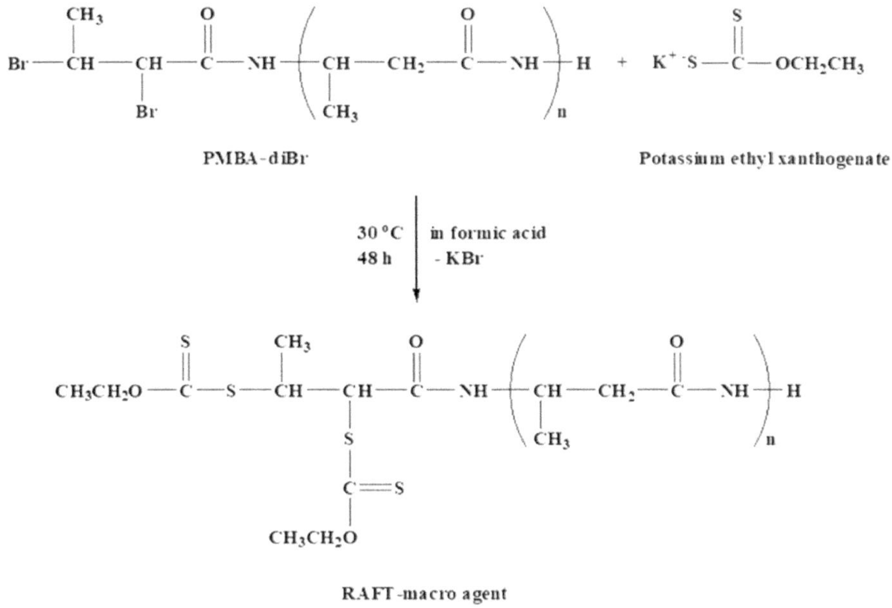


**SCHEME 23.2** Reaction outline in the synthesis of novel RAFT-macro agent.

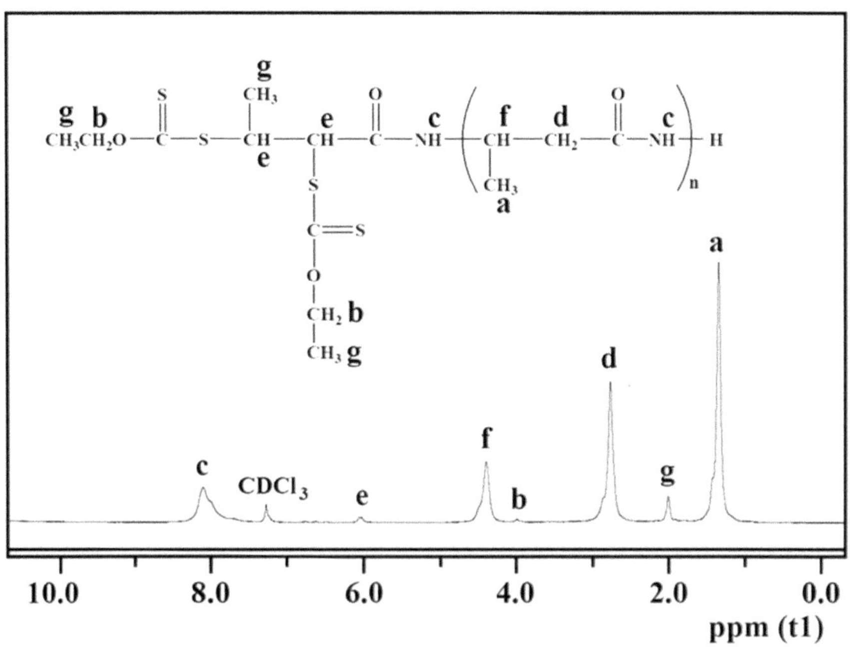


**FIGURE 23.1** $^1$H-NMR spectrum of RAFT-macro agent.

### 23.3.3 SYNTHESIS OF POLY(MBA-B-MMA) STAR-TYPE TRIARM BLOCK COPOLYMERS BY RAFT POLYMERIZATION

Poly(MBA-b-MMA) star-type triarm block copolymers were obtained by using RAFT polymerization of RAFT-macro agent and MMA. The polymer yield was between 41.35% (wt) and 24.80% (wt). Scheme 23.3 shows the pathway for synthesis of the star-type triarm block copolymer. The FT-IR spectrum of poly(MBA-b-MMA) star-type triarm block copolymer (HZM-1 in Table 23.1) in Figure 23.2 shows the signals at 3279 $cm^{-1}$ for -NH, 2500–3100 $cm^{-1}$ for aliphatic -CH, -$CH_2$, $CH_3$, 1728 $cm^{-1}$ for -C = O of PMMA, 1643 $cm^{-1}$ for -C = O of PMBA, and 1142 $cm^{-1}$ for -OC. The $^1$H-NMR spectrum of poly(MBA-b-MMA) star-type triarm block copolymer (HZM-1 in Table 23.1) in Figure 23.3 displayed at 1.0 ppm and 1.1 ppm for -C$\underline{H}_3$ of PMMA group, 1.4 ppm for -C$\underline{H}_3$ of PMBA group, 2.0 ppm for -C$\underline{H}_2$ of PMMA, 2.8 ppm for -C$\underline{H}_2$ of linked carbonyl unit of PMBA group, 3.8 ppm for OC$\underline{H}_3$ of PMMA group, 4.4 ppm for -NC$\underline{H}$ of PMBA group, and 8.1 ppm for -N$\underline{H}$ of PMBA group.

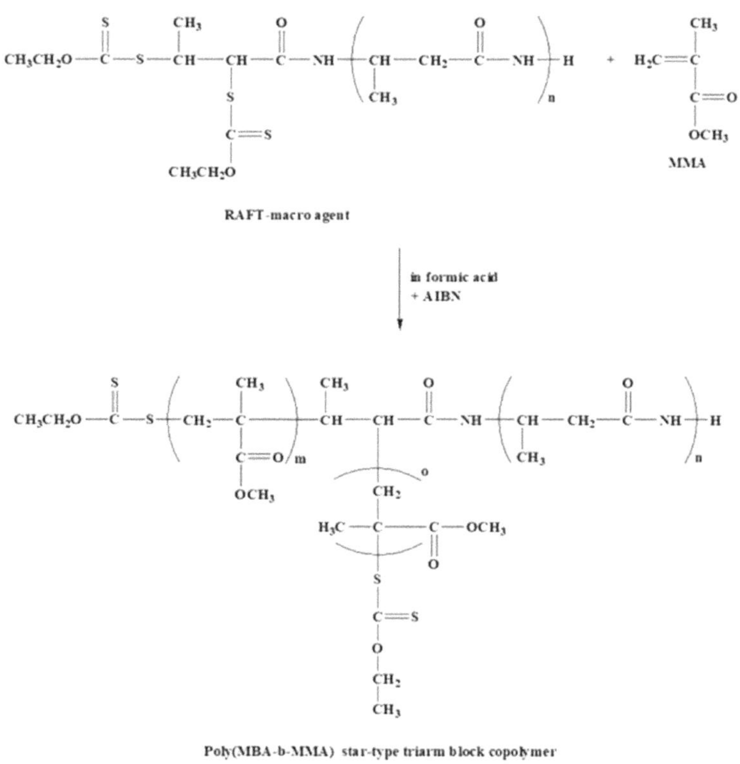


**SCHEME 23.3** Reaction pathway in the synthesis of novel poly(MBA-b-MMA) star-type triarm block copolymer.

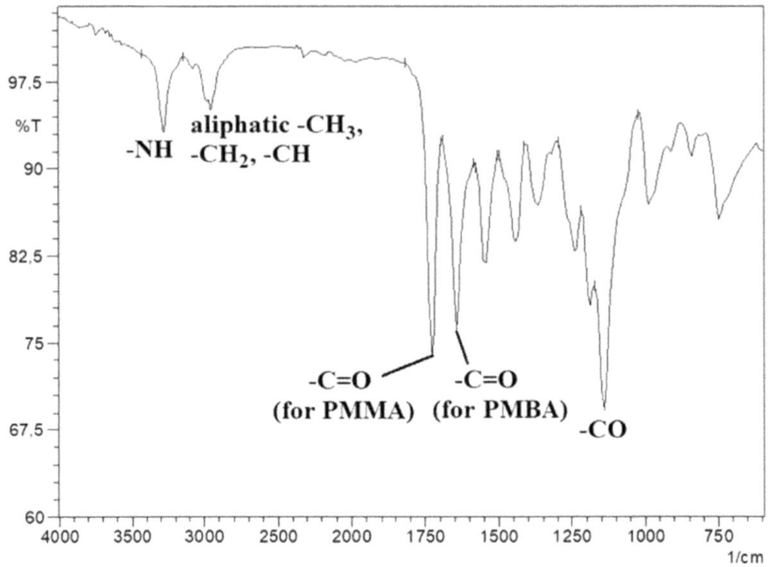


**FIGURE 23.2** FT-IR spectrum of poly(MBA-b-MMA) star-type triarm block copolymer (HZM-1 in Table 23.1).

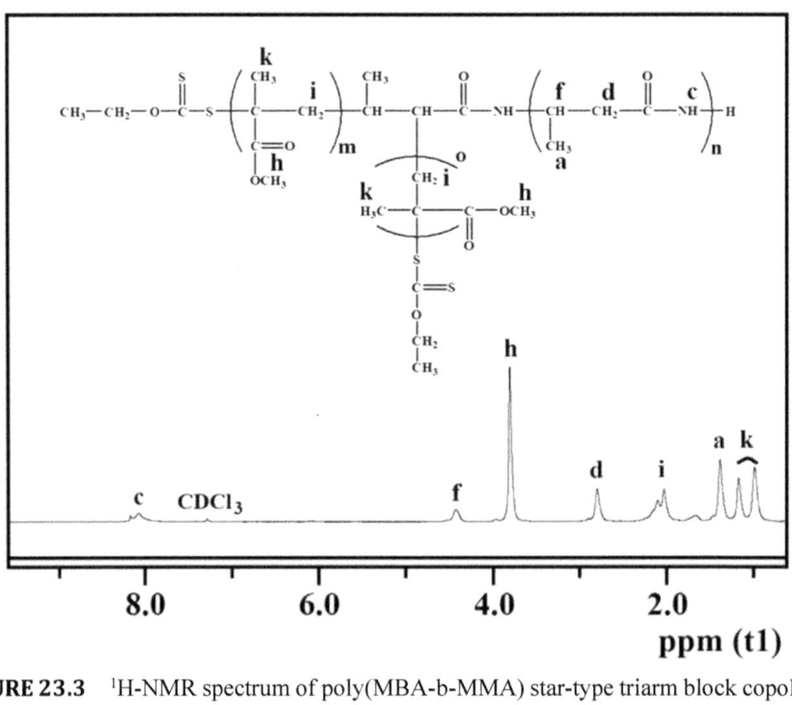


**FIGURE 23.3** $^{1}$H-NMR spectrum of poly(MBA-b-MMA) star-type triarm block copolymer (HZM-1 in Table 23.1).

Thermal analysis of poly(MBA-b-MMA) star-type triarm block copolymers was carried out by DSC and TGA. Decomposition temperatures (Td) of the copolymers were indicated in Table 23.2. In the case of poly(MBA-b-MMA) star-type triarm block copolymer, PMBA, and PMMA units have the individual decomposition temperatures (Td) (Figure 23.4). TGA has shown interesting properties of the copolymers indicating continuous weight loss starting from 200°C to nearly 500°C with two maxima at 329°C and 395°C (Figure 23.4(b)). The maximas at the derivative curve correspond to temperatures at which decomposition rates of amide and acrylate units are maximum. Tg shift in the DSC curve of HZM-2 sample was also determined as 110°C (Figure 23.5), which is compatible with the Tg values for PMMA (105°C) [36] and PBMA (105–125°C) [37]. The only one Tg value for the star-type triarm block copolymers may also be attributed to the miscible nature of the related polymers.

**TABLE 23.2** Thermal Properties of Poly(MBA-b-MMA) Star-Type Triarm Block Copolymers

| Code in Table 23.1 | Tg | Decomp. Temp. (°C) | |
|---|---|---|---|
| | | Td1 | Td2 |
| HZM | 139 | 294 | 460 |
| HZM-1 | n.d. | 329 | 403 |
| HZM-2 | 110 | 319 | 394 |
| HZM-3 | n.d. | 329 | 395 |
| HZM-4 | n.d. | 329 | 406 |

n.d.: not determinated.

MALDI-MS spectrum at the corner of Figure 23.6 has well-distributed signals between 1000 and 3000 Da. The expanded spectrum in Figure 23.6 was created, and some signal was labeled to reveal the copolymer formation. The signals separated by about 100 Da, which is the mass of MMA units, are also related with the signals between them. Because the mass difference between MMA and MBA units is equal to approximately 15 Da, mass differences between the sequential signals are also measured as about 15 Da. The observation is simple proof of the block copolymer formation.

## 23.4 CONCLUSIONS

The RAFT polymerization of novel poly(MBA-b-MMA) star-type triarm block copolymer from the two constituent monomers was acquired by using a novel RAFT-macro agent. The copolymers were obtained in high yield and high molecular weight. This method for the synthesis of the copolymer is simple and efficient. Products characterization was done using multi-instruments and fractional techniques. This study can provide a new well-characterized materials with wide biomedical application potential through the synergistic combination of the PMBA and PMMA.

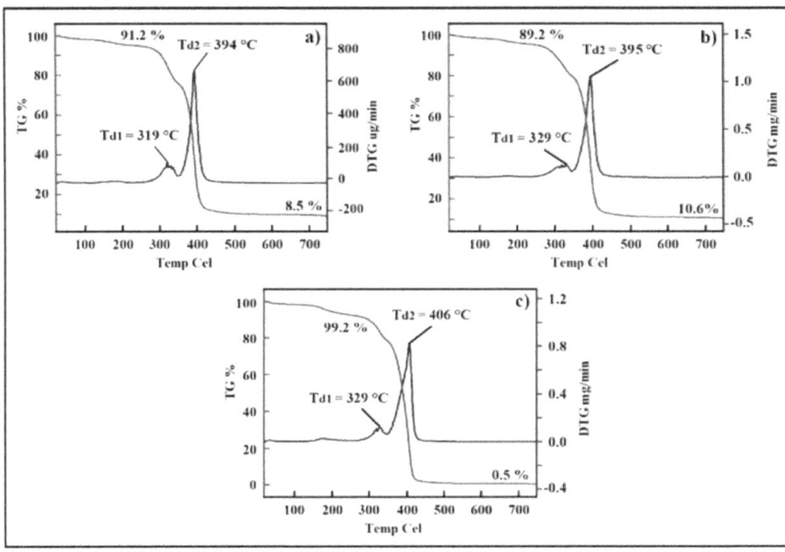


**FIGURE 23.4** TGA curves of poly(MBA-b-MMA) star-type triarm block copolymers: (a) HZM-2 in Table 23.2; (b) HZM-3 in Table 23.2; (c) HZM-4 in Table 23.2.

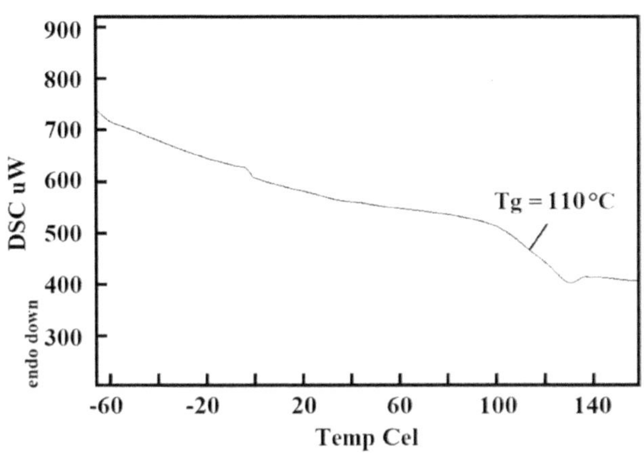


**FIGURE 23.5** DSC curve of poly(MBA-b-MMA) star-type triarm block copolymer (HZM-2 in Table 23.2).

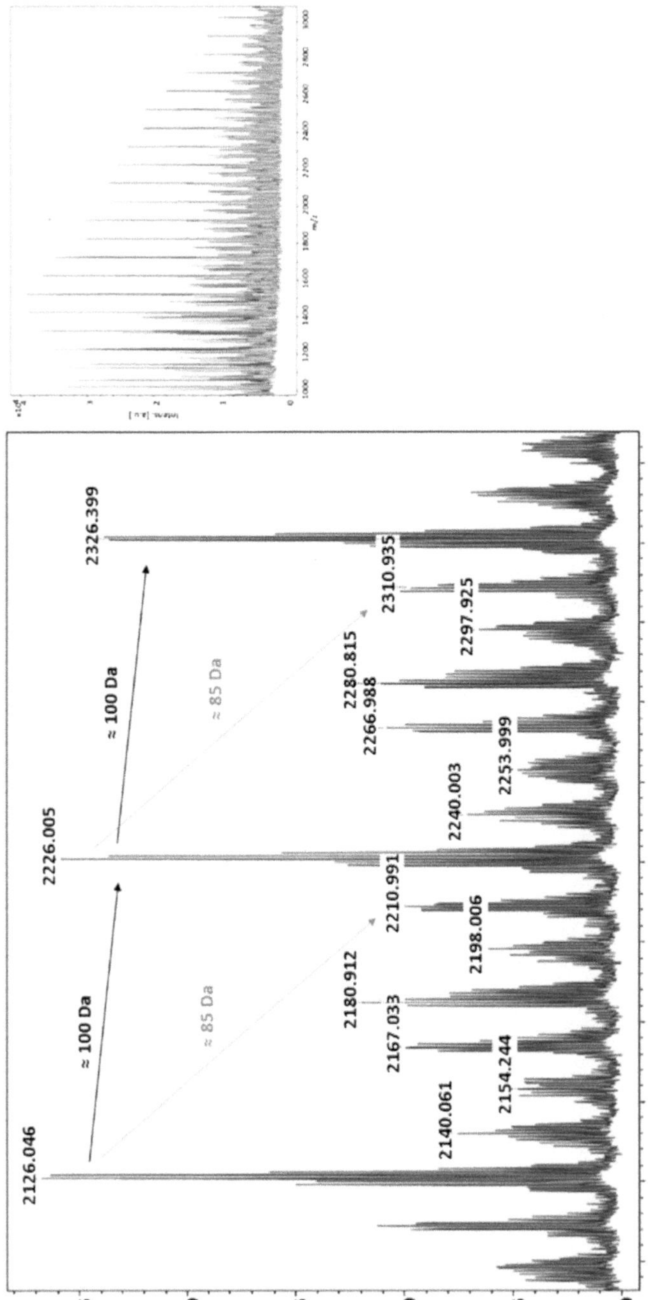


**FIGURE 23.6** Expanded MALDI-MS spectrum of HZM-1 sample.

## KEYWORDS

- **differential scanning calorimetry**
- **hydrogen transfer polymerization**
- **novel RAFT-macro agent**
- **poly(epichlorohydrin)**
- **reversible addition-fragmentation chain transfer (RAFT) polymerization**
- **star-type triarm block copolymer**

## REFERENCES

1. Chiefari, J., Chong, Y. K., Ercole, F., Krstina, J., Jeffery, J., Le, T. P. T., Mayadunne, R. T. A., et al., (1998). *Macromolecules, 31*(31), 5559.
2. Shimizu, K., & Kobatake, S., (2017). Chemistry Selec., 2(20), 5445.
3. Goto, A., Tsujii, Y., & Fukuda, T., (2008). *Polymer, 49*(24), 5177.
4. Robin, M. P., Wilson, P., Mabire, A. B., Kiviaho, J. K., Raymond, J. E., Haddleton, D. M., & O'Reilly, R. K., (2013). *J. Am. Chem. Soc., 135*(8), 2875.
5. Patton, D. L., Mullings, M., Fulghum, T., & Advincula, R. C., (2005). *Macromolecules, 38*(20), 8597.
6. Öztürk, T., Atalar, M. N., Göktaş, M., & Hazer, B., (2013). *J. Polym. Sci., Part A: Polym. Chem., 51*(12), 2651.
7. Öztürk, T., & Hazer, B., (2010). *J. Macromol. Sci., Part A, Pure Appl. Chem., 47*(3), 265.
8. Şanal, T., Oruç, O., Öztürk, T., & Hazer, B., (2015). *J. Polym. Res., 22*(2), 1.
9. Öztürk, T., Göktaş, M., & Hazer, B., (2011). *J. Macromol. Sci., Part A, Pure Appl. Chem., 48*(1), 65.
10. Öztürk, T., Kayğın, O., Göktaş, M., & Hazer, B., (2016). *J. Macromol. Sci., Part A, Pure Appl. Chem., 53*(6), 362.
11. Schmidt, B. V. K. J., & Barner-Kowollik, C., (2013). *Nat. Chem., 5*(12), 990.
12. Göktaş, M., Öztürk, T., Atalar, M. N., Tekeş, A. T., & Hazer, B., (2014). *J. Macromol. Sci., Part A, Pure Appl. Chem., 51*(11), 854.
13. Öztürk, T., Yavuz, M., Göktaş, M., & Hazer, B., (2016). *Polym. Bull., 73*(6), 1497.
14. Öztürk, T., Göktaş, M., Savaş, B., Işıklar, M., Atalar, M. N., & Hazer, B., (2014). *E-Polymers., 14*(1), 27.
15. Öztürk, T., Göktaş, M., & Hazer, B., (2010). *J. Appl. Polym. Sci., 117*(3), 1638.
16. Altintas, O., Tunca, U., & Barner-Kowollik, C., (2011). *Polym. Chem., 2*(5), 1146.
17. Altıntas, O., & Tunca, U., (2011). *Chem. Asian J., 6*(10), 2584.
18. Ruzette, A. V., & Leibler, L., (2005). *Nat. Mater, 4*(1), 19.
19. Çatıker, E., Öztürk, T., Atakay, M., & Salih, B., (2019). *J. Polym. Res., 26*(5), 123.
20. Çatıker, E., Meyvacı, E., Atakay, M., Salih, B., & Öztürk, T., (2019). *Polym. Bull., 76*(4), 2113.

21. Öztürk, T., & Cavicchi, C. A., (2018). *J. Polym. Mater., 35*(2), 209.
22. Bolton, J., & Rzayev, J., (2012). *ACS Macro Lett., 1*(1), 15.
23. Mitra, K., Hira, S. K., Singh, S., Vishwakarma, N. K., Vishwakarma, S., Gupta, U., Manna, P. P., & Ray, B., (2018). *Chemistry Select, 3*(31), 8833.
24. Johnson, J. A., Lu, Y. Y., Burts, A. O., Xia, Y., Durrell, A. C., Tirrell, D. A., & Grubbs, R. H., (2010). *Macromolecules, 43*(24), 10326.
25. Johnson, J. A., Lu, Y. Y., Burts, A. O., Lim, Y., Finn, M. G., Koberstein, J. T., Turro, N. J., Tirrell, D. A., & Grubbs, R. H., (2011). *J. Am. Chem. Soc., 133*(3), 559.
26. Zhang, M., Estournès, C., Bietsch, W., & Müller, A. H. E., (2004). *Adv. Funct. Mater., 14*(9), 871.
27. Djalali, R., Li, S. Y., & Schmidt, M., (2002). *Macromolecules, 35*(11), 4282.
28. Zhang, M., Drechsler, M., & Muller, A. H. E., (2004). *Chem. Mater., 16*(3), 537.
29. Li, C., Gunari, N., Fischer, K., Janshoff, A., & Schmidt, M., (2004). *Angew. Chem. Int. Ed., 43*(9), 1101.
30. Huang, K., & Rzayev, J., (2009). *J. Am. Chem. Soc., 131*(19), 6880.
31. Cheng, C., Qi, K., Khoshdel, E., & Wooley, K. L., (2006). *J. Am. Chem. Soc., 128*(21), 6808.
32. Al-Kaabi, K., & Van, R. A. J., (2008). *J. Appl. Polym. Sci., 108*(4), 2528.
33. Cho, B. S., & Noh, S. T., (2013). *Macromol. Res., 21*(2), 221.
34. Tang, T. T., Fan, X. S., Jin, Y., & Wang, G. W., (2014). *Polymer, 55*(16), 3680.
35. Çatıker, E., Güven, O., & Salih, B., (2018). *Polym. Bull., 75*(1), 47.
36. Brandrup, J., & Immergut, E. H., (1975). *Polymer Handbook* (2nd edn.). Wiley, New York.
37. Schmidt, V. E., (1970). *Angew. Makromol. Chem., 14*(1), 185.

# CHAPTER 24

# Using of Polymers Technologies in the Industry

L. V. TABATADZE,[1] V. V. SHVELIDZE,[2] E. J. CHURGULIA,[1] and N. G. SHENGELIA[1]

[1]*Department of Chemistry, Sokhumi State University, A. Politkovskaya, St., 12, Tbilisi–0186, Georgia*

[2]*Department of Physics, Tbilisi State University, I. Chavchavadze Ave., 1, 0179 Tbilisi, Georgia*

## ABSTRACT

The use of ozone in agriculture and in various fields of public health as a disinfection and sterilization method is very relevant today. The work describes the use of ozone generators "Samani 1" and "Samani 5" of different constructions developed by Sokhumi State University and TSU Scientific Group "Savelimisioni" in Georgia for disinfection and sterilization of wine necessary for wine production, also against microbial pollutants and disinfectants particular objects and the books in libraries.

## 24.1 INTRODUCTION

Agriculture is the oldest and most important field in Georgia. The history of vine culture is closely linked to the history of the Georgian nation. Georgia is considered one of the oldest horticulture-winemaking and one of the best countries in the world to produce high-quality wines.

The use of ozone in agriculture, in particular wine production, can completely change almost all of the previously used chemicals that will reduce the use of pesticides and environmental pollution. All operations require special preparation of necessary equipment for production because disinfection and sterilization of the internal surface of the equipment are determined by wine quality and organoleptic indexes. The use of ozone in winemaking is of great practical importance due to its high antibacterial properties.

## 24.2 MATERIAL AND METHODS

Today, ozone technology has an effective advantage. In small concentrations, ozone is the best disinfectant. It removes the outer layers of microorganisms, destroys staphylococci, streptococcus, sticks, hepatitis, and flu viruses, removes pesticides, herbicides, neutralizes salts of heavy metals, restores the homeostasis system in the body, and promotes the transportation of oxygen in the bloodstream. Ozone sharply reduces the bactericidal intake of the damaged skin of the living organism as well as on the surfaces of different tar, material, household items, inventory, and packaging materials. Its optimal concentration in case of short processing is about 0.5 g/m$^3$. Nowadays, industrial production of ozone is electrosynthesis based on the production of ozone molecules at the expense of energy allocated during electrical discharge. In 1985, at the Moldovan Research Institute, ozone was studied and researched for sterilization of the internal surface of the food industry technologies. Methods, controls, and technological schemes of the ozone process have been developed. It is also proven that a small amount of ozone is sufficient when processing the object. As well as the information of famous institutes of microbiology and epidemiology: *E. coli*, *St. Albus*, *Ps.* Fluorescent culture, also during the processing of various surface microflores, the concentration of 106–124 mg/m$^3$ ozone is high in disinfectants for 30–40 minutes.

Widespread use of ozone in our country hinders the absence of a reliable and perfect generator acceptable for ozone concentration required for practical use, the relatively low quality of processing, which is due to the fact that the processing of ozone does not occur in vacuum conditions, which results in a decrease in processing efficiency. For healthy and stable wine production, first of all, it is necessary to have special hygiene in winemaking. The current disinfectant filter cannot completely exclude ozone in the environment.

Against microbiological pollutants in wine production is often used steam, a saddle, acetic acid peroxide, salty water, and hot water. All these things have the appropriate antimicrobial effect, but negative impacts are also observed. Caustic soda and acetic acid are corrosive substances that can harm the personal in the wine industry. The advantage of the use of ozone in winemaking is to have very high antibacterial activity in minimal contact. Ozone concentration in aqueous solution is 0.02–2 mg/l.

Electricity expense is 5 watts/h, unlike the world analogs (3.5 times cheaper). The hybrid variant of new ozegenerators is tested and is underway. Due to the attractive technical and economic characteristics of the equipment, many customers in different spheres have been used in different enterprises in our country: in mineral water "Likani" enterprises, in Natakhtari meat factory, in 4 bread factory in Tbilisi and others, about 18 food stores in Georgia.

## 24.3 RESULTS AND DISCUSSION

Sokhumi State University and TSU Scientific-Technical Group "Savelimysioni" have been altering variants of different modification and purpose ozogenerators "Samani 1" and "Samani 5," which can be discharged at affordable prices for massive consumers.

Together with the Ministry of Agriculture and Vineculture Specialists in Kareli region, with ozogenerator "Samani 1" were processed the wineries and viticulture of the plantation. Ozone generated by the ozonegenerator in the mixing block is included with water. Ozone solutions are 11 times more than oxygen. The ozone aqueous solution is used to process under certain pressure. It is sprayed with grapes by a spray gun.

Such a mixture can also be used for disinfection of wine pitchers and storehouses in the wine factory. The test of ozone generator "Samani 1" was conducted for surface sterilization of the equipment of the wine industry and to clean the air. From three, the heat of the untreated plant in two was found the intestinal strain bacteria, during 20 minutes after the ozone layer, the intestinal strain bacteria were not observed in the factory.

The findings of the survey show that ozogenerator "Saman 1" can be successfully used in processing various volumes of wine production technologies. Works on viticulture plant are still underway. Preliminary data is positive and research-search continues (Figure 24.1 and Table 24.1).

Ozogenerator "Samani 5" was developed by Sokhumi State University and TSU Scientific Group "Savelimisioni," by which it is possible to disinfect and sterilize individual objects, consumer goods, and items in the available prices to the massive consumers. The presented ozogenerator is a device that contains a sterilization hermetic chamber (1), Sterilization agent generator (2), air heater (3), air cooler (4), vacuum pump (5), tap (6), and filter (7). Warm and filtered air in the ozonizer provides an efficient and complete treatment of objects in the disinfectant hermetic chamber. The device sterilizer eliminates the entry of ozone into the environment (Figure 24.2).

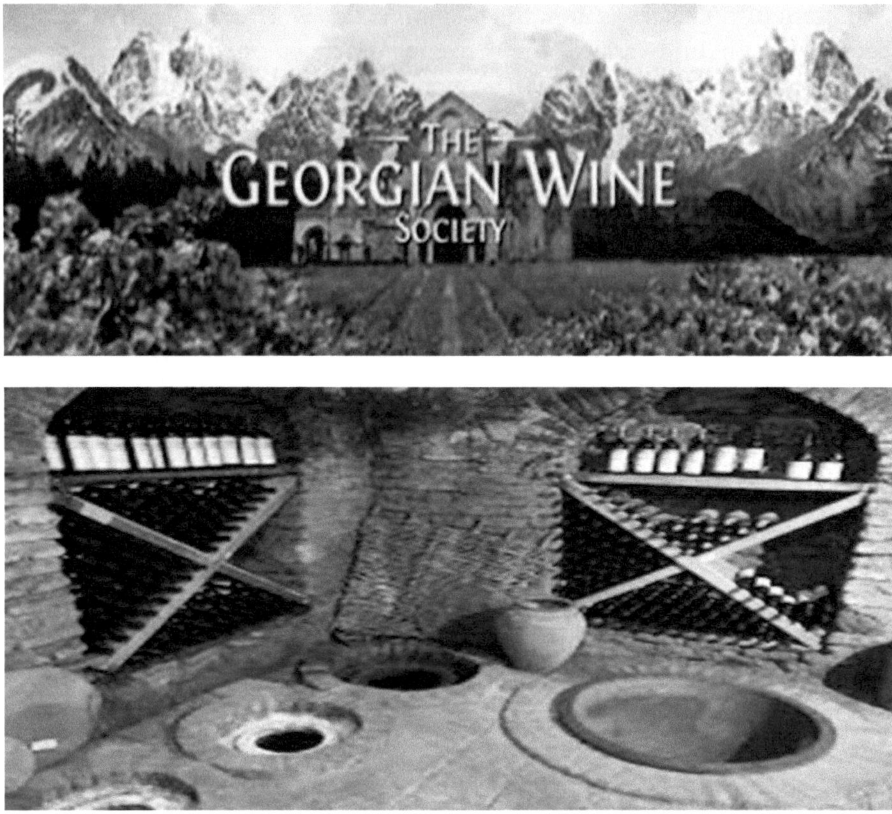


**FIGURE 24.1** Georgian wine.

**TABLE 24.1** The Time of the Processing Place

| Area of m² | Room Bulk Cube/m | Processing Time, Minute | Stuff Waiting Time, Minute |
|---|---|---|---|
| 20 | 60 | 20 | 15 |
| 40 | 120 | 30 | 20 |
| 60 | 180 | 50 | 30 |
| 100 | 300 | 70 | 50 |

At present, the preparatory research works are conducted in the central library of the TSU for conducting ozoneotherapy of books. Ozone allocated for disinfecting books is a mixture with the rest of the components of the air (water vapor, hydrogen peroxide, nitrogen, carbon dioxide, etc.). This ozone mixture is included in the upper part of the chamber, evenly distributed on the bottom of the disinfecting chamber on books, and then with gloves, it is necessary to flip through the books (Figure 24.3).

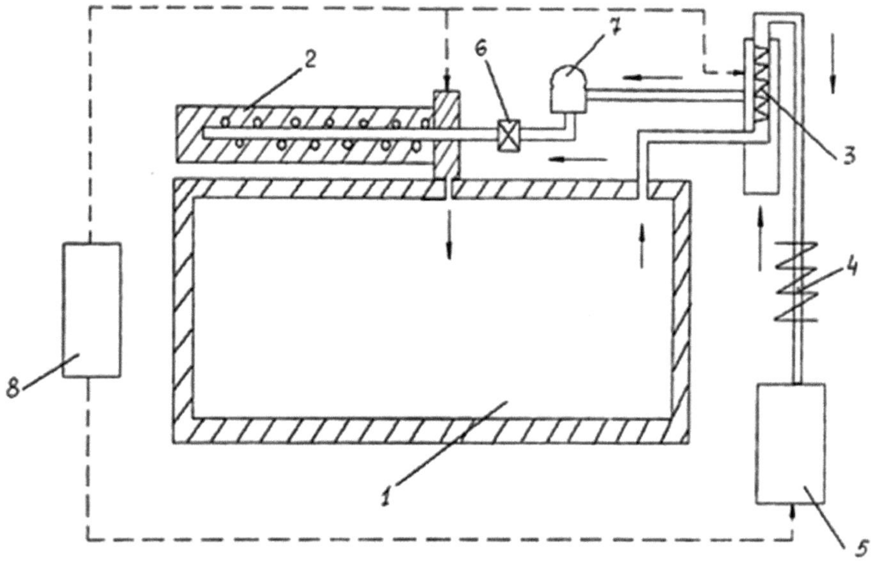


**FIGURE 24.2** Sterilization unit.

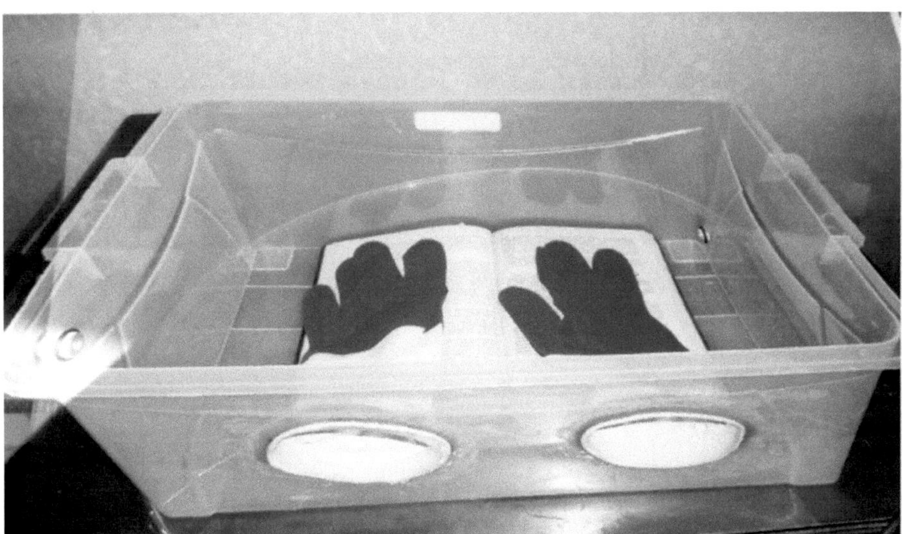

**FIGURE 24.3** Disinfecting chamber.

Laboratory of the National Center for Disease Control and Public Health held a bacteriological analysis of the books of the library. Identification numbers of sample material are 1GE012159 and 1GE012153. As a result of the research, both the samples were discovered by *Aspergillus* spp.; sticks and sample survey continues.

## 24.4 CONCLUSION

In the designed ozogenerators, low-pressure plasmas and spirally wound conductors are used as electrodes. Ozogenerator works on high frequency and low pressure. Ceramics are used as the dielectric. The device is performed on international technical terms: GOCT 26582-85.

Our studies show the ozone's apparent advantage over other disinfectants. The potential user of the use of ozone generators is the Institute of Manuscripts of Georgia and the National Museum.

## KEYWORDS

- **disinfectant hermetic chamber**
- **disinfection**
- **hydrogen peroxide**
- **microbes**
- **ozogenerators**
- **ozoneotherapy**

## REFERENCES

1. González-Barreiro, C., Rial-Otero, R., Cancho-Grande, B., & Simal-Gándara, J., (2015). Wine aroma compounds in grapes. *Critical Reviews in Food Science and Nutrition, 55*, 202–218.
2. Francis, I. L., & Newton, J. L., (2005). Determining wine aroma from compositional data. *Aust. J. Grape Wine Research, 11*, 114–126.
3. Ajikumar, P. K., et al., (2008). Terpenoids, opportunities for biosynthesis of natural product drugs using engineered microorganisms. *Molpharmaceut, 5*, 167–190.
4. Tabatadze, L., Shvelidze, V., & Gakhokidze, R., (2017). *Using an Ozogenerator "Samani-2" in Livestock* (Vol. 503, pp. 11–16). Works Georgian Technical University.

5. Tabatadze, L. V., & Shvelidze, V. V., (2018). *Developing Combined Ozone and Ultraviolet Light Solar Energy Devices* (Vol. 34, pp. 9–12). RS Global, Scientific Educational Center. Warshaw, Poland. World Science.
6. Rosenblum, J., Bohrerova, C. G. Z., Yousef, A., & Lee, J., (2012). Ozonation as a clean technology for fresh produce industry and environment, sanitizer efficiency and wastewater quality. *Journal of Applied Microbiology, 113*, 837–845.
7. Loreto, F., Pinelli, P., Manes, F., & Kollist, H., (2004). Impact of ozone on monoterpene emissions and evidences for an isoprene-like antioxidant action of monoterpenes emitted by *Quercus ilex* (L.) leaves. *Tree Physiology, 24*, 361–367.
8. Vahid, F., & Howarth, C. R., (1986). Synthesis of ozone in dielectrieless discharges. *Chemical Engineering Journal, 32*, 43–51.
9. Yang, P. P. W., & Chen, T. C., (1979). Effects of ozone treatment on microflora of poultry meat. *Journal of Food Processing and Preservation, 3*, 177–185.
10. Whistler, P. E., & Sheldon, B. W., (1989). Biocidal activity of ozone versus formaldehyde against poultry pathogens inoculated in a prototype setter. *Poultry Science Association, 68*, 1068–1073.

Geopolymers are synthesized out of reactive aluminosilicate powder (e.g., fly ash or calcined clay), which is activated with an alkaline activator (alkali hydroxide or alkali silicate solution) at ambient or slightly elevated temperature. They have a 3D dense amorphous microstructure, built out of interlinked $SiO_4$ and $AlO_4$ tetrahedra, and contain charge-balancing alkali [8].

The dissolution of aluminosilicate materials and the transition of silicate and aluminate anions to the liquid phase is a limiting stage in the kinetics of hardening of geopolymer materials. Therefore, the more active the aluminosilicate material in this regard, the higher the rate of curing. Compared with kaolinite, the product of its firing-metakaolin-sends silicate and aluminate ions to an alkaline solution much more actively; fly ash and slag are also inferior to metakaolin in the rate of dissolution [9]. The addition of metakaolin to fly ash or slag noticeably activates the alkaline hardening of these materials.

The widespread use of metakaolin in the production of building materials is hindered by the world's limited reserves of kaolin and kaolin clay. This also applies to Georgia, which does not possess these reserves. However, there are active searches on reception of metakaolin from clayey rocks by their thermal modification [10–15].

## 25.2 EXPERIMENTAL PART

### *25.2.1 MATERIALS*

For this study, clayey rocks of Georgia (shale, argillite, and low-melting clay) were used for the synthesis of geopolymer binders.

### *25.2.2 METHODS*

The mineral composition of clayey rocks was determined using an Optika B-383POL polarization microscope (Italy).

For thermogravimetric analysis (TGA), a NETZSCH derivatograph with STA-2500 REGULUS thermogravimetric and differential thermal analyzer (TG/DTA) was used. Samples were heated to 1000°C, in a ceramic crucible, heating rate 10°C/min—reference substance a-$Al_2O_3$.

The X-ray phase analysis was carried out using a Dron-4.0 diffractometer ("Burevestnik," St. Petersburg, Russia) with a Cu-anode and a Ni-filter. U = 35 kv. I = 20 mA. Intensity-2 degrees/min. $\lambda = 1.54178$ Å.

## 25.3 RESULTS AND DISCUSSIONS

Table 25.1 shows the chemical compositions of the studied materials.

**TABLE 25.1** The Chemical Composition of Clay Rocks, Mass. %*

| No. | LOI | $SiO_2$ | $TiO_2$ | $Al_2O_3$ | $Fe_2O_3$ | FeO | $Mn_2O_3$ | CaO | MgO | $SO_3$ | $Na_2O$ | $K_2O$ |
|---|---|---|---|---|---|---|---|---|---|---|---|---|
| 1. | 4.50 | 59.95 | 0.89 | 17.30 | 3.45 | 3.65 | 0.59 | 1.53 | 2.43 | 0.30 | 2.20 | 2.20 |
| 2. | 7.01 | 47.19 | – | 15.90 | 13.36 | – | 0.10 | 6.30 | 4.10 | 1.39 | 2.86 | 1.30 |
| 3. | 10.60 | 52.84 | – | 15.07 | 6.47 | – | – | 7.06 | 2.49 | 1.36 | 1.19 | 2.17 |

*No. 1-Shale; No. 2-Argillite; No. 3-Clay.

Figure 25.1 shows X-ray patterns of the studied clayey rocks where the presence of clay minerals is recorded (14.66–14.96, 7.14, 4.25, 3.66, 2.86, 2.327 Å); quartz (3.34 Å); feldspar (3.87 Å), carbonate (3.03 Å).

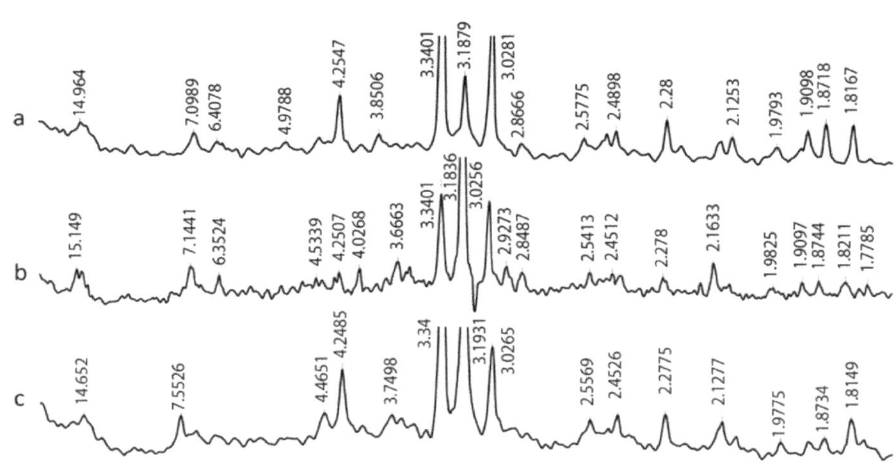


**FIGURE 25.1** X-ray patterns of clayey rocks: a-shale, b-argillite, c-clay.

To determine the temperature of the thermal modification of clayey rock, i.e., the temperature range in which the latter passes into the active-reactive modification, a TGA of clayey rocks was carried out. DTG curves are shown in Figure 25.2.

According to the data of differential thermal analysis (Figure 25.2), the endo-effect at 100–150°C is present on all DTG curves, which corresponds to the removal of physically bound water. In the temperature range of 650–850°C, an endo-effect is observed, which is obviously connected with the destruction of the crystal lattice of clay minerals and their transition to the active amorphous form (metakaolin). On this basis, the temperature of 700°C was chosen for the thermal treatment of clayey rocks.

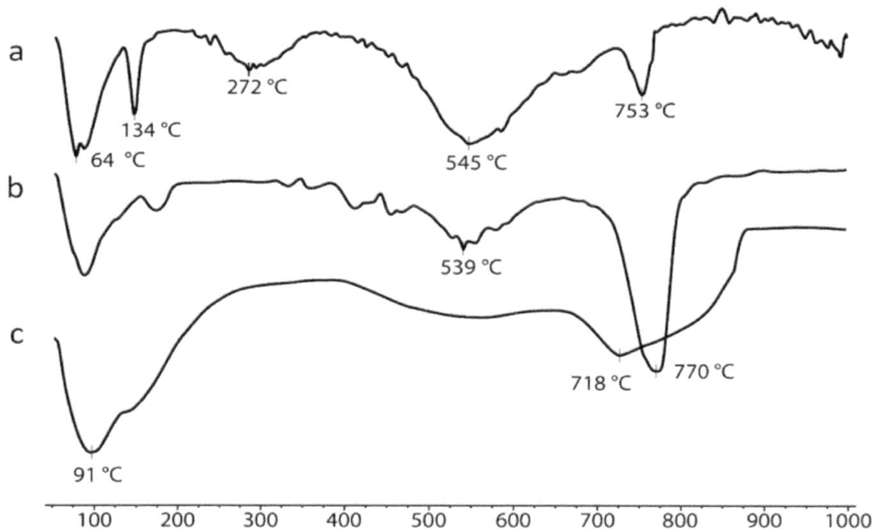


**FIGURE 25.2** DTG curves of clayey rocks: a-shale, b-argillite, c-clay.

Granulated blast furnace slag and modified at 700°C clayey rocks were used to produce geopolymer binder compositions, which were ground in a laboratory ball mill to a specific surface of 800–900 kg/m$^2$. The obtained mixtures were kneaded with alkaline component solution: sodium alkali (NaOH), sodium carbonate ($Na_2CO_3$) or alkali silicate $Na_2O{\cdot}(SiO_2)n$ to obtain normal density.

The resulting paste was used to mold 20 × 20 × 20 mm$^3$ samples, which were extracted from the mold on day 3. Some of the samples were stored in the air, some in water, and some in humid air conditions at +20°C for 28 days.

Some of the samples were heat-treated after being removed from the mold in the following way: withholding at 80°C for 20 hours. Compositions of geopolymer binders and the results of their physical-mechanical testing are shown in Table 25.2.

As the results of physical-mechanical testing have shown, after heat treatment of geopolymer binders at 80°C for 20 h, the mechanical strength is significantly increased compared to conventional curing samples. The strength of composition No. 11 on the basis of slag and modified clay with the addition of a mixture of: NaOH + $Na_2O{\cdot}(SiO_2)n$ differs especially. Obviously, further studies will be continued in this direction.

**TABLE 25.2** The Compositions of Geopolymer Binders and the Results of Their Physical-Mechanical Testing

| No. | Components (%) | | Alkaline Component (Dry Matter Over 100%), % | Compressive strength After 28 Days. Under Different Conditions of Hardening, MPa | | | Compressive Strength After Heat Treatment at 80°C for 20 h. |
|---|---|---|---|---|---|---|---|
| | | | | Air | Water | Humid | |
| 1. | Slag (80) | Shale (20) | NaOH (10) | 41.0 | 45.2 | 44.0 | 69.0 |
| 2. | Slag (80) | Shale (20) | $Na_2CO_3$ (10) | 21.0 | 24.5 | 24.0 | 53.7 |
| 3. | Slag (80) | Shale (20) | $Na_2O(SiO_2)n$ (10) | 18.7 | 33.4 | 21.2 | 48.8 |
| 4. | Slag (80) | Argillite (20) | NaOH (10) | 46.9 | 48.0 | 41.8 | 69.5 |
| 5. | Slag (80) | Argillite (20) | $Na_2CO_3$ (10) | 33.5 | 42.0 | 39.0 | 68.5 |
| 6. | Slag (80) | Argillite (20) | $Na_2O(SiO_2)n$ (10) | 53.6 | 47.2 | 45.0 | 85.6 |
| 7. | Slag (80) | Clay (20) | NaOH (10) | 46.0 | 51.0 | 47.8 | 63.0 |
| 8. | Slag (80) | Clay (20) | $Na_2CO_3$ (10) | 17.5 | 22.3 | 21.7 | 57.5 |
| 9. | Slag (80) | Clay (20) | $Na_2O(SiO_2)n$ (10) | 75.0 | 88.0 | 85.0 | 15.0 |
| 10. | Slag (80) | Clay (20) | NaOH (4) + $Na_2CO_3$ (6) | 21.5 | 254 | 23.0 | 266 |
| 11. | Slag (80) | Clay (20) | NaOH (4) + $Na_2O(SiO_2)n$ (10) | 85.0 | 94.0 | 93.5 | 102.5 |
| 12. | Slag (80) | Clay (20) | $Na_2CO_3$ (4) + $Na_2O(SiO_2)n$ (10) | 11.2 | 14.5 | 15.6 | 32.0 |

Heat treatment significantly accelerates the polymerization process and the rate of hydration of geopolymers. Temperature increase promotes the conversion of calcium hydrosilicates from gel-like phase to crystalline one. As the temperature rises, the solubility of ions increases: $Si^{4+}$, $Al^{3+}$, and $Ca^{2+}$, and consequently, the processes of geopolymer reactions are accelerated, which contributes to the increase of mechanical strength [16].

## 25.4 CONCLUSIONS

These studies have shown:

1. the possibility of obtaining geopolymer binders based on local clayey rocks modified at 700°C.
2. as the alkaline component, it is possible to use sodium alkali and sodium carbonate, alkali silicate, as well as combinations thereof.
3. after heat treatment of geopolymer binders at 80°C for 20 h, the mechanical strength is significantly increased compared to conventional curing samples.
4. it is possible to outline the prospects for the production of geopolymer binders based on local rocks as an alternative to OPC.

## ACKNOWLEDGMENTS

The authors are grateful to the Shota Rustaveli National Science Foundation of Georgia, with the financial support of which this work was carried out [Grant number FR-18-783].

## KEYWORDS

- **clayey rocks**
- **geopolymers**
- **metakaolin**
- **OPC**
- **sodium carbonate**
- **thermal modification**

## REFERENCES


1. Davidovits, J., (2017). Review: Geopolymers: Ceramic-like inorganic polymers. *J. Ceram. Sci. Technol., 8*(3), 335–350. doi: 10.4416/JCST2017-00038.
2. Davidovits, J., (1999). Chemistry of geopolymeric systems terminology. *Proceedings of Geopolymer* (pp. 9–40). Saint-Quentin, France.
3. Palomo, A., Marcias, A., Blanco, M. T., Puertas, F., Grutzeck, M. W., & Blanco, M. T., (1992). Physical, chemical and mechanical characterization of geopolymers. *9th Int. Con. On the Chem. Cem.*, 505–511.
4. Singh, B., Ishwarya, G., Gupta, M., & Bhattacharyya, S., (2015). Geopolymer concrete: A review of some recent developments. *Constr. Build. Mater., 85*, 78–90.
5. Robayo-Salazar, R. A., Mejia, J., & Mejia, D. G. R., (2017). Eco-efficient alkali-activated cement based on red clay brick wastes suitable for the manufacturing of building materials. *J. Clean. Prod., 166*, 242–252.
6. Davidovits, J., (1988). Soft mineralurgy and geopolymers. *Proceeding of Geopolymer 88 International Conference, the Université de Technologie* (pp. 49–56). Compiègne, France.
7. Korneev, V. I., & Brykov, A. S., (2010). Prospects of development of general binders. Geopolymers and their distinctive features. *Cement Appl., 2*, 51–55.
8. Duxson, P., Fernandez-Jimenez, A., & Provis, J., (2007). Geopolymer technology: The current state of the art. *J. Mater. Sci., 42,* 2917–2933.
9. Panagiotopoulou, C., Kontori, E., Perraki, T., & Glikeria, K., (2007). Dissolution of aluminosilicate minerals and by-products in alkaline media. *J. Mater. Sci., 42*(9), 2967–2973. doi: 10.1007/s10853-006-0531-8.
10. Gaifullil, A. R., Rakhimov, R. Z., & Rakhimova, N. R., (2015). The influence of clay additives in Portland cement on the compressive strength of the cement stone. *Mag. Civil Eng., 7*, 66–73. doi: 10.5862/MCE.59.7.
11. Fernandes, R., Martirena, F., & Scrivener, K. L., (2011). The origin of the pozzolanic activity of calcined clay minerals: A comparison between kaolinite, illiite and montmorillonite. *Cement and Concrete Research, 41*(1), 113–122.
12. Konovalov, V. M., Glikin, D. M., & Solomatova, S. S., (2015). Use of argillites in the production of mixed cements. *Modern Problems of Science and Education, 2*(2), 315–323.
13. Rakhimov, R. Z., Rakhimova, N. R., & Gaifullin, A. R., (2017). Influence of the addition of dispersed fine polymineral calcined clays on the properties of Portland cement paste. *Advances in Cement Research, 29*(1), 21–32.
14. Shapakidze, E., Nadirashvili, M., Maisuradze, V., Gejadze, I., Petriashvili, T., & Avaliani, M., (2018). Development of geopolymeric binding materials based on the calcined shales. *Ceram. Adv. Technol., 20*(2), 31–38.
15. Shapakidze, E., Nadirashvili, M., Maisuradze, V., Gejadze, I., Petriashvili, T., Avaliani, M., & Todradze, G., (2019). Elaboration of optimal mode for heat treatment of shales for obtaining metakaolin. *Eur. Chem. Bull., 8*(1), 31–33. doi: 10.17628/ecb.2019.8.31-33.
16. Hewlett, P. C., (1988). *Lea's Chemistry of Cement Concrete* (4th edn., p. 1087). John Wily and Sons, Inc., New York, USA.

## CHAPTER 26

# Investigation of Free Volume of Oriented Electrical Conducting Polymer Composites by Spin Probe Method

JIMSHER ANELI[1] and LEVAN NADAREISHVILI[2]

[1]*R. Dvali Institute of Machine Mechanics, Tbilisi, Georgia*

[2]*V. Chavchanidze Institute of Cybernetics, Georgian Technical University, Georgia*

## ABSTRACT

The spine probe method based on electron spin resonance (ESR) is used for the investigation of the free volume of the oriented polyvinyl alcohol (PVA) films filled with high dispersive carbon black. After the introduction of nitroxide stable radicals by diffusion to orient with different degree polymer films, the samples were tested by noted method via the definition of the measuring of correlation times of rotation of stable radicals around their axes. It is established that the free radicals concentration and their correlation time in local regions of the films are the lower, the higher is the orientation (stretching) degree of the films, which directly is connected with the volume of micro-empties in the polymer matrix. Noted processes correlate with analogical ones in the same polymer containing high dispersive electrical conducting filler (carbon black). In this case, the diffusion of free radicals to the polymer matrix is more difficult than in pure polymer–the significances of correlation time, and diffusion coefficients of stable radicals decrease because of additive interactions of these radicals with filler particles. It is made a conclusion about minimal empty of free volume in composites near the maximally oriented regions of the stretched films.

## 26.1 INTRODUCTION

ESR is a spectroscopic method able to detect paramagnetic centers, unpaired electrons, stable radicals, etc., in different low and high molecular substances of inorganic or organic media (particularly in polymers, biological bodies, etc.) [1, 2]. One important application of this method is the investigation of different structural-morphological changes in the polymer materials by ESR spectra of stable radicals introduced to the polymer matrix via diffusion (so called method of spin probes). Most commonly, free radicals of types of nitroxides are used in this method. Thanks to its structure, nitroxide radicals are rather stable in organic substances. Nitroxide radical possesses a free (unpaired) electron that belongs to the nitroxide group and occupies the $p_z$ orbital of the nitrogen atom. The nitrogen group is often surrounded by four methyl groups, substituted in **α** position, which sterically protect the paramagnetic center from a possible attacking reagent and hence contribute to the stability of a free nitroxide radical. The choice of an adequate nitroxide for the investigation of a given polymer system is particularly important. Therefore, it is necessary to take care not only of miscibility between the nitroxide and polymer matrix, but also of nitroxide size, shape, polarity, and flexibility [3, 4]. In other words, the chosen nitroxide radical must be compatible with the polymer chain segment in order to be able to follow its emotional behavior.

The spin probe method gives the possibility to introduce nitroxides into polymer systems by dissolving or swelling the polymer in a probe solution. After removal of the solvent, probes stay incorporated in a polymer matrix. It is more one method, firstly used by one of the authors-introduction of the stable radicals into polymers by dry diffusion (sublimation) at temperatures 50–70°C in the vacuum stove [5] (the last method was used in the presented work). Because of the necessary of a rather low concentration of nitroxides embedded in a polymer matrix (about 0.1 wt.% or lower) results in a good signal-to-noise ratio, an introduction of nitroxides into the polymer matrix-induced a slight perturbation of the matrix with no significant influence on its properties.

There are many investigations, where method ESR is successfully used for the study of the problems of the influence of the polymer materials micro-structure on their different physical and chemical properties [6, 7].

In the presented work, we have conducted the investigation of the effect of micro-empties (free volume), in which localize the stable radicals, and on the character of electrical conductivity of polymer composites based on polyvinyl alcohol (PVA) with the use of considered above spin probe method.

## 26.2 RESEARCH PROCEDURE

The film-composites based on polyvinyl alcohol (PVA) and high dispersive carbon black were obtained via the introduction of filler powder (40 wt.%) to the water solution of PVA with following evaporation of the solution at room temperature. The orientation of the films (length 40 mm, width 20 mm, thickness 0.3 mm) was conducted in an oven under conditions of temperature 90°C at the rate 0.1 cm/min with following slow cooling. There were obtained the films stretched on 50, 100, and 150%. The specimen is stretched as one whole along the entire length. In the film, the distribution of deformation in stretching direction is heterogeneous–the relative deformation increases and thickness decrease from clamps up to the central region, and the more, the more is the stretching degree. After this, we apply the rectangle greed (5 × 5 mm) to the stretched film. Then we cut the strip from the film in a stretching direction. It was obtained three types of samples with 5 mm width and lengths 60, 80, and 100 mm, which correspond to stretching degrees 50, 100, and 150%, respectively. Analogical manipulations were provided on obtaining of stretched films without filler (carbon black).

There were measured two parameters: electrical resistance and the stable radical concentrations in each elemental square of the central strip.

The electric conductivity of each elemental square of this strip was measured with the use of impedance spectroscopy. It was determined the value of electrical resistance R for each local square from one side up to the second side in terms of a unit volume of the squares with the use of two contact method.

At the next stage, it was conducted the doping of the same strip of the film with stable nitroxide radical (2,2,6,6-tetramethyl-4-carboximethylpiperidin-1-oxide) in the vacuum ($10^4$Pa) stove at 60°C during 30 min. The radical concentration was varied in the range of $10^{-3}$–$10^{-4}$ mol/l. After doping, the electric resistance of each square of the film was repeatedly measured.

In the next stage, the measuring of ESR spectra of the stable radicals localized in each elemental square of the strip was provided. With this aim, the probes from each square of the strip were cut, and then the ESR spectra of stable nitroxide radicals were recorded by standard method on the spectrometer of type Varian at room temperature. Values of the rotational diffusion coefficient of the radicals in the polymer matrix have been determined using formula $D_r = 1/6\tau$, where $\tau$ is a correlation time of radical rotation. With the use of the spatial atlas of ESR spectra obtained via theoretical calculation of all experimentally permissible ESR spectra of the nitroxide radicals, it is the possibility of obtaining of numerical estimation of the corresponding correlation time and diffusion coefficient of the stable radicals in the polymer medium [8].

## 26.3 EXPERIMENTAL RESULTS AND DISCUSSION

The measuring of the specific volumetric electrical resistance value of the separate regions of the film shows that this parameter depends on the coordinates of the elemental squares in all horizontal strips from one side of the film up to another side (Figure 26.1). The average resistance of squares increases exponentially in stretching direction with a maximum at the central square (N10). On the basis of the character of the curves in Figure 26.1, it may be made the following conclusions: (1) exponential increasing of the electrical resistance value is in accordance with well known low for the resistance of linearly elongated electric conducting polymer composites (exponential dependence of the resistance on the distance between conducting particles) [9]; (2) The higher is the degree of elongation the higher is a maximum of the electric resistance; (3) The resistances of the squares containing stable radicals are higher than for equivalent squares in the films without stable radicals. This result is explained by the phenomenon of the capture of charge particles by stable radicals, which leads to a decreasing of the amount of charge carriers in the polymer matrix and, consequently, to decreasing of the electric current to some extent [10].

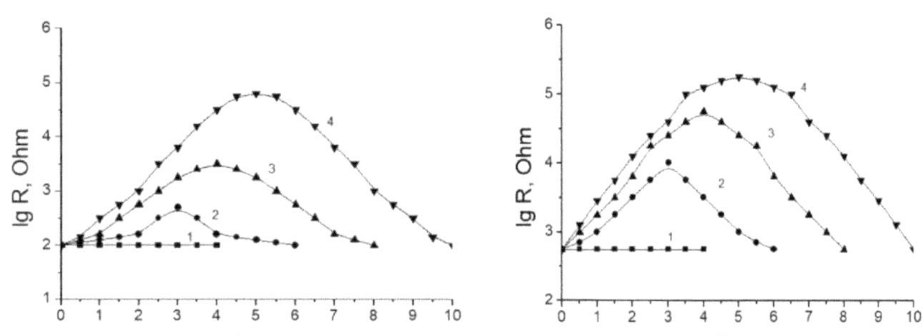


(a) (b)

**FIGURE 26.1** Dependence of the electrical resistance R of the PVA filled with CB without (a) and with nitroxide radicals (b) for unstretched (1) and stretched on 50 (2), 100 (3) and 150% (4) films (initial length of the sample 4 cm).

It is known that the spectrum of stable nitroxide radicals containing fragment > C-(N-O·)-C < with four methyl groups and unpaired electron in the liquid medium present asymmetric triplet (Figure 26.2(a)). However, this symmetry damages at increasing of viscosity of the medium-the more the less is the free volume around of the stable radical (Figure 26.2(b–d)).

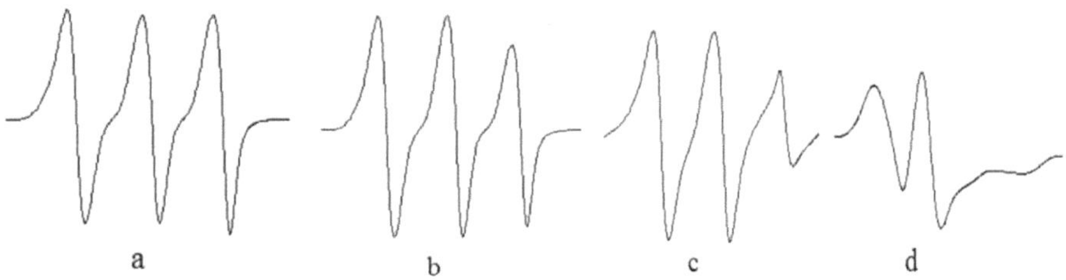


**FIGURE 26.2** The spectra ESR of stable radicals localized in different squares of the stretched strip of the film: a-in the liquid toluene medium; b-in the square N1; c) N6; d) N10.

By view of the shape of spectrum it will be possibility to estimate the viscosity or the density of the radical surrounding medium. Additively to note above it is known that the rate of rotation of radicals depends on segmental mobility of macromolecules [6]. The spectra ESR presented in Figure 26.2 correspond to stable radicals localized in the different regions with one and same sizes of the stripe of pure polymer films stretched on 150%. The obtained ESR spectra and calculated diffusion coefficients of rotating nitroxide radicals show that the values of these parameters decrease and reach to minimum near the middle of the stretched on 50, 100 and 150% film for both pure and filled with carbon black (Table 26.1). With the aim of comparison of the correlation times significances it was conducted the measuring of this parameter for radicals localized in each square from the central strip. It is clear that because of the density of filled polymer films is higher than for pure analogs this gradation of spectra ESR shape for composites is stronger than in pure ones.

Data in Table 26.1 show that the diffusion process of free stable radicals are fastest in polymers without fillers, which indicates on relatively weak interaction of these radicals with macromolecules. Another reason of this result may be the presence in the polymer microstructure of structural defects (empties), in which the radical movement proceeds with the least difficulties. The decreasing of both parameters enhances in the composites with carbon black because of increasing of the separate region's density and consequently decreasing of the free volume, as well as increasing of the interaction of radicals with active groups on the surface of carbon black particles (Figure 26.3).

**TABLE 26.1** Values of the Correlation Times of Nitroxide Radicals Localized in Unstretched and Stretched on 150% Polymer Film Strips of Four Regions with Equal Sizes from Left to Right up to Middle Region of the Strips*

| No. of Square from Strip (see Figure 26.1) | PVA | | PVA + CB | |
|---|---|---|---|---|
| | t,s | $D_r$, $s^{-1}$ | t, s | $D_r$, $s^{-1}$ |
| **Unstretch.** | $1 \times 10^{-10}$ | $1 \times 10^{10}$ | $3 \times 10^{-10}$ | $3 \times 10^{9}$ |
| 1 | $2 \times 10^{-10}$ | $5 \times 10^{9}$ | $6 \times 10^{-10}$ | $1 \times 10^{9}$ |
| 3 | $5 \times 10^{-10}$ | $2 \times 10^{9}$ | $2 \times 10^{-9}$ | $5 \times 10^{8}$ |
| 6 | $1 \times 10^{-9}$ | $1 \times 10^{9}$ | $1 \times 10^{-8}$ | $1 \times 10^{8}$ |
| 10 | $8 \times 10^{-9}$ | $1 \times 10^{8}$ | $9 \times 10^{-8}$ | $2 \times 10^{7}$ |

*The data for right part of the stretched films are not given, because the lasts reflect lefts.

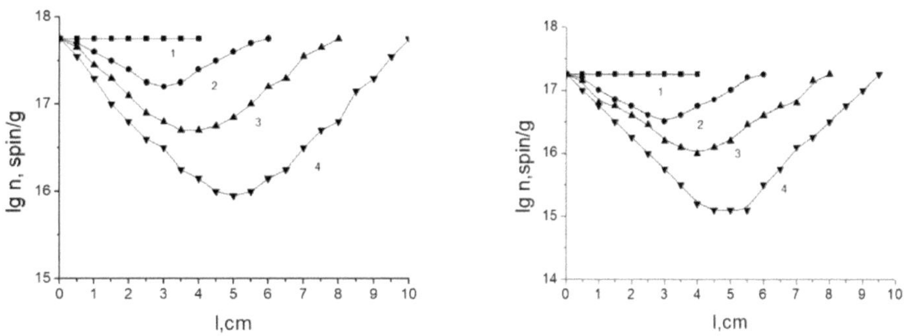


**FIGURE 26.3** Dependence of nitroxide radicals local concentration on the place of the corresponding region of pure PVA (a) and PVA filled with CB (b) for unstretched (1) and stretched on 50 (2), 100 (3), and 150% (4) films (initial length of the sample 4 cm).

## 26.4 CONCLUSIONS

With the use of ESR spectra of stable radicals, it is shown that as a result of the increasing of orientation degree, the correlation time of radical rotation of the local squares in the stretching direction from one side till middle square of the film correspondingly increases, which is due to decreasing of the free volume in the composite material.

By the study of electrical conductivity, it was established that as a result of the increasing orientation degree of the polymer composite, the specific electrical resistance of the materials increases, which is due to increasing of the average distance between conducting particles in stretching direction.

## KEYWORDS

- **correlation time**
- **diffusion of stable radicals**
- **electric conductivity**
- **electron spin resonance**
- **spin probe method**
- **stretched polymer composite**

## REFERENCES

1. Kovarski, A. L., & Sorokina, O. N., (2013). *Update on Paramagnetic Sensors for Polymers and Composites Research.* Rapra, UK.
2. Serda, M., Wu, Y. K., Barth, E. D., Halperin, H. J., & Rawal, V. H., (2016). *Chem. Res. Toxicology, 29*, 2153.
3. Didovic, M., Klepac, D., & Valic, S., (2008). *Macromolecular Symposia, 265*, 144.
4. Valic, S., (2010). *Rubber Nanocompostes* (p. 391). Academic Press, New York, USA, Chapter 15.
5. Aneli, J. N., & Khananashvili, L. M., (1994). *Intern. J. Polymeric Mater.*, (Vol. 27, pp. 125). New York, USA.
6. Veksil, Z., Andries, H., & Rakvin, B., (2000). ESR spectroscopy of the study of polymer heterogeneity [Review]. *Progress in Polymer Science, 25*(3), 949.
7. Jeschke, G., (2002). Determination of the nanostructure of polymer materials by ESR spectroscopy [Review]. *Macromolecular Rapid Communications, 23*, 227.
8. ESR Spectra of Spin Zondes and Labels, (1977). *Proceeding "Nauka."* Moscow (in Russian.).
9. Aneli, J. N., Khananashvili, L. M., & Zaikov, G. E., (1998). *Structuring and Conductivity of Polymer Composites*. New York, USA.
10. Masuda, K., & Susumu, N., (1974). Contribution of free radicals to electrical conductivity. In: *Energy and Charge Transfer in Organic Semiconductors*.

## CHAPTER 27

# Comparison of Modified Polyethylene Incubation Effects in Seawater and Composting Natural Environment

M. RUTKOWSKA, A. HEIMOWSKA, K. KRASOWSKA, and M. JASTRZĘBSKA

*Department of Industrial Commodity Science and Chemistry, Faculty of Entrepreneurship and Quality Science, Gdynia Maritime University, 83 Morska Str., 81-225 Gdynia, Poland, E-mail: m.jastrzebska@wpit.umg.edu.pl (M. Jastrzębska)*

### ABSTRACT

Polyethylene (LDPE MB), containing pro-degradant additive in the form of a master batch (MB) in amount of 20%, was used in this study. The MB consisted mainly of cornstarch, linear low-density polyethylene (LDPE) as the carrier resin, styrene-butadiene copolymer, and manganese stearate (the last two are referred to as the prooxidant system). The incubation of polymer samples took place in The Baltic seawater and in the living composting environment with plant treatment active sludge. The comparison of sample changes in the weight and tensile strength after incubation in seawater and composting natural environment is presented in this paper.

The very little microbial degradation was observed for modified LDPE in seawater and compost. This could be explained by low seawater temperature, moderate biological activity of microorganisms and low amounts of solar radiation reaching the films. There was also no defined trigger for the autooxidation process under ambient conditions in the dark as is in compost. In addition, the starch removal was limited because of the low content of moisture in compost, and this did not lead to disintegration of LDPE matrix. The changes of tensile strength were more visible. There was no correlation between small weights losses and tensile strength. They were continuously occur. Characteristic parameters of the compost with activated sludge used in the experiment are presented in Table 27.2.

**TABLE 27.2** The Characteristic Parameters of Compost with Active Sludge

| Parameter | Month | | | | | | |
|---|---|---|---|---|---|---|---|
| | July | August | October | November | December | February | April |
| Temperature [°C] | 19.0 | 22.6 | 15.0 | 12.0 | 8.0 | 6.0 | 8.3 |
| pH | 5.6 | 5.1 | 5.11 | 5.31 | 5.53 | 5.3 | 5.6 |
| Moisture content [%] | 45 | 47 | 41 | 41 | 40 | 40 | 44 |
| Activity of Dehydrogenases [mol/mg d.m.] | 0.0398 | 0.0331 | 0.0393 | 0.0216 | 0.0112 | 0.0121 | 0.0385 |

### 27.2.3 INVESTIGATION OF POLYMERIC MATERIALS

The films of LDPE were cut into 15 × 100 mm rectangles. After incubation in compost and seawater, the samples were taken from natural environments, washed thoroughly with distilled water, and dried at room temperature to constant weight.

The weight changes of degraded samples (%) were estimated using an electronic balance Gibertini E 42s. The results obtained from clean and dried samples of blends after biodegradation experiments were compared with those of respective samples before biodegradation.

Tensile strength of the blends in the form of films was measured using an Instron Model 4204 tensile tester at 20 mm/min tensile speed.

Fourier transform infrared spectroscopy (FTIR) analyses were performed on a Perkin-Elmer 1725x. In the infrared spectra, interest was focused on the carbonyl region (1715 $cm^{-1}$). Carbonyl absorbance at 1718 $cm^{-1}$ was measured relative to the $CH_2$ scissoring peak at 1463 $cm^{-1}$ [4]. This functional group was monitored for a change in order to establish the extent of biodegradation.

### 27.2.4 RESULTS AND DISCUSSION

The characteristic parameters of seawater in the natural environment presented in Table 27.1 indicate that the temperature of the Baltic seawater was lower than that preferred for enzymatic degradation, which is in the range of 20–60°C. The temperature of compost in the investigative period was the lowest in February 6°C and the highest in August 22.6°C. The activity of dehydratase was high in the summer months.

The results of weight changes of polymer samples after incubation in natural environments are presented in Figure 27.1.

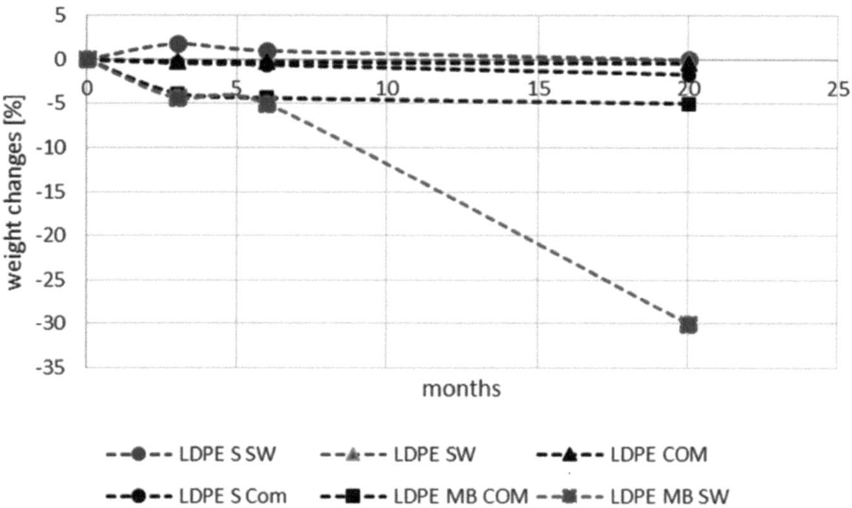


**FIGURE 27.1** Weight changes of modified polyethylene samples incubated in seawater (SW) and in compost (COM).

There were no visible weight changes of pure polyethylene. The degradation of polyethylene should be attributed primarily to photodegradation. The low amount of solar radiation reaching the film under the surface might be one of the factors responsible for the low extent of oxidation process of polyethylene. The degradation process should be sped up through microbiological consumption of the starch particles, producing a greater surface/volume ratio of the polyethylene matrix.

However, it has been demonstrated that LDPE modified with starch is also not very susceptible to microbial degradation in natural environments, although it has been known that starch is readily degraded by a wide variety of yeast, fungi, and bacteria. Moreover, the starch granulates have a tendency to absorb water. Therefore, the granules of starch swell first reversibly up to certain point after which the swelling becomes irreversible.

The changes in weight are connected not only with swelling of starch but also with growing of micro- and microorganism on polymer surface.

After 3 months of incubation starch had to be removed from the matrix by macroorganisms in order for it to contribute to the overall degradation.

During this process, free amylose molecules and some free amylopectine molecules should leave the granules by diffusion. The swollen granulates then slowly contracted and weight of samples started to decrease (after 3 months of incubation in seawater). However, the addition of starch to LDPE was not enough to enable weight decrease to be detected.

Therefore, there was a clear need for prooxidant in LDPE MB to initiate degradation. The oxygen content in seawater was in the range 10.8 ÷ 6.4 $cm^3/dm^3$. This probably helped to start degradation of LDPE MB. The small weight losses were observed for the system with prooxidant (polyethylene with MB) in the first period of the experiment (August–February). Then the degradation process sped up. Starch removal was much greater from films containing the prooxidant components, than that from those without them. The degradation of polyethylene with MB was due to hydroperoxide-catalyzed autooxidation of the prooxidant in synergistic combination with biodegradation of the starch particles [2]. Because of the synergistic process, the weight loss was much higher than the amount of the starch in added MB.

Analysis of the results shows that the temperature of the environment has an influence on the rate of degradation.

During winter (November–February), no weight loss was observed for blends of polyethylene with starch and for polyethylene with pro-degradant additive. Higher temperature in summer months has visible influence on the weight loss of polyethylene with the MB additive. The degradation process in this period was considerable, and after 20 months weight loss of the samples reached 26%.

There was no correlation between weights losses and tensile strength but the changes of tensile strength after incubation in the Baltic Sea were more visible than changes of weight (Figure 27.2). We could observed that the tensile strength decreased continuously for all samples (including pure polyethylene) incubated in the natural marine environment, which could be explained by some shear stresses (the fluctuation off water). Mechanical damage to the polyethylene macrochains may also be caused by the swelling and bursting of growing cells of the invading microorganisms in the seawater. Owing to mechanical damage of microchains, the value of the tensile strength of modified polyethylene decreased about 30% although the weight decreased about 0.6%.

The unusual change of tensile strength after 12 months of the experiment for polyethylene with 8% starch additive was probably caused by some release of starch (from the samples). Reduction of starch weakened the polymer sample by creating holes on the polymer surface after starch removal.

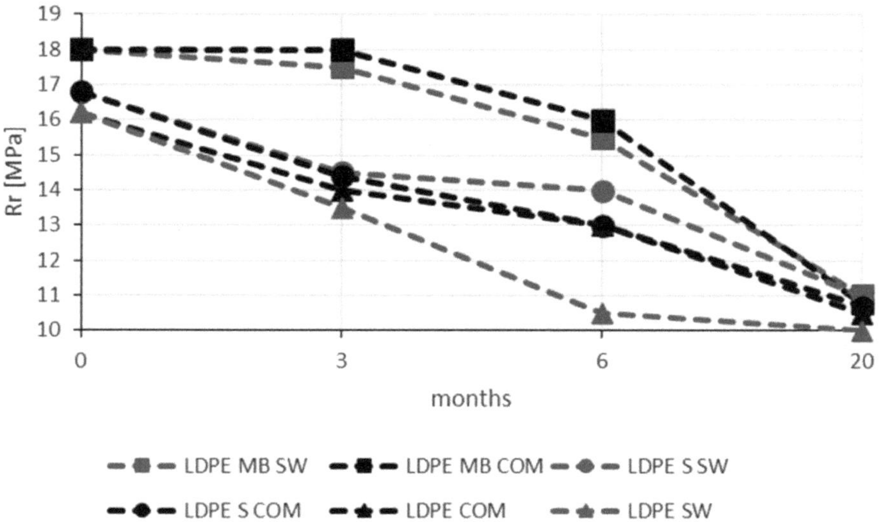


**FIGURE 27.2** Tensile strength of modified polyethylene samples incubated in seawater (SW) and in compost (COM).

FTIR analysis of biotically aged modified LDPE confirmed the observed degradation process (Figures 27.3 and 27.4). Attenuated peaks in the range of 1790–1750 $cm^{-1}$ are due to the formation of ketone or aldehyde C=O groups, which can be assigned to aldehydes, ketones, and acid groups-various types of oxidation product formed during polyethylene biodegradation [4], even initiating the breakage of LDPE chains is the longest and difficult step in its degradation. There were no visible changes of carbonyl index for LDPE MB samples after 6 months of incubation in compost, but that index change after incubation in seawater. This feature was due to the presence of transition metal within the MB of starch and prooxidant addition. The control index was not observed at the beginning of experiment and appeared very small after 6 months of incubation and what was the evidence of the first step of degradation process. Long incubation times produce a sufficient quantity of carbonyl groups to continue the decomposition process.

The little changes in the concentration of the functional groups (carbonyls) at the surface at LDPE observed after incubation in two natural environments with microorganisms because of their either consumption or production (accumulation). This is depending on the balance of rates of oxidation and degradation, which in turn depends on the nature of the microorganisms present. Oxidized groups are more easily degraded by microorganisms because they modulate microtrial attachment by increasing the hydrophilicity of the surface.

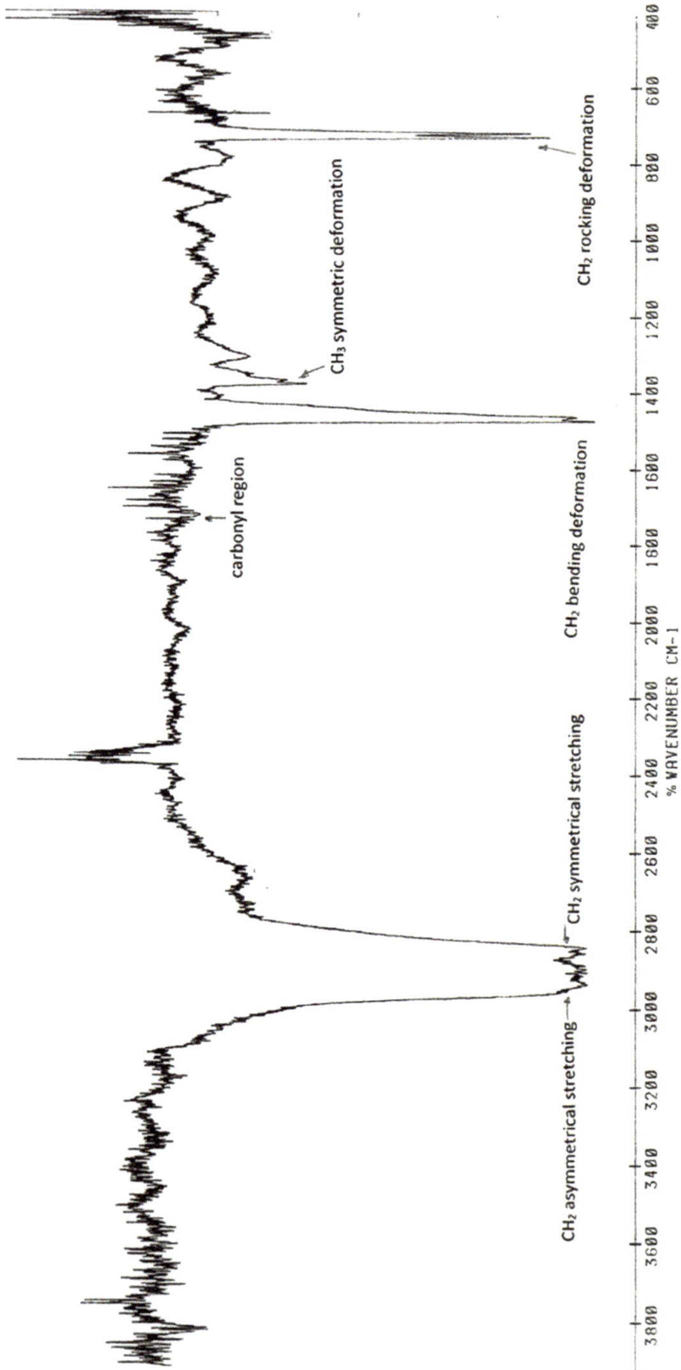


**FIGURE 27.3** FTIR spectra of LDPE before incubation.

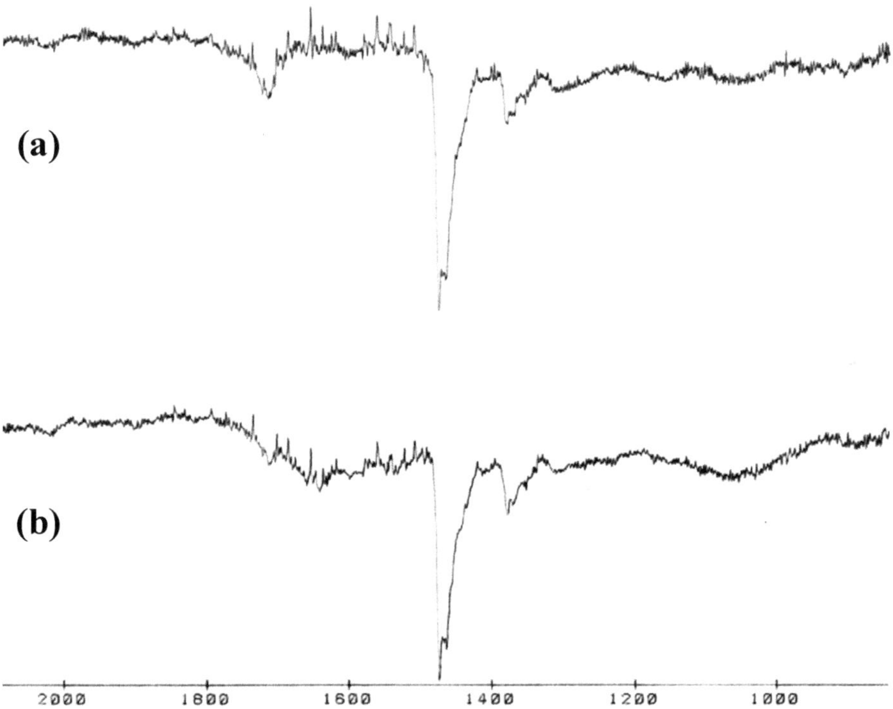


**FIGURE 27.4** FTIR spectra of LDPE MB before (a) and after 6 months incubation in seawater (b).

The obtained results confirm good resistance of pure LDPE to natural environments-seawater and compost and indicate that the introduction of the pro-degradant additive into the LDPE improves biodegradability.

## 27.3 CONCLUSIONS

LDPE has demonstrated good resistance to natural environments-seawater and compost. The addition of starch to LDPE was not enough to enable a weight decrease to be detected because of swelling of starch granulates and growing of micro- and macroorganisms on polymer surface.

The incorporation of prodegradant additives in LDPE matrix in the form of a MB improves the biodegradability of polymer systems due to hydroperoxide-catalyzed autooxidation of perooxidant in synergistic combination with biodegradation of starch particles.

## KEYWORDS

- **biodegradation**
- **Fourier transform infrared spectroscopy**
- **low-density polyethylene**
- **masterbatch**
- **natural environment**
- **polyethylene**


## REFERENCES


1. Restrepo-Florez, J. M., Bossi, A., & Thompson, M. R., (2014). *International Biodeterioration and Biodegradation, 88*, 83–90.
2. Albertsson, A. C., Barenstedt, C., Karlsson, S., & Lindberg, T., (1995). *Polymer, 36*, 3075–3083.
3. Rutkowska, M., Heimowska, A., Krasowska, K., & Janik, H., (2002). *Polish Journal of Environmental Studies, 11*(3), 267–274.
4. Nowak, B., Pajak, J., Drozd-Bartkowicz, M., & Rymarz, G., (2011). *International Biodeterioration and Biodegradation, 65*, 757–767.

# CHAPTER 28

# Obtaining of Biodegradable Polymers by Targeted Polycondensation Reaction

GIVI PAPAVA, NORA DOKHTURISHVILI, MARINA GURGENISHVILI, IA CHITREKASHVILI, ETER GAVASHELIDZE, NAZI GELASHVILI, RIVA LIPARTELIANI, and KETEVAN ARCHVADZE

*TSU Petre Melikishvili Institute of Physical and Organic Chemistry 31, A. Politkovskaia Str., Tbilisi–0186, Georgia,*
*E-mail: marina.gurgenishvili@yahoo.com (M. Gurgenishvili)*

## ABSTRACT

The intense growth of the population demands an increase in the production of grain crops. One of the ways to resolve the problem is the application of nitrogen-containing fertilizers in increased doses. According to the current data, annually, more than 200 million tons of nitrogenous fertilizer is introduced into soil, but because of its good water solubility its major part is lost due to its volatility and washing-off, which results in substantial economic losses. To elevate the yield of grain crops the ecologically pure and economically efficient bio-composites which contain linear structure polymerized nitrogenous fertilizers acting by the prolongation mechanism and microorganisms able to destruct such fertilizers were developed. The process of creation of polymerized fertilizers was studied. It was shown that in the range of 110–125°C, up to deep conversion of fertilizers, reaction rate constants keep constant values, when they are computed according to Arrhenius's second-order equations.

Results of I.R. spectroscopy studies showed that at the first stage of the reaction, of carbamide-formaldehyde interaction, when methylol-derivatives are formed (specter wave 1030 $cm^{-1}$) we also observe conversion of methylol groups into dimethylene ether groups (specter wave 1085 and 1110 $cm^{-1}$). Rectilinear dependence of on the reaction rate constant logarithm alteration to inverse absolute temperature refers to the fact that reaction rate constants undergo change according to the Arrhenius equation.

At the application of polymerized nitrogenous fertilizers, the fixed hectare norm of nitrogenous fertilizers decreases minimum by 40%, productivity increases by 15–20%, and the environment is protected from pollution by nitrogenous fertilizers. The technology of obtaining polymerized nitrogenous fertilizers and polymer biodegradable biocomposites has been developed.

## 28.1 INTRODUCTION

An increase in the production of grain crops is due to the intense growth of the population demand. Annually arable lands decrease because of intense urbanization and industry intensification. One of the ways to resolve the problem is the assimilation of intense technologies, especially the application of nitrogen-containing fertilizers in increased doses. According to the current data, annually, more than 200 million ton nitrogenous fertilizers are introduced into soil, but because of their good water solubility, the major part (more than) and volatility and washing-off 50%, which results in substantial economic losses. At the same time, it contributes to global pollution, causing many diseases.

To resolve the problem, a new type of so-called "exchange fertilizers" was developed in the United States of America. They consist of hardly soluble ingredients, which are released in the soil as a result of the buffer effect of chemical reactions going on there and are assimilated by plants [1–4].

Our research aimed to receive a chemical synthesis of a polymer biodegradable in soil and containing amide and peptide groups of prolongation action, to isolate its destructor from the soil, inclusive urease activity microorganisms, and to create nitrogen-containing biocomposites easily biodegradable in soil.

## 28.2 EXPERIMENTAL PART

The subject of our study was the linear structure of nitrogenous fertilizer of prolongation action synthesized by us, which was obtained in the solution as well as in the melt, via carbamide and formaldehyde polycondensation [5].

To study biodegradation of nitrogenous fertilizers with prolongation effect, we isolated freely residing nitrogen-fixing microorganisms (Azotobacter sp., Clostridium sp.) and various strains of urobacteria (Urobacillus sp.) from soil [6].

Isolation of microorganisms from soil and their identification was realized according to the methods of M. Burger and T. Dobrovolskaya [7]. Nitrogen was determined by I. Kjeldal method.

The process of the creation of a polymer in the melt was investigated. To simplify the process of penetration into the polymer, we synthesized linear polymers. High efficiency was achieved at a definite molar ratio of starting components. Study of the process of creation of carbamide based on linear structure polymer was implemented at the following molar ratios of the initial components-carbamide and formaldehyde: 1:0.9; 1:1; 1:1.1 and 1:1.2, correspondingly. Reaction temperature was 60, 70, 80, 90, 100, 135°C. Macromolecule formula is expressed as follows:

$$HOH_2C[-NH-CO-NH-CH_2]_n-NHCONH_2$$

## 28.3 RESULTS AND DISCUSSION

In the FTIR spectra of the synthesized polymers with methylol groups One can see absorption bands characteristic for valence oscillation of C-H, -NH-, $-NH_2$, - OH, C-N, bonds at 3000–2840, 3450, 3600–3400, 850 $cm^{-1}$. It should be noted that valence oscillation of –NH–, $-N_{H2}$ and –OH groups overlaps to eachother. In the spectra oscillation of C=O carbonyl groups distinctly is observed at 1720 $cm^{-1}$.

In the process of synthesis of carbamide-formaldehyde oligomers we consider expedient to study some kinetic regularities of the process within the range of 110–125°C, at the carbamide-formaldehyde molar ratio 1:1.2. Control over reaction progress was carried out according to the changes in formaldehyde concentration. It turned out that in the 110–125°C temperature range, till deep conversion, the reaction rate constants during the reaction keep constant values, when they are computed by the second order equation.

Rectilinear dependence of the reaction rate constant of logarithm alteration to the inverse absolute temperature refers to the fact that reaction rate constants undergo change according to the Arrhenius equation.

The second order is also proved by the ratio of 1/a-x and reaction duration. In this temperature range, the rectilinear relation is preserved. Activation energy computed according to the Arrhenius equation equals to 1554 kcal/mol (Table 28.1).

The process of degradation of linear structure polymers, at the impact of bacteria, is simplified because of the linear structure of molecules. Urease ferments easily penetrate into macromolecules and biodegrade them.

**TABLE 28.1** Values of Reaction Rates and Activation Energy[x] at 110–125°C Range at the Interaction of Formaldehyde and Carbamide–[x]

| № | Reaction Duration, sec. | Conversion Degree,% | | | Reaction Rate Constant К. 10–2, l.mol–1, sec.–1 | | |
|---|---|---|---|---|---|---|---|
| | | 110°C | 120°C | 125°C | 110°C | 120°C | 125°C |
| 1. | 30 | 71.66 | 74.50 | 89.16 | 2.81 | 3.24 | 3.38 |
| 2. | 60 | 83.60 | 85.48 | 94.25 | 2.83 | 3.27 | 3.34 |
| 3. | 90 | 88.39 | 89.83 | 96.00 | 2.80 | 3.27 | 3.38 |
| 4. | 120 | 91.14 | 92.15 | 97.01 | 2.85 | 3.26 | 3.34 |
| 5. | 150 | 92.83 | 93.53 | 97.60 | 2.87 | 3.21 | 3.38 |
| 6. | 180 | 93.97 | 94.60 | 98.01 | 2.88 | 3.24 | 3.37 |
| 7. | 240 | 95.33 | 95.86 | 98.50 | 2.80 | 3.22 | 3.36 |
| 8. | 300 | 96.25 | 96.66 | 98.27 | 2.85 | 3.22 | 3.37 |

[x]Çarbamide and formaldehyde molar ratio 1:1.2.

Nitrogen concentration was determined in carbamide and fertilizers with prolongation action. Besides, the dynamics of changes in nitrogen content was defined in soil. The transition of nitrogen into soluble form in the case of the fertilizer characterized by prolongation action takes place purposively slowly.

We have studied the degradation of linear structure polymer (fertilizer) by the emission of ammonia. With this in view, microorganisms apt to destruct the tested polymer were isolated from definite type soils (chestnut, alluvium-acidic, brown-carbonaceous) of east Georgia.

As a source of nitrogen, various quantities of the polymer (to adapt microorganism) were added to the medium of cultivation of microorganisms (Christiansen medium, where initially starting quantity of peptone was used as nitrogen sources); when the adapted microorganisms were obtained experiments were carried out according to the following scheme: only polymer, without inoculation (control) only with urea and polymer with inoculation (in the nutrient medium starting quantity of peptone and adapted microorganism).

Analysis of the obtained results showed that at the $4^{th}$ hour of cultivation of microorganisms, ammonia content in the medium was approximately 0.6 mg/ml and by the $10^{th}$ hour of the cultivation it fell to 0. At this period, source of ammonia in cultivation medium is peptone (starting dose); by the $10^{th}$ h of cultivation, peptone reserve is exhausted and microorganism starts to use carbamide as nitrogen sources (it destructs the application of nitrogen fixers on the seed material (*Azospirillum brasilense*) together with fertilizer characterized by the prolongation action, showed that hectare norms of nitrogen compared with the control (obtained by agro-rules) were reduced by more than 40–45%, while productivity was increased by 15–20%, correspondingly. This effect was achieved by joint application of fertilizers of prolongation action and nitrogen-fixers. Such technology practically ensures reduction of nitrogenous fertilizers wash-off and evaporation to the minimum, while the plant is provided with nitrogen along the whole vegetation period, which ensures the growth of productivity and environment protection from pollution. Finally, it will give a substantial economy and ecological effects.

## 28.4 CONCLUSIONS

New nitrogenous fertilizer of linear polymer structure was synthesized, and kinetic regularities of its creation were studied. Active strains of urobacteria able to destruct nitrogen-containing polymer of the linear structure were isolated from soil. Biocomposite containing polymerized carbamide and its destructor microorganisms were obtained. Biodegradable nitrogenous fertilizer of linear polymer structure increases the production of grain crops by 15–20% and decreases hectare norms of nitrogenous fertilizers by 40–45%. The technology of obtaining the polymerized nitrogenous fertilizers and polymer biodegradable biocomposites is developed.

## KEYWORDS

- **biocomposites**
- **fertilizers**
- **microorganisms**
- **nitrogen-fixers**
- **polymer**
- **prolongation mechanism**

## REFERENCES

1. Papava, G., Lomtatidze, Z., Gugava, E., Gurgenishvili, M., Chitrekashvili, I., Gorozia, I., & Molodinashvili, Z., (2012). Prolonged-release nitrogen fertilizers for reducing

global environmental pollution. *Journal of Chemistry and Chemical Engineering, 6*, 520–525. ISSN: 1934–7375, USA. Chicago, IL, USA.

2. Papava, G., Gugava, E., Lomtatidze, Z., Gorozia, I., Gurgenishvili, M., & Chitrekashvili, I., (2014). New type bio-composites containing nitrogenous fertilizers of prolonged action and nitrogen-fixing microorganisms. *Journal of Chemistry and Chemical Engineering, 5*, 581–586. ISSN: 1934-7375, USA. Chicago, IL, USA.
3. Ming, D. W., & Allen, E. R., (2001). Use of natural zeolites in agronomy, horticulture and environmental soil remediation. *Reviews in Mineralogy and Geochemistry, 45*(3), 649–654.
4. Eberl, D. D., Barbarick, K. A., & Lai, T. M., (1995). Influence of $NH_4$-exchanged clinoptiolite on nutrient concentrations in sorghum-Sudangrass. In: Ming, D. W., & Mumpton, F. A., (eds.), *Natural Zeolites 93 Int. Comm. Natural Zeolites, 8*, 491–504. Brockport, NY.
5. Papava, G., Ebralidze, K., Dolidze, A., & Gugava, E., (2001). Biodegrading homogeneous and mixed matrixes of amidoaldehyde type for micro capsulation of multicomponent composition containing biologically active substances. *Bull. Georg. Acad. Sci., 164*(3), 495–496.
6. Lomtatidze, Z., Papava, G., Gugava, E., & Gorozia, I., (2013). New technology to obtain bio-composites of prolonged effect on the basis of synthetic fertilizers and nitrogen-fixing microorganisms for the elevation of cereal yield. *Journal Microbiology and Biotechnology, 4*, 63–68.
7. Dobrovolskaya, T., Skvortsova, I., & Lysak, (1989). Methods of identification and isolation of soil bacteria. *Textbook for Students of Biology and Soil Faculties*, Moscow University Publishing House, 21–35 (in Russian).

# CHAPTER 29

# Enhancing of the Properties of Polymer Composites with Mineral Fillers Modified by Tetraethoxysilane

O. MUKBANIANI,[1,2] J. ANELI,[2,3] and E. MARKARASHVILI[1,2]

[1]*I. Javakhishvili Tbilisi State University, I. Chavchavadze Ave., 1 Tbilisi–0127, Georgia, E-mail: omarimu@yahoo.com (O. Mukbaniani)*

[2]*Institute of Macromolecular Chemistry and Polymeric Materials, Ivane Javakhishvili University, Ilia Chavchavadze Blvd. 13, Tbilisi–0179, Georgia*

[3]*R. Dvali Institute of Machine Mechanics, 10 Mindeli Str. Tbilisi–0186, Georgia*

## ABSTRACT

Ultimate strength and softening temperature of the polymer composites based on diane epoxy resin (ER) (type ED-20) and low-density polyethylene (PE) with unmodified and modified by tetraethoxysilane (TEOS) mineral basalt are described. Comparison of experimental results obtained for investigated composites shows that they containing modified filler have more high values of mentioned above parameters than composites with the unmodified filler at definite loading. The maximum of ultimate strength for composites with modified filler is higher than that for composites with unmodified filler. First composites have also higher softening temperatures. Dependence of the properties of composites has an extreme character-maximums of the technical parameters appear the composites with definite concentrations of filler. Experimental results are explained in terms of structural peculiarities of polymer composites.

## 29.1 INTRODUCTION

Today the polymer composites attract great attention from scientists and engineers. High monitoring of the exploitation properties in the wide interval, high durability, stability to aggressive media, lightness, easy technology of obtaining, and finally low cost are the fundamental characteristics of these materials. The noted factors led to the high competitiveness of such traditional materials like metals, ceramics, wood, and skin. Currently, the polymer (synthetic or natural) composites with different mineral fillers are very spread [1–7]. Thanks to these fillers, many properties of the composites are improved – increases the durability and rigidity, decrease the shrinkage during hardening process and water absorption, improves thermal stability, fire proof and dielectric properties, and finally, the price of composites becomes cheaper [3–5]. At the same time, it must be noted that the mineral fillers at high content lead to some impairment of different physical properties of composites. Therefore, the attention of the scientists is attracted to substances, which would remove the mentioned leaks. It is known that silicon-organic substances (both low and high molecular) as "soft materials" reveal hydrophobic properties, high elasticity, and durability in a wide range of filling and temperatures, and consequently, introduction of these materials to the polymer blends, in general, can increase the compatibility of ingredients and respectively increase of the mineral filler concentration in the composites [6, 7]. Since the price of mineral fillers is commonly very low than that for polymer (binder), the main investigation in the field of composite technology is directed to the creation of high filled composite materials. If we foresee that the composites may be obtained on the basis of different natural or artificial remains, the application of these materials will be connected with great economical effect.

Application of the method of so-called polymerization filling (polymerization on the filler particles) as a rule leads to obtaining of composites with high mechanical properties due to the formation of the chemical bonds on the filler surfaces between the polymer and active groups on the particle's surfaces [3, 4]. However, this method consists of several stages of processes and is relatively expensive.

At this time, the polymer layered silicate nanocomposites are rather popular [5–7]. There are three main methods of their synthesis: *in situ* polymerization, solution intercalation, and melt processing (or melt blending). A fourth method, using sol-gel techniques, can in principle be applied but is not utilized often, because of the higher temperatures needed and the tendency of the inorganic component to aggregate. Each of the three main approaches may yield exfoliated, intercalated, or a mixed of exfoliated and intercalated structures. The degree of exfoliation versus intercalation depends on a host of experimental conditions, such as monomer type, solvents, temperatures, etc.

The purpose of presented work is the investigation of effect of modify by tetraethoxysilane (TEOS) of the mineral basalt on mechanical strengthening and thermal stability of polymer composites containing basalt powders.

## 29.2 EXPERIMENTAL

The initial TEOS was supplied from Sigma-Aldrich. Mineral-bentonite as a filler was used. The organic solvents were purified by drying and distillation. FTIR spectra were recorded on a Jasco FT/IR-4200 device. The KBr pellets of samples were prepared by mixing (1.5–2.00) mg of samples, finely grounded, with 200 mg KBr (FT-IR grade) in a vibratory ball mixer for 20s. The physical-mechanical and thermal properties of obtained composite polymer materials based on activated basalt were testified by standard methods.

Reaction of silanization of basalt with TEOS was provided by following manipulations: in a three-necked flask supplied with a mechanical mixer, thermometer, and dropping funnel, to a solution of 50 g basalt in 80 ml anhydrous toluene the toluene solution of 1.5 g (0.0072 mol) in 5 ml toluene was added. The reaction mixture heated up to boiling point of used solvent during 5 h. After this the reaction mixture was filtered and dried under vacuum at 100°C up to constant weight.

High temperature condensation reaction between basalt and TEOS was carried out in anhydrous toluene solution (~35%). The masses of TEOS were 5 and 10% from the mass of filler. The reaction system was heated at the solvent boiling temperature (~110°C) during 5–6 hours by stirring. After that solvent ethyl alcohol was separate from precipitate and remained TEOS on the rotor-evaporator and work thermal the modified basalt as at the first stage pure basalt up. The reaction goes by Scheme 29.1 (the direction of the reaction was improved by FTIR spectra analysis):

On the basis of modified basalt and ER (of type ED-20) the polymer composites with different content of filler were obtained after careful wet mixing of components in mixer. After the blends with hardening agent (polyethylene-polyamine) were placed to the cylindrical forms (in accordance with standards ISO) for hardening at room temperature during 24 h. The samples hardened later were undergo to temperature treatment at 120°C during 4 h.

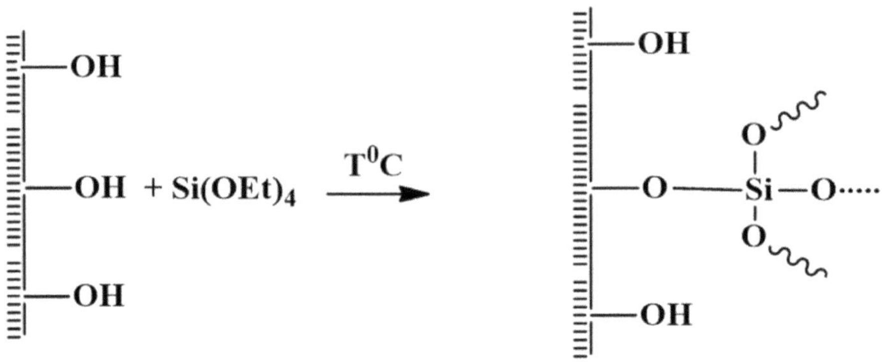


**SCHEME 29.1** Silanization of basalt surface.

It was shown that the modification goes more successfully when the masses of TEOS were near 5% from the masses of the filler. The concentration of powder basalt (average diameter up to 50 micron) was changed in the range 20–60 mass %. Filler was modified with 5 and 10 mass % of TEOS. Analogical procedures were conducted in the case of composites based on PE.

## 29.3 RESULTS AND DISCUSSION

In Figures 29.1 and 29.2, the curves of the dependence of ultimate strength on the filler concentration of the composites based on dian ER and Polyethylene (PE) filled by high dispersed basalt with average diameter of particles < 50 microns are presented.

The forms of all presented in Figures 29.1 and 29.2 curves have an extreme character with maximums of mechanical strength. However, the place on the diagrams and the height of these maximums essentially depend on the amount of modifier TEOS.

The curves presented in Figure 29.1 allow us to make a conception that mechanical properties of the composites increases till definite maximal significances after which decreases. The curve 1 corresponds to composite containing the filler without modifier. The shape of this curve reflects well-known dependence for filled polymer composites. The mechanical strengthening of this composite with unmodified basalt has the maximum at relatively low concentration of the filler, while this maximum for analogical composite with modified with TEOS (5 and 10 wt.%) basalt the noted maximum shares to more high concentrations of this filler (Figure 29.1, curves 2 and 3). This result has practically important meaning-the higher is mineral filler concentration in the composite, the lower is its cost.

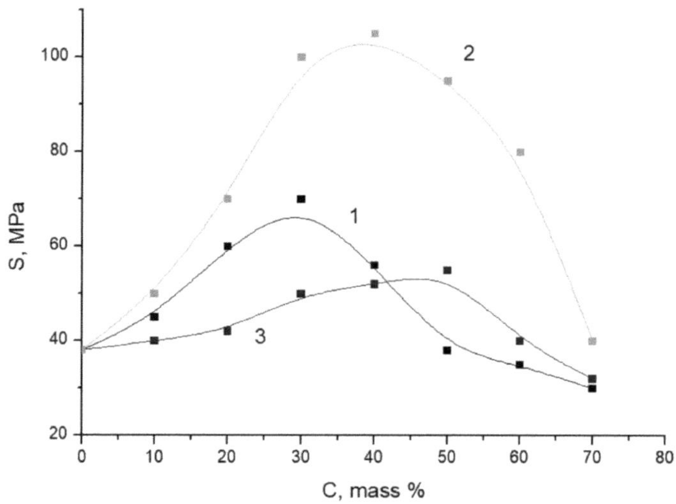


**FIGURE 29.1** Dependence of the ultimate strength at compression on the filler concentration for composites based on epoxy resin and basalt. 1-with unmodified basalt, 2 and 3-with modified by 5 and 10 mass % of tetraethoxysilane, respectively.

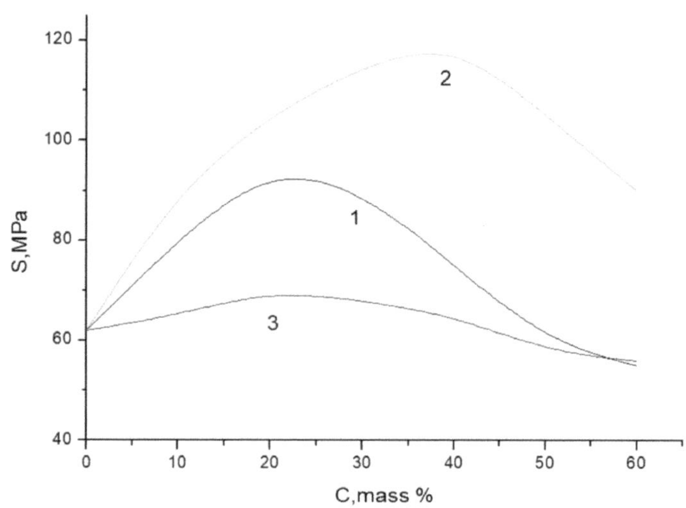


**FIGURE 29.2** Dependence of the ultimate strength at compression on the filler concentration for composites based on polyethylene and basalt. 1-with unmodified basalt, 2 and 3-with modified by 5 and 10 mass % of tetraethoxysilane, respectively.

The curves of the dependence of mechanical strengthening of the composites with modified and unmodified bentonite show that modifier molecules displaced on the filler particle surface increases its activity (expressed with enhancing of composite material as a result of chemical reactions between active groups of ingredients) till definite concentrations of the filler, higher which the mechanical characteristics of composite decreases. One of the main reasons of this phenomenon is the formation of number of structural defects (mechanical cracks, empties) in result of formation of the surfaces non reacted with organic part of the binder, amount of which increases with increasing of filler concentration. It is due to so-called effect of high filling (Rebinder effect).

In the composites with modified fillers the characteristic maximums on the curve of dependence ultimate strengthening on the filler concentration modifier molecules enhance the interaction between heterogeneous phases from one side and absorb the mechanical stresses in the composite body at hardening from another one. Therefore, the maximums on the curves for the composites with modified fillers share to more high concentrations of the filler. However, at following increasing of filling in the composite the modifier phase increases, which is usually soft phase and plays the role of structural defects. Increasing of this phase leads to softening or decreasing of composite mechanical strengthening.

It is interesting that the mechanical properties of the composites, from the other side, are better for the composites, filler in which is modified by 5% of TEOS. The composites with fillers modified with a more high concentration of the modifier (10%) are characterized with less strength. It may be proposed that a reason of this phenomenon is non-full using modifiers in the reactions of filler modifications. In this case, the modifier molecules can play the role of plasticizer and therefore weaken the strength of the material. The decreasing of this parameter for composites takes place also at relatively high concentrations of the filler, which can be attributed as appearing of the regions with dry filler particles in the polymer matrix, creating the defect places in it and, consequently, lead to the weakening of the composites. Analogical discussion can be provided in the case of composites based on PE and basalt.

The results of investigations of thermal stability of composites obtained by method Vicat are presented as the dependences of the value of deepening of the indenter in the body of the composite sample at increasing temperature (Figures 29.3 and 29.4).

In accordance with Figures 29.3 and 29.4, the change of thermal stability of composites based on ER and different concentration reflex the character of the dependence of composite mechanical strength on the filler concentration. It can be supposed that the high mechanical properties composites are characterized with perfect structure due to homogeneous distribution of the filler particles in the polymer matrix at relatively low concentrations of the structural defects ensuring a creating homogeneous temperature field (without gradients) in the composites.

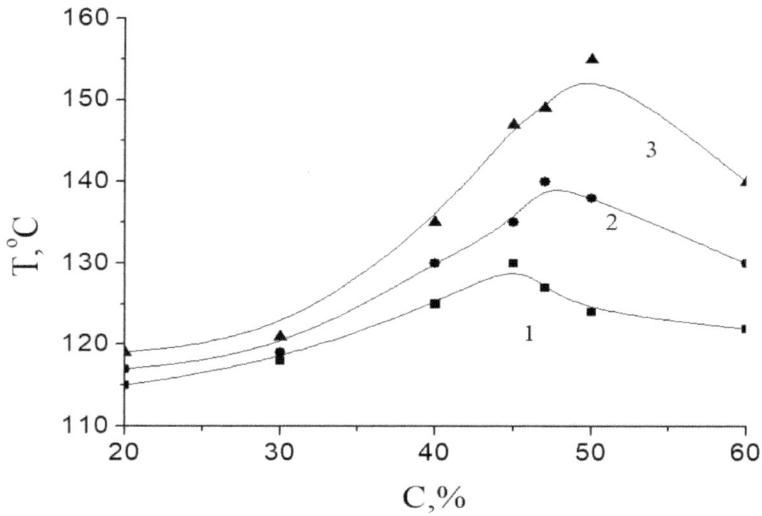


**FIGURE 29.3** Dependence of the softening temperature of composites based on epoxy resin and basalt (measured by method Vicat) on the filler concentration.

From the curves presented in Figures 29.3 and 29.4, it is seen that modified with TEOS basalt as filler in the composite effects to a definite extent on the thermal stability of the composite. Namely, the softening of the composites containing modified by TEOS basalt begins at relatively high temperatures than in the case of ones containing the unmodified filler. This result is in good agreement with the ones obtained at the investigation of the mechanical properties of corresponding materials.

Enhancing the compatibility of the ingredients in the composites containing modified by TEOS also reflected on the characteristic of water absorption of the composites. In accordance with Figures 29.5 and 29.6, the composites with modified filler show more high water stability than analogs with unmodified fillers. It can be proposed that the composites containing this filler modified with a more high concentration of the same modifier will be more hydrophobic than presented samples, because the named composites with a high concentration of modifier as a high hydrophobic component will be more waterproof.

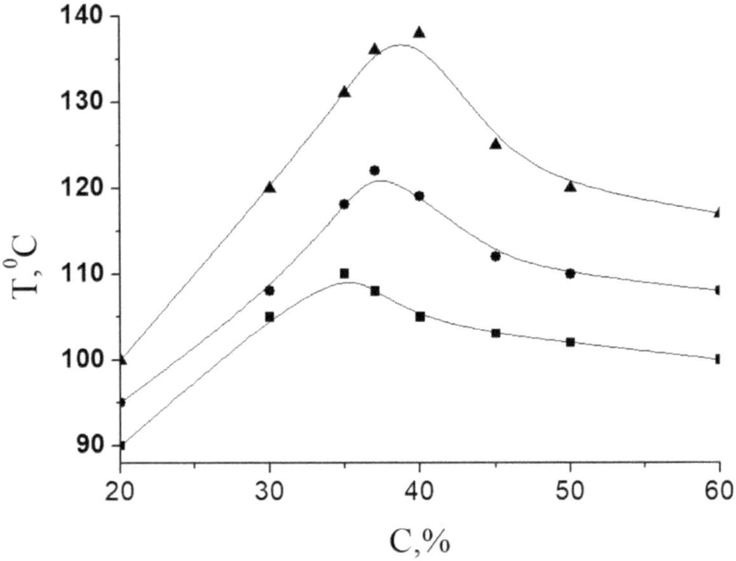


**FIGURE 29.4** Dependence of the softening temperature of composites based on polyethylene and basalt (measured by Vicat method) on the filler concentration.

The obtained experimental results may be explained in terms of composite structure peculiarities.

Silane molecules displaced on the surface of basalt particles lead to activation of them and participate in chemical reactions between active groups of ES (hydroxyl) and homopolymer (epoxy group). Silane molecules create the "buffer" zones between the filler and the homopolymer. This phenomenon may be one of the reasons for the increasing strengthening of composites in comparison with composites containing unmodified fillers. The composites with modified basalt display more high compatibility of the components than in the case of the same composites with unmodified filler. The modified filler has more strong contact with a polymer matrix (due to silane modifier) than unmodified basalt. Therefore, mechanical stresses formed in composites by stretching or compressing forces absorb effectively by relatively soft silane phases, i.e., the development of micro defects in carbon chain polymer matrix of composite districts and finishes in silane part of the material the rigidity of which decreases.

The structural peculiarities of composites also display in the thermo-mechanical properties of the materials. It is clear that the softening of composites with modified by TEOS composites begins at relatively high temperatures. This phenomenon is in good correlation with corresponding composite mechanical strength. Of course, the modified filler has more strong interactions (due to modifier) with epoxy polymer molecules than unmodified filler.

## 29.4 CONCLUSIONS

Investigation of mechanical strengthening and softening temperature for polymer composites based on ER and polyethylene with mineral filler basalt unmodified and modified by the low amount (5 and 10 mass%) TEOS leads to the conclusion that modifies agent stipulates the formation of heterogeneous structures with higher compatibility of ingredients and consequently to enhancing of noted technical characteristics. The increasing of mechanical and thermal properties of the composites with a relatively high content of modified by TEOS basalt filler has practical aspect-saving of the flexibility of polymer materials after filling and improvement of their properties.

## KEYWORDS

- **basalt**
- **epoxy resin**
- **mechanical and thermal properties**
- **polyethylene**
- **polymer composite**
- **water absorption**

## REFERENCES

1. Achilias, D. S., Roupakias, C., Megalokonomos, P., & Lappas, A., (2007). *Journal of Hazardous Materials, 149*, 536.
2. Jassim, A. K., (2017). *Procedia Manufacturing, 8*, 635.

3. Gorrasi, J., Tortora, M., Vittoria, V., Pollet, E., Lepoittevin, B., Alexandre, M., & Dubois, P., (2003). *Polymer, 44*(8), 2271.
4. Chen, H., Giannelis, E. P., & Sogah, D. Y., (1999). *Journal of the American Chemical Society., 121,* 1615.
5. Pavlović D., Linhardt J. G., Künzler J. F., & Shipp D. A., (2010). Synthesis and characterization of PDMS-, PVP-, and PS-containing ABCBA pentablock copolymers. *Macromolecular Chemistry and Physics*. 211: 1482-1487. DOI: 10.1002/macp.200900523.
6. Vo, L. T., & Giannelis, E. P., (2007). *Macromolecules, 40*, 8271.
7. Pavlidou, S., & Papaspyrides, C. D., (2008). *Progress in Polymer Science, 33*, 1119.

CHAPTER 30

# Peculiarities of Copolymerization of Vinyloxycyclopropanes with Maleic Anhydride and Synthesis of Photosensitive Polymers

RITA SHAHNAZARLI

*Institute of Polymer Materials of Azerbaijan National Academy of Sciences, S. Vurgun Str., 124, AZ5004, Azerbaijan,*
*E-mail: shahnazarli@mail.ru*

## ABSTRACT

The radical joint polymerization of a number of functionally substituted vinyloxycyclopropanes (VOCP) with maleic anhydride (MA) has been carried out, and the copolymers of equimolar composition and alternating structure have been obtained. The copolymerization constants and *Q-e*-parameters of the investigated copolymers have been determined. The formation of complexes between comonomers has been revealed and the complex-formation constants values and thermodynamic parameters of the process have been calculated. The modification of the obtained copolymers with various carbinols has been carried out and the modifiers containing reactive functional groups in macromolecules have been made. The photo structuring of the synthesized modifiers has been studied and their photosensitivity has been determined.

## 30.1 INTRODUCTION

The chemical reactions of the reactive groups presenting in macromolecules of functional polymers allow obtaining the various practically useful multipurpose materials. In this connection, the copolymers of maleic anhydride

(MA) undergoing the various chemical reactions with the opening of the anhydride cycle are of some practical interest. The easiness of opening the anhydride cycle in macromolecules of MA copolymers makes them convenient initial products for synthesis of the film-forming coatings, binders for reinforced plastics and other important materials [1]. The joint polymerization of vinyloxycyclopropanes (VOCP) with MA is of interest from the point of view of preparation of new copolymers with specific properties and allows regulating the content of various functional groups in the macromolecule within a wide range, which is very important in using them as film-forming materials.

Most of the investigations described in the literature on chemical reactions of the anhydride links in MA copolymers are referred to the reactions of hydrolysis and salt formation, esterification, amidation, and condensation with epoxide compounds [2]. The products obtained as a result of such polymer-analogous reactions have the higher adhesion and cohesion and easily pass into a cross-linked form in a case of action of radiation or temperature on them, which allow making various photo- and electrono-resists used in microelectronics on the basis of such materials [3, 4].

Taking into account the actuality of the problem of introduction of functionally active groups into macromolecular chain, we have carried out the modification of the copolymers of cyclopropane-containing vinyl esters (compounds **1–9**) with MA, studied the regularity of their interaction with various carbinols (vinyl carbinol, vinyloxycyclopropyl carbinol and phenylcyclopropyl carbinol) and investigated the structuring process of the modified products.

## 30.2 EXPERIMENTAL

The IR spectra of the synthesized monomers and copolymers were taken on apparatus "Specord" M 80 in the field of prisms KBr, NaCl, LiF as thin films.

The PMR spectra were taken on spectrometer of mark "Fourier" of firm "Bruker" (300 MHz) in $CDCl_3$, internal standard-hexamethyldisiloxane. The chemical shifts of signals have been presented in a scale δ.

The copolymerization of VOCP with MA was carried out both in solution (benzene) and in mass. The obtained copolymers were purified by precipitation of their acetone solution to petroleum ether and dried at 50–60°C to a constant mass. The modified copolymers were isolated from solutions in methyl ethyl ketone (MEK) by precipitation with 3–4 fold quantity of petroleum ether, the precipitated powdered products were dried in a vacuum at 30–40°C.

MW of the obtained copolymers were found on values of the characteristic viscosities of their solutions determined on Ubbelohde viscometer at temperature of 20C, using the Mark-Kuhn-Houwink equation (taking К = $4.11 \cdot 10^{-4}$ and $\alpha = 0.89$ [5].

MWD was determined by a method of turbidimetric titration of the copolymer solutions (VOCP: MA = 1:1, polymer concentration in solution-0.25 g/dl.), finding "precipitation threshold" (hexane served as the precipitator) on device FEK-56M (cuvette size-18 × 30 mm, light filter № 6). According to the results of turbidimetric titration, the integral and differential MWD curves were constructed.

The photosensitive characteristics of the polymers were investigated under the following conditions: the light source-mercury lamp DRT 220 (or DRSH-250), the current strength on the lamp-2.2 A, the mobile shutter rate-360 mm/h, the distance between sample and lamp-25 cm, the sample temperature-30–40°C. The light intensity was measured by radiation thermoelement RTN-10C. The photostructuring process of the polymers was studied by irradiation of their films for the required time (depending on degree of esterification of the copolymers) at room temperature. The polymer solution in MEK or acetone with concentration 8–16 mass % was applied to glass substrates by centrifugation and the films with thickness of 0.15–0.70 mcm (thickness of the films was measured by Linnik MII-4 microinterferometer) was obtained. Cyclohexanone was used as a developer. Development time is up to 15 seconds. Drying of photoresist layers was carried out at 120°C for 30 min.

## 30.3 RESULTS AND DISCUSSION

It was known that an introduction of the reactive functional group into macromolecule by chemical modification leads to a change of its structure and properties. Consequently, the copolymers modified in various conditions are differed from each other both on degree of modification and on other characteristics.

The choice of MA as a comonomer has been stipulated by the fact that it has a deficit of the electron density of the double bond, which is important at its copolymerization with electron-donor monomers, which are VOCP. In addition, MA shows a tendency to form the alternating copolymers [6, 7]. At copolymerization of vinyl ethers with electron-acceptor monomers, for example, MA, a donor-acceptor complexes occur, which extremely increases the reaction polar effect [8].

The spectral analysis showed that the copolymers obtained both in mass and in solution is identical to composition, intramolecular distribution of links and chemical structure:

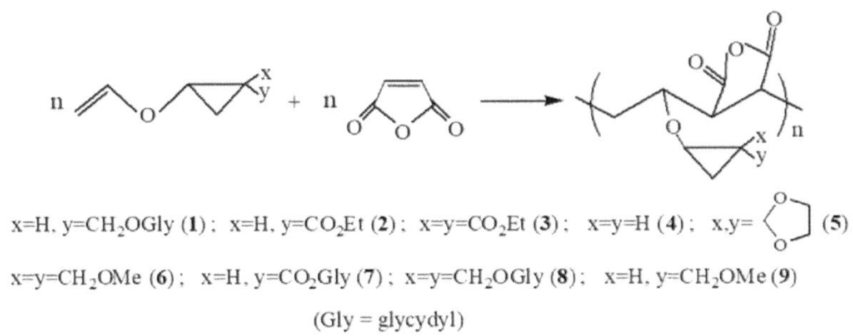


It has been established, according to the data of IR spectroscopy and elemental analysis that the composition of the obtained copolymers doesn't practically depend on ratio of the initial monomers and close to equimolar one (Table 30.1).

**TABLE 30.1** Copolymerization of Substituted Vinyloxycyclopropanes with Maleic Anhydride (Concentration AIBN 0.02 mol/l, Temperature-70°C)

| Code of VOCP | Composition of the Initial Mixture, mol.% VOCP | Copolymerization Time, min. | Conversion, % | Copolymer Composition, mol.% VOCP |
|---|---|---|---|---|
| 3 | 90.0 | 20 | 8.2 | 50.7 |
| | 75.0 | 20 | 9.6 | 50.0 |
| | 50.0 | 20 | 14.1 | 49.5 |
| | 25.0 | 30 | 12.4 | 48.9 |
| | 10.0 | 60 | 11.0 | 48.6 |
| 5 | 50.0 | 20 | 15.7 | 52.8 |
| 7 | 50.0 | 20 | 14.1 | 50.1 |
| 1 | 50.0 | 20 | 12.4 | 50.2 |
| 6 | 50.0 | 20 | 13.9 | 50.1 |
| 8 | 50.0 | 20 | 13.7 | 50.3 |

The copolymer maximum yield corresponds to a ratio of monomers in the initial mixture 1:1. As a result of copolymerization, the copolymers insoluble in aromatic hydrocarbons, but well soluble in polar organic solvents have been obtained.

The results of turbidimetric titration of the copolymer solutions show the homogeneity of their composition and a sufficiently wide MWD. The maximum one of the differential curve corresponds to the copolymer obtained at ratio of comonomers 1:1. A displacement towards lower MW values was observed at ratio of monomers in the initial mixture differing from equimolar one. Under these conditions, the obtained copolymers are less homogeneous on MWD, as evidenced by the broadening of the differential curve. The calculated MW copolymers values obtained under various conditions indicate to their relatively low MW, the characteristic viscosity of solutions, which vibrated from 0.47 to 0.60 dl/g.

The IR spectra of the obtained copolymers showed the availability of the absorption bands in the field of 1765–1845 $cm^{-1}$ characteristic for carbonyl groups in anhydride fragments. The availability of the absorption at 1020–1050 $cm^{-1}$ characterizes the skeleton vibrations of three-membered carbon cycle. In the case of copolymerization with participation of compounds **1** and **7** the copolymers, in the IR-spectra of which was observed the availability of the absorption bands at 830–840 $cm^{-1}$ and 930 $cm^{-1}$ characteristic for epoxide groups have been obtained. The same one was also observed in the IR spectrum of the copolymer obtained from monomer **8**. In the spectra of the copolymers on the basis of compounds **7** and **8** there were the absorption bands at 1720 $cm^{-1}$ characteristic for carbonyl group. At the same time, the absorption in the field of 1635–1645 $cm^{-1}$, corresponding to valence vibrations of double C = C-bond, in the IR spectra were absent. All these data allow to conclude that, firstly, only the double bond of VOCP (in this case, the three-membered cycle is not affected) takes part in the copolymerization process; secondly, the copolymerization proceeds with the participation of both comonomers with the formation of copolymers of equimolar composition; and thirdly, the ratio of the initial reagents does not influence on the composition of the obtained copolymers. After determining a quantity of anhydride groups in the copolymers, obtained at various ratios of comonomer, the copolymerization constants of some copolymers have been calculated (Table 30.2). Using the values $r_1$ and $r_2$, *Q-e* parameters for these monomers characterizing the specific activity of the initial monomers and the polarity of the radicals formed from them have been calculated. The observed large difference between *e* values for MA and VOCP, as well as $r_1$ and $r_2$ values show the predominance of the cross-growth chain reaction and the copolymers formation of the alternating structure.

It was known that the complex-formation between comonomers is a characteristic peculiarity of the copolymerization process with the participation of MA. For elucidation of this and the reason for alternation of the monomer links, the PMR spectra in $CDCl_3$ solution of both pure MA and MA in the presence of VOCP were taken.

**TABLE 30.2** Copolymerization Constant Values ($r_1$ and $r_2$) and *Q-e* Parameters of Copolymers of Some VOCP with Ma

| Monomer | $r_1 \cdot 10$ | $r_2 \cdot 10$ | $r_1 – r_2 \text{-} 10^3$ | $-e_1$ | $\Delta e^*$ | $Q_1 \text{-} 10^2$ |
|---|---|---|---|---|---|---|
| 1 | 0.27 | 0.29 | 0.78 | 0.43 | 2.68 | 1.9 |
| 2 | 0.30 | 0.25 | 0.75 | 0.43 | 2.68 | 2.2 |
| 4 | 0.14 | 0.25 | 0.35 | 0.57 | 2.82 | 1.7 |
| 5 | 0.32 | 0.22 | 0.70 | 0.45 | 2.70 | 1.76 |
| 7 | 0.31 | 0.26 | 0.81 | 0.45 | 2.74 | 2.2 |

*values Q and e for MA are equal to 0.23 and 2.25, respectively.

As the data of PMR spectroscopy showed, the chemical shift of protons of the double bond of MA in the presence of VOCP comonomer undergoes the displacement from field of $\delta = 7.25$ ppm to the field of $\delta = 7.13$ ppm. Such change of the chemical shift of MA protons has been connected with formation of the donor-acceptor complexes in the system, i.e., a charge transfer from donor (VOCP) to acceptor (MA) occurs. From obtained data of PMR investigations, carried out similarly to the method [9], using the Beneschi-Hildebrant formula, the complex-formation constants $K_p$ for VOCP-MA system have been calculated (Table 30.3):

$$\frac{1}{\Delta} = \frac{1}{K_p \cdot \Delta_0} \cdot \frac{1}{[Д_0]} + \frac{1}{\Delta_0};$$

where; $\Delta = \delta_H^A - \delta_0^A$; $\Delta_0 = \delta_{АД}^A - \delta_0^A$.

The formation of complexes between comonomers is also evidenced by IR spectroscopy data: a careful consideration of the IR spectra of both MA and its mixture with VOCP showed that the absorption band at 1640 $cm^{-1}$, characterizing the valence vibrations of the double bond, is subjected to some displacement, which indicates the formation of complexes. It can be assumed that the complex between VOCP and MA should be considered as a monomer with reactivity considerably greater than in the initial monomers.

$K_p$ values presented in Table 30.3 show that as a result of the donor-acceptor interaction between comonomers, there is a weak charge transfer from the highest occupied molecular orbital of the vinyloxy group of VOCP to the lowest vacant MA orbital. This assumption well explains the reason for the structure regularity of the obtained copolymers.

**TABLE 30.3** Temperature Dependence of the Complex-Formation Constants between VOCP and MA and Enthalpy and Entropy Values of the Complexes

| Complex | Temperature, °C | $K_p$, L/mol | –ΔH, kcal/mol | –ΔS, cal/mol·F |
|---|---|---|---|---|
| 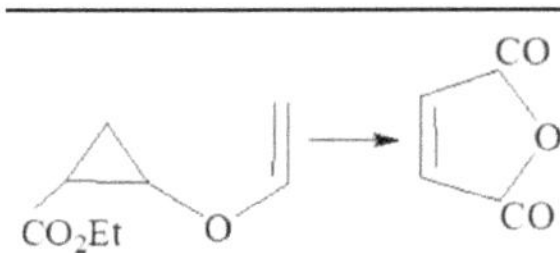 | 25 | 0.45 | 1.99 | 8.25 |
| | 45 | 0. 38 | | |
| | 65 | 0.30 | | |
| 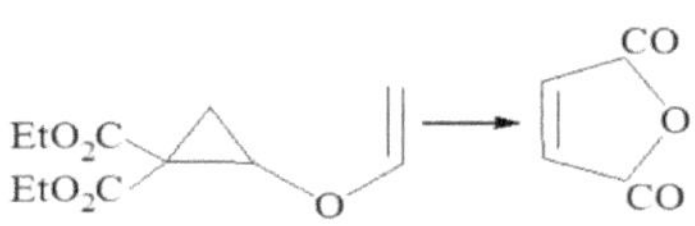 | 25 | 0.37 | 1.39 | 6.76 |
| | 45 | 0.32 | | |
| | 65 | 0.27 | | |
| 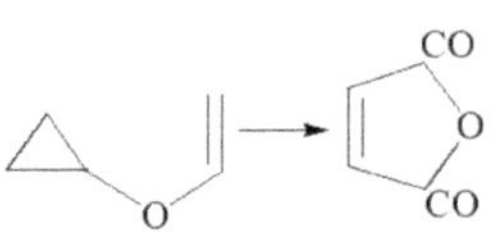 | 25 | 0.25 | 2.03 | 9.74 |
| | 45 | 0.20 | | |
| | 65 | 0.16 | | |
| 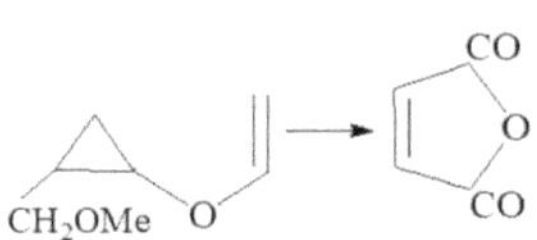 | 25 | 0.45 | 1.23 | 5.28 |
| | 45 | 0.41 | | |
| | 65 | 0.38 | | |

The investigation of the temperature dependence of $K_p$ complex-formation constant values also enables to evaluate the thermodynamic parameters of the complex-formation. Using the Vant-Hoff equation, there have been calculated the enthalpy and entropy values of the complex-formation (Table 30.3), which are in good agreement with the available data in the literature. As follows from the data of Table 30.3, the obtained values for ΔH show the very weak donor-acceptor interactions in the investigated complexes. An introduction of any functional group into cyclopropane ring of monomer 4 leads to a change of the enthalpy and entropy values of the complex-formation, although these changes are insignificant, nevertheless, in a case of availability of substituent with a +J-effect (compound 9) in cyclopropane ring, the complex formation is somewhat easier in comparison with compounds with-J-effect (compounds 2 and 3).

With the aim of the preparation of new photosensitive materials, we have carried out the esterification of copolymers of VOCP with Ma [10–13]. Vinyl-(A), vinyloxycyclopropyl-(B) and phenylcyclopropyl-(C) carbinol were taken as modifying compounds. As an example, one can present the esterification of the copolymer of ethoxy-carbonyl cyclopropyl vinyl ether with MA:

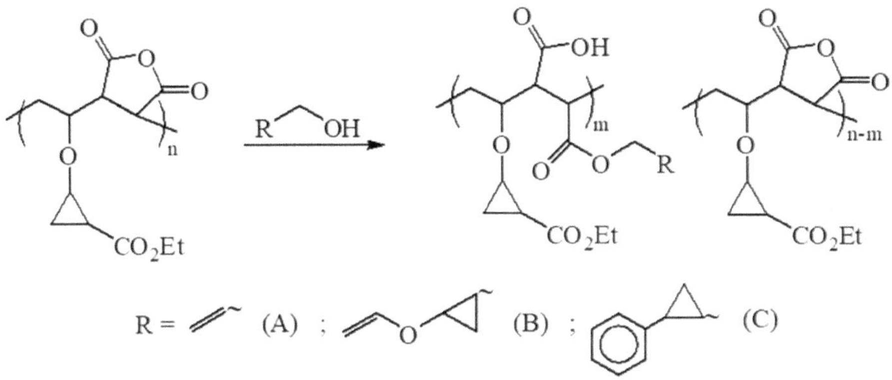


The reaction was carried out in a solution of cyclohexanone or DMFA at 80°C for 2 h in the absence of the catalyst. The initial reagents were taken in an equimolar ratio. The degree of modification was determined on quantity of free carboxyl group forming as a result of the reaction. It has been established in this case that the degree of esterification of the anhydride groups depends, mainly, on ratio modifier: copolymer.

The composition and structure of the obtained modifiers were established by methods of IR spectroscopy, elemental analysis, and potentiomertric titration. In the IR spectra of the modified copolymers there are the absorption in the field of 1720 $cm^{-1}$ and 1015 $cm^{-1}$, corresponding to C=O and C–O–C bonds. The appearance of new absorption bands has been easily detected by a comparison of the IR spectra of the copolymer itself and its modified analog. So, along with absorption bands at 1770 $cm^{-1}$, 1860 $cm^{-1}$, and 1035 $cm^{-1}$, corresponding to carbonyl groups in anhydride fragment and cyclopropane, in the spectrum of copolymer of compound **2** with MA esterified by vinyl carbinol, the absorption bands at 1640 $cm^{-1}$ and 980 $cm^{-1}$ corresponding to vinyl group are also appeared. The availability of the absorption bands in the field of 1770 $cm^{-1}$ and 1860 $cm^{-1}$ has been connected with the presence of the anhydride groups remaining unsubstituted in the macromolecule.

The absorption bands at 1030–1150 $cm^{-1}$ and 1700–1745 $cm^{-1}$, characteristic for the ester group, indirectly confirm the availability of vinyloxycyclopropyl group in the modified copolymer, since the used carbinol, opening the anhydride cycle, favors the formation of a fragment containing the ester group. The absorption bands at 3450 $cm^{-1}$ can be attributed to the unplanar deformation vibrations of carboxyl group, and the bands characteristic for valence vibrations of carbonyl group of carboxyl are overlapped with strong absorption bands of the ester group in the field of 1700–1745 $cm^{-1}$.

For determination of the degree of esterification of the copolymers with vinyl carbinol by IR spectroscopic method, two absorption bands have been selected: the band at 1035 $cm^{-1}$, characteristic for cyclopropane fragment and the band at 1770 $cm^{-1}$, characteristic for C-O–C fragment of the anhydride cycle. As expected, with an increase of a degree of opening of the anhydride cycle and its conversion into the ether group, the intensity of the absorption band at 1770 $cm^{-1}$ is decreased. A degree of conversion (α) was calculated on formula:

$$\alpha = (1 - A_t/A_0) \times 100$$

where; $A_o$ and $A_t$ are the initial and during esterification *t* ratio between intensities of the absorption bands at 1770 $cm^{-1}$ and 1035 $cm^{-1}$.

In Table 30.4, the results of IR spectroscopic investigations of esterification of the copolymer with compounds **4** and MA with vinyl carbinol are presented.

**TABLE 30.4** Change of Intensity of the Absorption Bands in the IR Spectrum of the Copolymer of Vinylcyclopropyl Ether and MA Esterified by Vinyl Carbinol

| **Absorption Band, $cm^{-1}$** | **$A_0$** | **$A_t$ 1 h** | **$A_t$ 2 h** | **$A_t$ 3 h** | **$A_t$ 4 h** | **$A_t$ 5 h** | **$A_t$ 6 h** |
|---|---|---|---|---|---|---|---|
| 1770 (-C–O–C-) MA | 0.363 | 0.269 | 0.202 | 0.151 | 0.124 | 0.115 | 0.103 |
| 1035 (-C–C-) Δ | 0.119 | 0.119 | 0.122 | 0.118 | 0.116 | 0.120 | 0.121 |
| Ratio 1770/1035 | 3.050 | 2.261 | 1.656 | 1.280 | 1.069 | 0.958 | 0.851 |
| $A_t/A_o$ | – | 0.741 | 0.557 | 0.416 | 0.342 | 0.317 | 0.284 |
| α, % | – | 25.9 | 44.3 | 58.4 | 65.8 | 68.3 | 71.6 |

For revealing the optimal conditions of esterification of VOCP-MA copolymers with vinyl carbinol, the reaction was carried out in various temperature-time regimes.

Since the esterification reaction is carried out without catalyst and the carboxyl groups formed as a result of the opening of the anhydride cycle are in a sterically hindered position, their reaction with carbinol becomes unlikely. Consequently, a determination of a quantity of free carboxyl groups can allow judging the degree of conversion of anhydride links during esterification reaction.

It has been revealed that with four-fold molar excess of carbinol in relation to the copolymer of compound **4** with MA, at temperatures of 80–90°C, the degree of conversion of the anhydride links in the macromolecular chain for 180 min reaches 80–87%. Under similar conditions, the degree of conversion of the anhydride links of the copolymer of compound **9** with MA is 80–83%. At the temperature rise from 70 to 90°C, a quantity of free carboxyl groups forming for 50 min. increases from 28 to 46%.

Under optimal conditions (ratio vinylcarbinol: copolymer = 4:1, temperature 80°C, reaction duration-180 min), a degree of conversion of the anhydride links in the molecular chain of the copolymer is sufficiently high (70–80%). It should be emphasized that a used quantity of carbinol is completely consumed for esterification reaction and does not remain in the system in free form. The obtained data have also been confirmed by elemental analysis, which completely coincided with the results of potentiometric titration of the copolymer samples. The obtained results allow concluding that the copolymers are quite easily esterified by vinyl carbinol without catalyst with a sufficiently high degree of conversion of the anhydride links in the polymer chain.

It was known that the polymers, whose macromolecules contain unsaturated or cyclic groups in the main or side chains, during heat treatment or in the action of radiation or UV-irradiation on them, easily pass into three-dimensional net structures [14]. VOCP and MA copolymers after esterification by carbinols have been subjected to structuring by the influence of the film of UV-irradiation on them.

For checking the possibility of application of the obtained modified copolymers as photoresists, their solutions in acetone with various concentrations (8–16 mass %) without addition and with the addition of a sensitizer-benzophenone (in quantity ~10 mass %) have been prepared under identical conditions and the photoresist films with a thickness of 0.15–0.70 μm have been obtained. After certain periods of time of UV irradiation with light, the optical density of the films was determined and compared with the optical density of the films before irradiation. The changes occurring in the copolymer films subjected to the irradiation were determined by weight method-determination of a quantity of the soluble fraction by extraction with acetone. It has been detected that the cross-linking process proceeds differently depending on the nature of the chromophore groups (the intensity of the absorption bands varied to various degrees) included in the polymer macromolecule [15, 16].

At the investigation of the photochemical properties of the modified cyclopropane-containing anhydride copolymers in their irradiation with monochromatic light with $\lambda$ = 365 nm, we have noticed that in the IR spectra of irradiated copolymers, the intensity of the absorption band of the double bond at 1645 $cm^{-1}$ is changed. Taking the absorption band at 1035 $cm^{-1}$ characteristic for cyclopropane group as an internal standard, we have evaluated the film photosensitivity made from modified copolymer, which was 70–80 $mJ/cm^2$. This radiation dose necessary for exposure allows forming the topological structures with minimum possible element sizes for this system in the film of the modified copolymer [17, 18]. In irradiation of film with dose to 80 $mJ/cm^2$ a degree of conversion of the side double bonds was 40–45%. The irradiation of film higher 120 $mJ/cm^2$ increases the depth of conversion of the vinyl groups up to 80%, but such exposure is accompanied by side phenomena-postradiation conversions, accompanied by deterioration of resolving capacity of resist.

The change of relative intensity of the absorption band for modifiers ($M_1$–$M_4$), obtained from copolymer of compound **2** with MA with various degree of substitution, depending on irradiation time presented in Figure 30.1, was determined by means of IR spectroscopy, taking into account the change of intensity of the absorption band at 1645 and 980 $cm^{-1}$ (for C=C double bond).

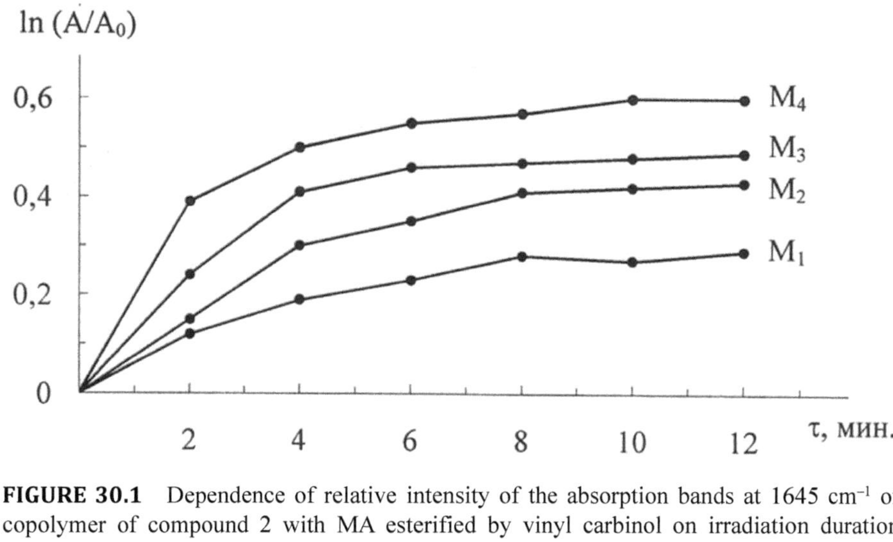


**FIGURE 30.1** Dependence of relative intensity of the absorption bands at 1645 $cm^{-1}$ of copolymer of compound 2 with MA esterified by vinyl carbinol on irradiation duration (degree of esterification $M_1$-14%, $M_2$-44%, $M_3$-60%, $M_4$-78%).

It should be noted that after irradiation of the polymer films for 2 min, it was detected a fall of the solubility of the modified copolymers, which is probably connected with the appearance of cross-linking due to double bonds during the irradiation process. It is seen from the graph that during the first minutes of irradiation, a fall of the intensity of the absorption bands occurs with a high rate. One can say that 30%-s conversion is completed within 3–4 min. The subsequent decrease of the intensity of the absorption bands occurs with a lower rate on the reason that the dense-net polymer particles (microgels) formed at the initial stage of the polymerization screen the vinyl groups not undergoing the reaction, considerably decreasing their activity. In addition, the formed spatially cross-linked structure complicates the relaxation processes (translational and segmental motion of the particles). Therefore, the inhibited movement (or displacement) of the side groups makes it less probable the participation of the vinyl fragments in the photochemical structuring.

The degree of cross-linking of the irradiated copolymer films was determined by holding the polymer films for 3 min. in an acetone solution at room temperature and by their drying to a constant mass. The uncross-linked part of the films completely dissolved. With the increase of the degree of cross-linking of the copolymer films, their solubility in acetone solution decreased [19, 20]. It has been found that after 15 min. of irradiation, the mass losses remained unchanged. The spectroscopic investigations showed that the maximum degree of cross-linking with the participation of the vinyl groups reached 5–6 min, which is in accordance with the calculated content of gel-fractions.

## 30.4 CONCLUSIONS

Under the influence of UV irradiation, as shown by IR and PMR spectroscopic investigations, the structuring occurs in the macromolecular chain of the modified copolymers, and it proceeds only due to the opening of the double bond of the vinyl group, in this case, the other groups and bonds do not participate in this process. As a result of photo-action, a three-dimensional cross-linked structure is formed, which leads to a loss of solubility of the modified copolymers. Thus, the high polymerization activity of the modified copolymers under the action of UV irradiation enables to use of them as a basis for the creation of the photo-and electron-resists.

## KEYWORDS


- **complex-formation**
- **copolymer**
- **esterification**
- **photosensitivity**
- **reactivity**
- **vinyloxycyclopropane**


## REFERENCES


1. Ghosh, A., Carran, R. S., Grosvenor, A. J., Deb-Choudhury, S., Haines, S. R., & Dyer, J. M., (2016). Feather meal-based thermoplastics: Methyl vinyl ether/maleic anhydride copolymer improves material properties. *Fibers and Polymers, 17*(1), 9–14.
2. Atıcı, O. G., Akar, A., & Rahimian, R., (2001). Modification of polymaleic anhydride-co-styrene) with hydroxyl containing compounds. *Turk. J. Chem., 25*, 259–266.
3. Houlihan, F. M., Wallow, T. I., Nalamasu, O., & Reichmanis, E., (1997). Synthesis of cycloolefin-maleic anhydride alternating copolymers for 193 nm imaging. *Macromol., 30*(21), 6517–6524.
4. Hyun-Sang, J., Dong, C. S., Chang, M. K., Young, T. L., Seong, D. C., et al., (2003). Ring opened maleic anhydride and norbornene copolymers (ROMA) have a good character in resist flow process for 193-nm resist technology. *Proc. SPIE*, V. 5039, *Advances in Resist Technology and Processing XX*, 725.
5. Antonovich, A. A., Kruglova, V. A., Skobeeva, N. I., et al., (1980). Copolymerization and complex-formation in the system of 2-тtrichloromethyl-methylene-1,3-dioxalane-maleic anhydride. *Vysokomolek. Soyed., t.22*(A), (10), 2273–2278.
6. Rzayev, Z. M. O., (2011). Graft copolymers of maleic anhydride and its isostructural analogues: High performance engineering materials. *Intern. Review of Chemical Engineering, 3,* 153–215.
7. Shahnazarova, R. Z., & Guliyev, A. M., (2004). Copolymerization of vinylcyclopropyl ethers with maleic anhydride. *Azerb. Chem. J.,* (2), 25–30.
8. Shahnazarli, R. Z., Aliyeva, S. G., & Guliyev, A. M., (2014). Vinylcyclopropyl ethers: Synthesis, radical reactions of addition and polymerization. *Coll. of Scientific of IPM, ANAS,* 32–52.
9. Emmanuel, N. M., (1971). *Experimental Methods of Chemical Kinetics* (p. 176). M.: Vysshaya Shkola.
10. Shahnazarli, R. Z., Ramazanov, G. A., & Guliyev, A. M., (2012). Esterification of anhydride-containing copolymers by vinyl(phenyl)cyclopropyl carbinols and preparation of photosensitive materials. *Materials of VII Intern., Symposium of Fundamental and Applied Problems of a Science, 3,* 15–22. Moscow.

11. Shahnazarli, R. Z., (2012). Modification of copolymer of vinyl ethoxy carbonyl cyclopropyl ether with maleic anhydride by vinyl cyclopropyl carbinol. Functional monomers and polymers. *Coll. of Scientific Works of IPM, ANAS,* 205–218.
12. Shahnazarli, R. Z., & Guliyev, A. M., (2013). Esterification of polyvinyl alcohol and synthesis of photosensitive polymers. *Coll. of Scientific Works of XXVI Intern., Scientific-Techn., Conf. Reactive-2012* (pp. 119–127). Minsk.
13. Shahnazarli, R. Z., Garayeva, S. H., & Guliyev, A. M., (2019). Photosensitive copolymers on the basis of *gem*-disubstituted vinyloxycyclopropanes. In *Applied Chemistry and Chemical Engineering. V.3- Interdisciplinary Approaches to Theory and Modeling with Application.* Ch. 12, pp. 157–170.
14. Moro, U., (1990). Microlitography: Principles, methods, materials. In: Timerova, R. K., (eds.), *Transl. from Engl.,* (pp. 1, 606).
15. Shahnazarli, R. Z., Aliyeva, A. A., Nazaraliyev, K. G., & Guliyev, A. M., (2008). Photosensitive materials on the basis of copolymers of vinyl cyclopropyl ethers with maleic anhydride. *J. Appl. Chemistry, 81*(2), 304–307 (In Russia).
16. Shahnazarli, R. Z., (2013). Cyclopropane-containing photosensitive resists on the basis of modified PVC. *IX Intern. Scient. Pract. Conf. of Advanced Scientific Developments* (pp. 65–74). Praga.
17. Garayeva, S. H., Shahnazarli, R. Z., Ramazanov, G. A., & Guliyev, A. M., (2018). Synthesis of cyclic acetals of vinyloxycyclopropanes and photosensitive polymers on their basis. *Eur. Science Review,* (3, 4), 313–321.
18. Shahnazarli, R. Z., Garayeva, S. H., Ramazanov, G. A., & Guliyev, A. M., (2018). Advanced copolymers on the basis of vinylcyclopropyl ethers. *Poly. Char., 26,* p. 36. World Forum on Advanced Materials. Georgia, Tbilisi.
19. Shahnazarli, R. Z., (2014). Polymer photoresists on the basis of acylated polyvinyl alcohol. *European Applied Science,* (2), 142–146.
20. Ramazanov, G. A., Shahnazarli, R. Z., & Guliyev, A. M., (2010). Joint polymerization of cyclopropyl-substituted dioxolanyl methacrylates with styrene and synthesis of photosensitive copolymers. *J. Appl. Chemistry, 83*(3), 484–488 (In Russia).

# CHAPTER 31

# About of the Rotation Mechanisms of the Molecular "Motors"

N. S. VASSILIEVA-VASHAKMADZE,[1] R. A. GAKHOKIDZE,[2]
T. S. VASHAKMADZE,[3] M. Z. GORGOSHIDZE,[4] and P. L. TOIDZE[5]

[1]*Georgian Academy of Engineering, Tbilisi, Georgia,*
*E-mail: nonavas@rambler.ru*

[2]*Iv. Javakhishvili Tbilisi State University, Department of Bioorganic Chemistry, Tbilisi, Georgia*

[3]*Iv. Javakhishvili Tbilisi State University, Vekua Institute of Applied Mathematics, Tbilisi, Georgia*

[4]*E. Andronikashvili Institute of Physics, Iv. Javakhishvili Tbilisi State University, Tbilisi, Georgia*

[5]*Georgian Technical University, Department of Chemical and Biological Technologies, Tbilisi, Georgia*

## ABSTRACT

The rotation mechanisms of the molecular "motors" cannot be observed directly, but it is interesting to study them with some theoretical methods. In this paper, a proposed mechanism of molecular rotor rotation is theoretically justified by the example of FOF1 $Na^+$, $K^+$ ATP-synthase of *Propionigenium modestum*, as well as the torque mechanism of bacterial flagella. Made is an assumption: a regular link between the gate current and appearance of a magnetic field and the ability of the Lorentz force to create a torque that acts on moving charges.

## 31.1 INTRODUCTION

A considerable experimental material has been accumulated during the last years that can serve as a base for evaluation of the nature of forces that build the torque of molecular rotors [1–5]. It is known that the proton and sodium FOF1 ATP-synthases of the most diverse range of organisms have almost the same structure and configuration-they consist of two large compounds FO and F1, and small/minor structures. The mandatory elements are the stator and the rotor, and the operation is carried out by rotating the rotor with the participation of the stator using the energy from external sources, which is converted to potential, in particular, into a proton or sodium potential. Therefore, it is sometimes acceptable to mention a type of ATP-synthase without mentioning the organism from which this enzyme is isolated, e.g., "FOF1 $Na^+$, $K^+$ ATP-synthase." It is suggested that the basis of the rotor rotation mechanism of the FOF1 $Na^+$, $K^+$ ATP-synthase, as well as $H^+$, $K^+$ ATP-synthase and also the basal bodies of bacterial flagella and other molecular motors is a combination of some factors: a change in transmembrane potential when cells are activated, which creates gate currents in the protein components of the ion channels of the stator; a regular link between the charge current and appearance of a magnetic field; and the ability of the Lorentz force to create a torque that acts on moving charges. ATP-synthase, $Na^+$, $K^+$ ATP-synthase FOF1, which has been found in *Propionigenium modestum* and *Ilyobacter tartaricus* are very similar to the proton ATP-synthase. Its FO compound is located in the cytoplasm membrane and is in contact with exoplasm. F1 is fully localized in the cytoplasm [6, 7] (Figure 31.1). This allows us to suppose that the study results of bacterial ATP-synthase working mechanism can be applied to other types of ATP-synthases..

## 31.2 RESULTS AND DISCUSSION

### *31.2.1 ELECTRIC AND MAGNETIC FIELDS OF LIVING ORGANISMS*

It is known that the natural sources of magnetic and electromagnetic fields can be moving charges, including ionic and gate currents. Those fields disappear when the relevant charges stop moving [8–10]. Functioning of living organisms, including humans, transmission of nerve impulses, contraction of the heart muscle, etc., are performed by "molecular machines"-protein complexes, including ATF-synthase, ATP-ase-which is accompanied by the appearance of "gate" and "ion" currents-sources of electric and magnetic fields that disappear after the death of animals [11–14]. Special devices such as electronic and magnetic encephalographs, SQUIDs, electronic, and magnetic myographs, electronic, and magnetic cardiographs, etc., can record lifetime electric and magnetic fields [15, 16]. $ab_2$ FOF1 ATP-synthase subunits as well as the basal bodies of bacterial flagella penetrate through the cytoplasmic membrane. The mutually opposite gates of ion channels located inside the subunit fall into zones with different ionic composition and, accordingly, with the following potential difference [17]:

$$\varphi = \frac{RT}{F} \ln \frac{\sum_k c_k^e p_k^e}{\sum_k c_k^i p_k^i} \tag{1}$$

where; φ is transmembrane difference of potentials, *R* is universal gas constant, *T* is absolute temperature, *F* is Faraday constant, $C_k^e, C_k^i$ is concentration of *k*-type ions in exo- and cytoplasm, respectively, and $P_k^e, P_k^i$ is *k*-type ion permeability in exo- and cytoplasm, respectively.

Any changes in ion composition in the local area around the external or internal gate of the ion channel are connected with changes in the transmembrane difference of potentials, which is accompanied by emerging of the gate current [11, 12].

### *31.2.2 FOF1 $Na^+$, $K^+$ ATP-SYNTHASE OF PROPIONIGENIUM MODESTUM*

The rotor or C-ring of $Na^+$, $K^+$-ATP-synthase is spatially close, but not connected by a chemical bond with the fixed subunit $ab_2$ (Figure 31.1).

Starting from the peak of F1, the component $ab_2$ goes around the surface of F1 localized in the cytoplasm and, after crossing the membrane, it goes to the exoplasm. That is why the peak of the stator $ab_2$ and its base appear in zones with different ionic compositions, between which there is a transmembrane potential difference (Figure 31.1).

The magnetic field of the gate current in the α-helical subunit of the wall protein component of the $Na^+$, $K^+$ ATP-ase ion channel can be described with the proportion: [8, 9, 18].

#### *31.2.2.1 THE $Na^+$, $K^+$-ATP-SYNTHASE*

According to the image taken with electron microscopy, C-disc-rotor $Na^+$, $K^+$-ATP-synthases is a protein complex in a near to cylindrical shape formed by 8–15 packs of α-helical subunits. The C-disk is immersed in the thickness of the cytoplasmic membrane and does not come into contact with the cytoplasm. The lower base of the C-disk is in contact with the exoplasm. In the center of the upper base of the C-disk, there is anchored the γ subunit oriented perpendicular to its surface. The γ subunit crosses the cytoplasmic membrane and continues in the cytoplasm, passes through the lumen in the lower part of the compound and reaches its apex (Figure 31.1). Therefore, between the center and the periphery of the C-disk, the same potential difference and intensity are created as in the transmembrane:

$$\vec{E} = -grad\varphi \tag{2}$$

A change in the ionic composition in the cytoplasm or exoplasm causes a change in the transmembrane potential difference $\varphi$ and intensity, and hence the potential difference between the center and periphery of the C-disk, as well as the electric field intensity $\vec{E}$. In case of sufficiently large values of $\vec{E}$ in the electric field, a movement of charges occurs under the action of the following force:

$$\vec{F} = q\vec{E} \tag{3}$$

In this case, this leads to the appearance of a radial current between the center and the periphery of the C-disk.

### 31.2.2.2 THE FOF1 $Na^+$, $K^+$ ATP-SYNTHASE STATOR

During the operation of FOF1 $Na^+$, $K^+$ATP-synthase, a change in the ionic composition in cyto- or exoplasm initiates a change in the transmembrane potential difference, and therefore the potential difference between the lower and upper gates of the ion channel passing through the $ab_2$ complex, penetrating the cytoplasmic membrane. When this occurs, the gate current appears. According to the laws of electromagnetism, a magnetic field is formed around the current. In the case of gate current we have [8, 9]:

$$\vec{B} = \mu_0 \mu \frac{N}{l} J \cdot \vec{n} \tag{4}$$

where; $B$ is the magnetic field induction, $\mu_0$ is the magnetic constant, $\mu$ is the magnetic capacity, $N$ is the number of turns, $l$ is the length of a α-spiral fragment, $J$ is the gate current, and $\vec{n}$ is the unit vector directed along the axis of α-helix.

The magnetic field is not local, and therefore this magnetic field has an influence on the rotor of ATP-synthase (C-ring) (3). Therefore, the Lorentz force acts on the rotor charges (C-disk) [8, 9]:

$$\vec{F} = q\vec{E} + q\left[\vec{v} \cdot \vec{B}\right]; \tag{5}$$

$$\vec{F} = \vec{F}_1 + \vec{F}_2; \ \vec{F}_1 = q\vec{E}; \ \vec{F}_2 = q\left[\vec{v} \cdot \vec{B}\right] \tag{6}$$

where; $q$, $v$ are charges and their speeds, accordingly, and $\vec{B}$ is the magnetic field created by the stator (3).

The $\vec{F}_1$ component (2) gives radial velocity to $\vec{v}$ charges. The force $\vec{F}_2$ (5) acts perpendicular to the direction of movement of charges in the plane of the C-disk and therefore creates a torque $\vec{N}$ that acts on radially moving charges:

$$\vec{N} = q\left[\vec{R} \cdot \vec{F}_2\right]; \tag{7}$$

$$\left|\vec{F}\right| = qv\ B; \quad v = \omega R; \quad qvB = \frac{mv^2}{R}; \ \rightarrow \ \omega = \frac{qB}{m} \tag{8}$$

According to the works of Hodjkin, Huxley, Armstrong, Bezanilla, Nikolls, and other authors, the potential-dependent ion channels of different systems are structurally similar to each other, the gate currents are about 3 orders of magnitude weaker than the ionic currents. Given this, we will conduct an approximate numerical estimate of the rotor speed of ATP synthase.

$$\omega = \frac{q}{m}\left|\vec{B}\right|; \ \left|\vec{B}\right| = B = \mu_0 \mu \frac{N}{l} J_g; \ \mu_0 = 4\pi \cdot 10^{-7}\ \frac{H}{m}; \mu \sim 1; N \sim 10^3; l \sim 10^{-9}\ m; \tag{9}$$

$$J_g \approx 10^{-14}\ A; \text{Therefore, } B \sim 10^{-8}\ Tesla; \ \frac{q}{m} \sim 1{,}8 \cdot 10^{11}\ \frac{C}{kg}.$$

$$\text{Consequently } \omega = \frac{q}{m}\left|\vec{B}\right|; \sim 10^3; \ \nu = \frac{\omega}{2\pi} \sim 10^2\ s^{-1} \tag{10}$$

Under the action of the torque $\vec{N}$ (6), the radial current of the C-disk charges begins to rotate and involve the entire rotor system, including the γ subunit rigidly connected with it. When the base of the γ subunit that is anchored in the center of the ATP-synthase rotor rotates, it's opposite end rotates in the lumen of the F1 compound between the α3β3 components, which makes possible the enzyme function. Rotation of the rotor does not change the mutually perpendicular orientation of the radial direction of the C-ring charge speed $\vec{v}$ relatively to the magnetic field induction $\vec{B}$ direction of the stator. Therefore, the force $\vec{F}_2$ (5) that is responsible for the appearance of torque (6), continues to act for some time and causes multiple rotations of the ATP-synthase rotor.

## 31.3 THE BASAL BODY OF THE BACTERIAL FLAGELLUM

The processes in the basal body of bacterial flagella are somewhat different but still similar in many respects to the mechanism of ATP-synthase [19–22].

The rotor of the bacterial motor (MS-disk or ring) contacts the cell cytoplasm with its internal (cytoplasmic) base and the rest of the disk is immersed into the thickness of the cytoplasmic membrane (Figure 31.2).

Unlike the γ subunit that rotates together with the C-ring of the rotor in FOF1 $Na^+$, $K^+$ ATP-ase, passes through the lumen along the symmetry axis of the F1 compound and enters the cytoplasm, the bacterial flagellum leaves the rotor-the MS-ring immersed in the cytoplasmic membrane-and enters its surrounding medium-the exoplasm-and then ensures the movement of bacteria in the external environment by rotating with the rotor [20–22].

## 31.4 THE STATOR OF THE BASAL BODY OF THE BACTERIAL FLAGELLUM

The basal body stator-MotA, MotB-is a system of several α-helical protein bundles that penetrate the cytoplasmic membrane. It is important that their intracellular terminals are immersed in the cytoplasm, and the external ones in the exoplasm. Therefore, when the bundles of α-helical protein subunits that form the stator are immersed with their negative charge ends into the zones with differing ionic compositions and potentials, they have an initial electric dipole moment $\vec{D}_0$.

When a change in the transmembrane potential is caused by some external influences, as well as by the approaching of charged particles to the ion channel gate according to the concept of Hodjkin and Huxley [11] and experiments of Bezanilla and Armstrong [12], a gate current appears in the protein components of the ion channel walls. As is known, any charge current creates not only electric but also magnetic field. As it was shown for $Na^+$, $K^+$ATP-synthase, the direction of the magnetic field of the gate current is perpendicular to the surface of the cytoplasmic membrane and is determined the same way (3). The magnetic field of the stator is not localized and it can influence on the rotor system located near to the stator.

## 31.5 THE ROTOR OF THE BASAL BODY OF BACTERIAL FLAGELLA

The rotor of the basal body of the flagellated "motor," which rotates the flagellum, is an MS-ring consisting of 8 to 12 bundles with α-helical protein subunits (Figure 31.2). The MS ring is located in the cytoplasmic membrane so that the "lower" regions of the α-helical protein subunits of the MS ring are in contact with the cytoplasm, and the rest part is immersed in the thickness of the cytoplasmic membrane.

In the center of the "upper" base of the rotor, there is anchored a flagella, the end of which goes out into the external environment, and its "lower" base is in contact with the cytoplasm. Therefore, between the center of the MS-ring, where the flagellum is anchored-and its periphery, there should be a potential difference φ that coincides with the transmembrane potential difference. Accordingly, there is electric field strength $\vec{E}$ between the center and the periphery of the rotor (1). A sufficiently large transmembrane potential difference creates an electric field $\vec{E}$ capable of inducing a radial charge current between the center and the periphery of the MS disk, because the force $\vec{F}$ (2) acts on each charge in the electric field.

The stator of the basal body of the bacterial flagellated motor (MotA, MotB) is constructed similarly to the FOF1 $Na^+$, $K^+$ ATP-synthase stator subunit $ab_2$ (Figures 31.1 and 31.2). In both cases, the gate current in the α-helical subunits of the protein component of the ion channel, which occurs when the transmembrane potential changes, creates a magnetic field $\vec{B}$ (3).

A change in the transmembrane potential (and hence the potential difference between the center and periphery of the MS-ring) causes a radial movement of charges in the rotor. In this case, the rotor-the system of MS-rings together with the base of the flagellum-gets into the magnetic field of the stator. In the magnetic field, the Lorentz force (4, 5) acts on moving charges. The magnetic component of the Lorentz force is on the plane of the MS-ring, it is directed perpendicular to the velocity of the charges and the magnetic field. Therefore, a torque N is generated that acts on radially moving charges (6).

Using the well-known equations, we can obtain the following:

$$\left|\vec{F}\right| = qv\ B;\quad v = \omega R;\quad qvB = \frac{mv^2}{R} \ \rightarrow\ \omega = \frac{qB}{m} \tag{11}$$

where; $B$ is the magnetic field induction of stator, $R$ is the radius of the $MS$ -disc, $q,m$ are the charge size and its mass, respectively, $v$ is the charge rate of $MS$ -disc, $\omega$ is the cyclic charge frequency in the $MS$ -disc.

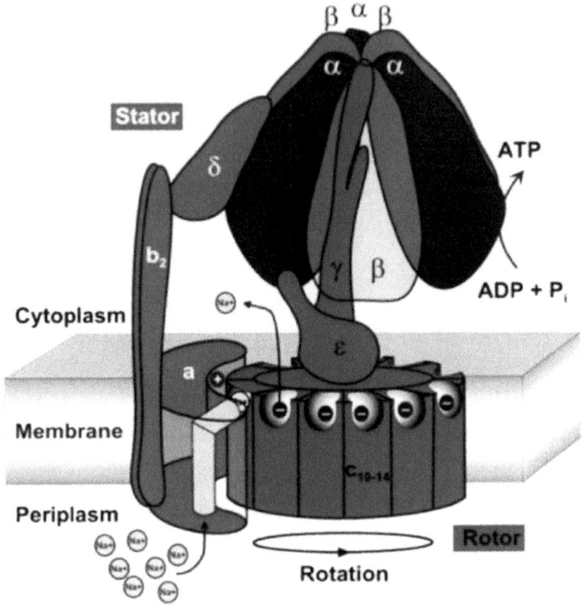


**FIGURE 31.1** FOF1 $Na^+$ ATP-synthase of *Propionigenium modestum* [4].

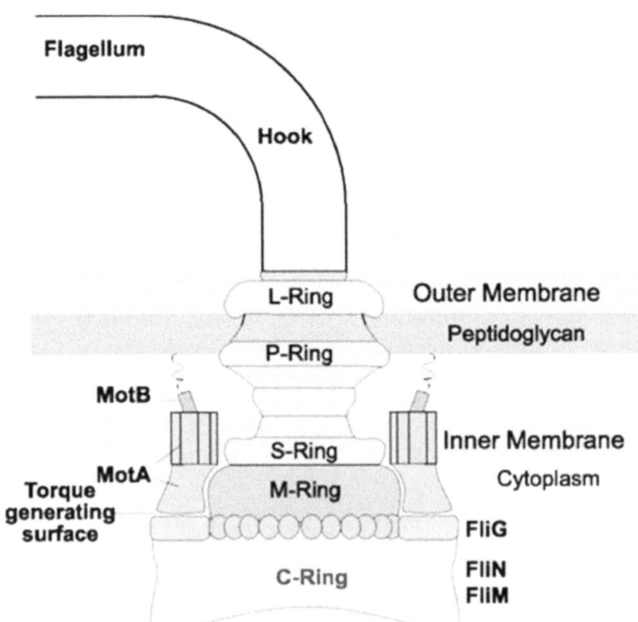


**FIGURE 31.2** Architecture of the bacterial flagella [23].

A numerical estimate of the frequency of rotation of the rotor of bacterial flagella, carried out similarly to (6), (7), (8), (9), gives: $\nu \sim 10^2\ s^{-1}$

## 31.6 CONCLUSION

1. Rotation of the rotor FOF1 $Na^+$, $K^+$ ATP-synthase is ensured by the Lorentz force-torque, which is created by the magnetic field of the gate current of the stator. The possibility of multiple rotations can be explained by the fact that the angle between the direction of movement of the rotor charges and the Lorentz force does not change during the rotation, which stipulates a long-lasting effect of the torque.
2. In the bacterial flagella, the Lorentz force $\vec{F}$ (5) creates the torque $\vec{N}$ (6), which acts on the charges that rotate at the cyclic frequency $\omega$ (7), and as the charges are connected with their molecules, they involve the entire system in the rotation, including the base of the flagellum. Therefore, the flagellum rotates with the cyclic frequency $\omega$ (7).

The FO compound (indicated as the rotor in the picture) resembles a mushroom stipe that is located in the cytoplasmic membrane and goes into the exoplasm. F1 (indicated as the stator) resembles a mushroom cap and is located in the cytoplasm.

The ion channels of the stator pass through the compound $ab_2$ that permeates through the cytoplasmic membrane. The outer gate of the ion channel opens into the exoplasm, and the inner gate opens into the cytoplasm. The rotating subunit $\gamma$ that is anchored in the center of the rotor forms a connecting isthmus between the FO and the central zone of the F1 complex, which is completely in the cytoplasm. The rotation mechanism of the rotor is described in the text.

In Figure 31.2, MotA and Motb are stator components with ion channels inside them that go from the cytoplasm to the peptidoglycan layer. M and S rings form the rotor. It is localized in the cytoplasmic membrane, and its lower surface in contact with the cytoplasm. In the central zone of the upper surface of the rotor, which is immersed into the cytoplasmic membrane, it is penetrated by the base of the flagellum, along the major axis of which passes a channel that is connected with exoplasm. Between the exoplasm and cytoplasm, and hence between the base of the flagellar channel and the periphery of the rotor (MS-ring), there is a potential difference and an electric field strength in which the force acts on the charges and initiates their radial motion between the center and the periphery of the rotor. The magnetic field of the stators creates the Lorentz force, which initiates the torque that causes rotation of the charges, and the SM-ring with an anchored bacterial flagellum rotates with them.

## KEYWORDS


- **cytoplasmic membrane**
- **molecular "motors"**
- **molecular rotating mechanisms**
- **motility protein A**
- **motility protein B**
- **symmetry mismatch**


## REFERENCES


1. Skulachev, V. P., (1997). *Laws of Bioenergy* (pp. 9–14). Lomonosov Moscow State University.
2. Romanovsky, Y. M., & Tikhonov, A. N., (2010). Molecular energy transducers of a living cell. *Proton ATP-Aza-Rotating Molecular Motor. UFN., 180*, pp. 931–956.
3. Kinosita, K., Yasuda, Jr. R., Noji, H., & Adachi, K., (2000). A rotary molecular motor that can work at near 100% efficiency. *Philos. Trans R. Soc. Lond. B. Biol. Sci., 355*, 473–489.
4. Georg, K., (2001). *$Na^+$-Translationing $F_1$ $F_0$ ATP Synthase Propionigene Mechano-chemical Understanding of Motor $F_0$ that Controls ATP synthesis Biochemistry and Biophysics Acta (BBA)-Bioenergy, 1505*(1), 94–107.
5. Jonna, K. H., Adriana, L. K., Jan, H., Eckhardt-Strelau, L., Bernd, B., Janet, V., & Thomas, M., (2012). Structural study on the bacterial ATP synthase $F_0$ motor. *Proc. Natl. Acad. Sci. USA, 109*(30), pnas.1203971109.
6. Hiroyuki, N., Yasuda, R., Yoshida, M., & Kinosita, K. Jr., (1997). Direct observation of the rotation of $F_1$ ATPase. *Nature, 386*(6622), 299–302.
7. Wu, C., (1997). Molecular motor spins out energy for cells. *Science News, 151*(12), 173.
8. Parcell, E., (1965). *Electricity and Magnetism* (p. 444).
9. Matveev, A. N., (2010). *Electricity and Magnetism* (p. 464). ISBN: 978-5-8114-1008-8.
10. Landau, L. D., & Lifshits, E. M., (1982). Electrodynamics of continuous media. *M. Science*, 240.
11. Hodjkin, A. L., & Huxly, F. A., (1974). *Physiolojy, 117*(4), 500–544.
12. Besanilla, F., & Armstrong, C. M., (1977). Inactivation of the sodium channel. Sodium current experiments. *J. Gen Physiol., 70*(5), 549–66.
13. Wilson, J., & Hunt, T., (1994). *Molecular Biology of the Cell* (p. 520).

14. ATP Synthase, (2017). Wikipedia.
15. Shestakova, A. N., Butorina, A. V., Osadchy, A. E., & Shtyrov, Y. Y., (2012). Magnetoencephalography is the newest method of functional mapping of the human brain. *Jour. Experimental Psychology, 5*(2), 119–134.
16. Vvedensky, V. L., & Ozhogin, V. I., (1986). Supersensitive magnetometry and biomagnetism. *M. Science,* 199.
17. *Derivation of the Nernst Equation.* http://www.helpiks.org/6-49371.html; https://studopedia.info/2-52448.html (accessed on 19 October 2020).
18. Besanilla, F., (2005). *Voltage-Gated Ion Channels, 4*(1).
19. Chen, S., Beeby, M., Murphy, G. E., Leadbetter, J. R., Hendrixson, D. R., Briegel, A., Lim Z., et al., (2011). Structural diversity of bacterial flagellar motors. *MBO J., 30*(14), 2972–2981.
20. Renui, L., (2017). Stepwise formation of the bacterial flagellar system. *PNAS, 104*(17), 7116–7121.
21. Carlotte, K. O., Gibbsons, I. R., & Dagger, R. K., (1999). Rotation of the micro pair of eucaryotic flagella. *Molec. Biology of the Cell., 10*, 14–44.
22. Wikipedia, (2016). The structure of the bacterial flagella. *Flagella of Prokaryotes*, http://ru.wikipedia.org (accessed on 19 October 2020).
23. Steve, D. V., (2004). *The Amazing Motorized Germ* (Vol. 27, No. 1). AIG.

# Index

## D

## N

## T

## U



## V